T0257786

Genetic Engineering of DNA and Protein

Volume II

Genetic Engineering of DNA and Protein Volume II

Edited by **Tom Lee**

New York

Published by Callisto Reference,
106 Park Avenue, Suite 200,
New York, NY 10016, USA
www.callistoreference.com

Genetic Engineering of DNA and Protein
Volume II
Edited by Tom Lee

© 2015 Callisto Reference

International Standard Book Number: 978-1-63239-352-4 (Hardback)

This book contains information obtained from authentic and highly regarded sources. Copyright for all individual chapters remain with the respective authors as indicated. A wide variety of references are listed. Permission and sources are indicated; for detailed attributions, please refer to the permissions page. Reasonable efforts have been made to publish reliable data and information, but the authors, editors and publisher cannot assume any responsibility for the validity of all materials or the consequences of their use.

The publisher's policy is to use permanent paper from mills that operate a sustainable forestry policy. Furthermore, the publisher ensures that the text paper and cover boards used have met acceptable environmental accreditation standards.

Trademark Notice: Registered trademark of products or corporate names are used only for explanation and identification without intent to infringe.

Printed in the United States of America.

Contents

Preface

Every book is a source of knowledge and this one is no exception. The idea that led to the conceptualization of this book was the fact that the world is advancing rapidly; which makes it crucial to document the progress in every field. I am aware that a lot of data is already available, yet, there is a lot more to learn. Hence, I accepted the responsibility of editing this book and contributing my knowledge to the community.

This book provides readers with research works in discussion worldwide on the topic of latest molecular genetics. There are two approaches of every research work published in this book; first, to make the research chapters understandable to majority of readers and second, to describe the genetic tools and pathways used in research. The one fact mostly highlighted is the necessity of genetic insight in solving an issue. This book will prove to be an interesting read to those interested in genetic discoveries because of its structure, which has been made with a point of view for attracting readers and familiarizing them with genetic approaches in disease - related research, applied research and new tools for molecular genetics.

While editing this book, I had multiple visions for it. Then I finally narrowed down to make every chapter a sole standing text explaining a particular topic, so that they can be used independently. However, the umbrella subject sinews them into a common theme. This makes the book a unique platform of knowledge.

I would like to give the major credit of this book to the experts from every corner of the world, who took the time to share their expertise with us. Also, I owe the completion of this book to the never-ending support of my family, who supported me throughout the project.

Editor

Molecular Genetics in Disease-Related Research

New Insights into the Epithelial Sodium Channel Using Directed Mutagenesis

Ahmed Chraibi and Stéphane Renauld
University of Sherbrooke, Québec
Canada

1. Introduction

Directed mutagenesis is a fundamentally important DNA technology that seeks to change the base sequence of DNA and test the effect of the change on gene or DNA function. It can be accomplished using the polymerase chain reaction (PCR). For more than 20 years, many applications in both basic and clinical research have been revolutionized by PCR. The development of this technique allowed the substitution, addition or deletion of single or multiple nucleotides in DNA (Mullis and Faloona, 1987). Because of redundancy in the genetic code, such mutations do not always alter the primary structure of proteins. In this chapter, we will review the contribution of PCR-directed mutagenesis in the determination of the structure-function relationship of the epithelial sodium channel (ENaC), particularly with respect to the domains involved in proteolytic activation and ligand-induced stimulation of the channel.

2. Physiological role and structure of ENaC

ENaC is a key component of the transepithelial sodium transport. It is expressed at the apical membrane of a variety of tissues, such as the distal nephron of the kidney, lungs, exocrine glands (e.g., sweat and salivary glands) (Brouard et al., 1999; Duc et al., 1994; Perucca et al., 2008; Roudier-Pujol et al., 1996) and distal colon (Kunzelmann and Mall, 2002). In aldosterone-sensitive distal nephron (ASDN) and distal colon, this channel plays a major role in the control of sodium balance and blood pressure (Frindt and Palmer 2003; Garty and Palmer 1997). In lungs, ENaC regulates mucus secretion and aids in the protection of the airway surface (Randell and Boucher, 2006). Its role was clearly demonstrated in mice in which the ENaC gene was inactivated by homologous recombination (Hummler et al., 1996). ENaC belongs to a gene family with members found throughout the animal kingdom, the so-called ENaC/degenerin family, including the acid sensing ion channel (ACIC) and the Phe-Arg-Met-Phe amide-gated ion channel (FaNaCh), (Kellenberger and Schild. L, 2002). Using the *Xenopus* oocyte expression system and a distal colon cDNA library, the primary structure of ENaC was identified; and electrophysiologic characteristics of ENaC channel were determined (Canessa et al., 1993; Canessa et al., 1994; Lingueglia et al., 1993). ENaC is a heteromeric channel made of three subunits (α, β and γ) encoded by 3 different genes SCNN1a, SCNN1b and SCNN1g, respectively. Each subunit exhibits ~30% identity at the amino acid level and shares highly conserved domains. The membrane topology of each subunit predicts the presence of two transmembrane domains (M1 and M2), a large extracellular loop (~70% of the size of the channel) and relatively short amino and carboxyl termini. The stoichiometry of ENaC was much discussed: several examples of biochemical and functional evidence are consistent with a heterotetrameric structure (2α, 1β, 1γ) (Anantharam A, 2007; Dijkink et al., 2002; Firsov et al., 1998), but octameric or nonameric structures have also been suggested (Eskandari et al., 1999; Snyder et al., 1998). Recent crystallographic data obtained on the related ASIC1 channel suggest ENaC most likely exists functionally as an $\alpha\beta\gamma$ heterotrimer complex (Jasti et al., 2007; Stockand et al., 2008). ENaC is characterized by high sodium selectivity ($P_{Na+}/P_{K+} > 100$), a low single-channel conductance (4-5 pS), gating kinetics characterized by long opening and closing times, and a specific block by amiloride (Ki: 100 -200 nM).

Sodium homeostasis requires that the entry of sodium through the apical membrane of epithelial cells is tightly controlled. This control may be realized by regulation of ENaC activity and expression. The role of different domains involved in this regulation has been determined by directed mutagenesis.

3. Mutations in ENaC subunits cause hereditary human disease

The role of ENaC in the regulation of blood pressure and regulation of extracellular fluid volume has been highlighted by the discovery of two severe human diseases. The diseases are due to loss or gain of function of ENaC. Homozygous inactivating mutations in the α, β or γ ENaC subunits cause pseudohypoaldosteronism type 1 (PHA-1), characterized by hypotension and severe hyperkalemic acidosis (Chang et al., 1996). Activating mutations in the genes for the β or γ ENaC subunits lead to Liddle's syndrome, characterized by autosomal-dominant hypertension accompanied by hypokalemic Alkalosis and volume expansion (Shimkets et al., 1994).

The mutations causing PHA-1 have been identified, and the mechanisms by which they led to a hypofunction of ENaC have been addressed. See (Kellenberger and Schild. L, 2002) for review.

In particular Chang et al. (1996) showed that a single point mutation (G37S) in the coding region for a highly conserved motif in the amino-terminal domain of the β subunit induces PHA-1. Grunder and co-authors (1997) showed that this domain is involved in the gating of ENaC. They identified that the mutation G37S in the gene for the β subunit and homologous mutations in the other subunit genes reduce channel function by changing the open probability.

Liddle syndrome has been linked genetically to mutations that delete or alter a conserved PY (proline-tyrosine) motif located in the carboxy-terminal domain of either β or γENaC (Hansson et al., 1995; Hansson et al., 1995; Shimkets et al., 1994; Tamura et al., 1996). Such deletions or point mutations lead to elevated channel function after expression in Xenopus oocytes, suggesting that the PY motif is involved in the regulation of activity and the density of ENaC channels at the cell surface (Firsov et al., 1996; Kellenberger et al., 1998; Schild et al., 1995; Schild et al., 1996; Shimkets et al., 1997). Mutations within the coding region for the PY motif were generated *in vitro* by directed mutagenesis. They have been widely studied to investigate the role of Nedd4-2 in the regulation of the number of ENaCs at the cell surface (Abriel and Horisberger, 1999; Debonneville et al., 2001; Kamynina and Staub, 2002;

Renauld et al., 2010; Staub et al., 1997). Thus, this has allowed the identification of tyrosine-carrying ubiquitin residues involved in Nedd4-2 dependent-internalization of the channel.

Intracellular C termini also harbor multiple phosphorylation sites and participate in the activity of the channel, suggesting that aldosterone, insulin, SGK1, PKA and PKC modulate the activity of ENaC by phosphorylation (Renauld et al., 2010; Shimkets et al., 1997).

4. Directed mutagenesis and regulation of ENaC by extracellular factors

Several members of the ENaC/degenerin family are clearly extracellular-ligand-gated channels. Numerous studies suggest that ENaC may also be a ligand-gated channel (Horisberger and Chraibi, 2004). A number of extracellular factors of various types have been shown to activate or inhibit ENaC. Amongst these factors, there are serine proteases, sodium itself, other inorganic cations, organic cations, and small molecules.

4.1 Activation by Serine proteases

In 1997 we cloned a serine protease that acts as a channel-activating protease, called CAP1 (Vallet et al., 1997); and we explored the mechanism by which it stimulates ENaC (Chraibi et al., 1998). We showed that the effect of CAP1 is done on the extracellular part of the channel, and it can be mimicked by trypsin or chymotrypsin. Ion selectivity, single channel conductance and channel density are not modified, which suggests that the serine proteases increase the open probability. During the last ten years, many studies showed that ENaC can be activated by other proteases, such as prostasin or furin (Hughey et al., 2004; Vuagniaux et al., 2000). Further progress in the understanding of the mechanism by which serine proteases activate ENaC has been made by functional investigation in heterologous expression systems combined with directed mutagenesis. Mutation of the CAP1 GPI-anchored consensus motif completely abolishes ENaC activation. However, catalytic mutants of CAP1 do not fully stimulate ENaC, suggesting that a noncatalytic mechanism is partly involved in this regulation pathway (Vallet et al., 2002). Thus, a putative site for CAP1 and trypsin action has been identified. However, there is no clear evidence of their role in the proteolytic activation of ENaC. Masilamani and co-authors (1999) first provided evidence for a possible cleavage of the γENaC subunit. These authors were able to show that the aldosterone infusion, or salt restriction, induced a shift in molecular weight of the gamma subunit from 85 to 70 KDa. Subsequently, it was shown that the serine proteases, including prostasin, plasmin, elastase and furin, cleave the extracellular domain of the α and γ subunits (Bruns et al., 2007; Caldwell et al., 2005; Hughey et al., 2004; Passero et al., 2008; Rossier, 2004; Vuagniaux et al., 2002). A basic motif (RKRK[186]) has been identified as a cleavage site for CAP1/Prostasin in the extracellular loop of γENaC (Bruns et al., 2007; Diakov et al., 2008). Additional cleavage sites within extracellular loop of α and γ subunits have been described (Garcia-Caballero et al., 2008; Myerburg et al., 2006). However, no site for furin was described in βENaC (see Figure 1).

4.2. Effects of extracellular sodium and other small molecules

4.2.1 Self-inhibition

We have shown that the external sodium exerts a fast inhibitory effect on ENaC activity, a phenomenon called sodium self-inhibition (Chraibi and Horisberger, 2002). We observed that the apparent affinity constant for the site responsible for self-inhibition was significantly lower, with a K½ of 100-200 mM. The kinetics of this phenomenon strongly depended on temperature and the extent of proteolytic processing of the ENaC subunits. We demonstrated that the effect of temperature was due to a large decrease in the probability of channel opening at high temperatures, while the unitary current increased with temperature (Chraibi and Horisberger, 2003). Later Sheng et al. (2004, 2002) showed that the mutation of His282 in the α subunit or His239 in the γ subunit (these amino acids reside in close proximity to the defined sites for furin cleavage) enhanced and eliminated the sodium self-inhibition response, respectively.

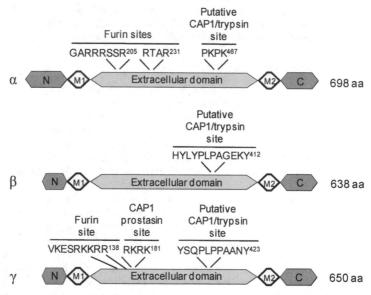

Fig. 1. Schematic representation of the rat ENaC subunits and their identified and putative sites for furin, CAP1, trypsin and prostasin. M1, M2: transmembrane domains; N, C: intracellular amino- and carboxy-termini, respectively.

4.2.2 Effects of cpt-cAMP and cpt-cGMP

cpt-cAMP, a membrane permeant cAMP analogue, has been described to be a species-dependent extracellular activator of ENaC. Rat and *Xenopus laevis* ENaC expressed in *Xenopus* oocytes are not sensitive to cpt-cAMP (Awayda et al., 1996). However, guinea pig (gp) channels could be activated by cpt-cAMP perfusion in the oocyte expression system (Liebold et al., 1996). The gp αENaC has been shown to be essential for this stimulation (Schnizler et al., 2000). However, the mechanism leading to ENaC stimulation did not exclude the possibility of an intracellular pathway involving protein kinase A (PKA). Further experiments demonstrated that PKA inhibitor PKI 6-22 did not prevent cpt-cAMP stimulation of gpENaC expressed in *Xenopus* oocytes. Furthermore, the α subunit containing the gp extracellular loop with rat intracellular C and/or N termini expressed in *Xenopus* oocytes together with rat βγ ENaC were sensitive to cpt-cAMP (Chraibi et al., 2001). This chimeric channel demonstrated that the extracellular domain of the gp α subunit was the determinant for ENaC stimulation by cpt-cAMP. Thus, the molecule can be considered to be a ligand for the channel. Moreover, the outside-out configuration of the patch clamp showed an increase of the open probability and the number of open channels (N.Po) exposed to cpt-cAMP, confirming a direct interaction with the extracellular domain of the gpα, ratβγ chimera expressed in *Xenopus* oocytes. To determine which part of the extracellular domain of αENaC is involved in this regulation, we made four chimeric constructions of that subunit (Figure 2).

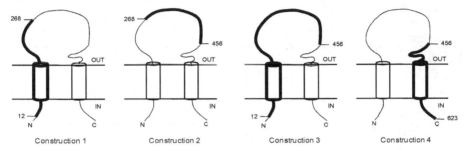

Construction 1   Construction 2   Construction 3   Construction 4

Fig. 2. Schematic representation of the ENaC subunits. Chimeric constructions of α subunits by fusion of the coding region for the αgp part (bold line) with the αrat part (thin line). Numbers indicate residues at corresponding positions on the αgp sequence.

To do so, two restriction sites were generated in guinea pig and rat αENaC cDNAs at homologous positions using a PCR technique. Then the appropriate fragment of gp cDNA was inserted into the rat cDNA between the restriction sites. Amiloride-sensitive current was measured in the presence and absence of 10 μM cpt-cAMP. We generated eleven swapping mutants of rat and gp αENaC using PCR-directed mutagenesis and expressed each of these mutants with the rat β and γ subunits in *Xenopus* oocytes. Among the eleven substitutions, Ile481 in the gp αENaC extracellular domain plays a major role in cpt-cAMP-induced ENaC activation. The *I481N* mutation in the gene for the αgp subunit completely abolished stimulation of ENaC. The *N510I* mutation in the gene for the αrat subunit caused intermediate sensitivity to cpt-cAMP. All other mutations or combination of mutations, including *N510I* in the αrat gene, did not increase the cpt-cAMP effect (Renauld et al., 2008).

Similarly to what we described with cpt-cAMP, Hong-Guan and coworkers (Nie et al., 2009) suggested that cpt-cGMP stimulates human, rat and mouse ENaC through direct interaction and not through the intracellular pathway. Indeed directed mutagenesis of the coding regions for potential phosphorylation sites for the cGMP-dependent kinases on ENaC did not affect cpt-cGMP-induced activation in *Xenopus* oocytes. Furthermore, knockdown of PKG isoforms did not prevent cpt-cGMP-dependent activation. Han and colleagues (2011) confirmed that cpt-cGMP-induced ENaC activation was mediated through direct interaction and an increases of N.Po. By directed mutagenesis, these authors were able to show that the mutations abolishing self-inhibition (β∆V348 and γH233R) lost their responses to cpt-cGMP. The mutations augmenting this phenomenon (αY458A and γM432G) facilitated the stimulatory effects of this compound. Thus, these data suggest that the elimination of self-inhibition may be a novel mechanism for cpt-cGMP to stimulate ENaC.

α subunit	Icpt/Ictl	SEM	unpaired t-Test VS α rat wt
α gp wt	2.28	0.05	P<0.001
α rat wt	1.13	0.01	
construction 1	1.14	0.01	NS
construction 2	1.28	0.03	P<0.001
construction 3	1.34	0.03	P<0.001
construction 4	1.85	0.06	P<0.001
αr L493S	1.06	0.02	NS
αr S500N	1.08	0.03	NS
αr S507N	1.06	0.03	NS
αr I509T	1.05	0.02	NS
αr N510I	1.45	0.03	NS
αr K524T	1.02	0.03	NS
αr E531Q	1.12	0.02	NS
αr N542S	1.05	0.04	NS
αr K550N	1.12	0.02	NS
αr F554Y	1.16	0.02	NS
αr K561R	1.12	0.02	NS
αgp I481N	1.58	0.07	P<0.001

Table 1. Effect of cpt-cAMP on different constructions and mutants of the αENaC subunit expressed in *Xenopus* oocytes together with the β and γrat subunits. Results are presented as a ratio of amiloride-sensitive current measured after and before cpt-cAMP perfusion (Icpt/Ictl). gp, guinea pig; r, rat; wt, wild type; NS, not significant relative to αrat wt

4.2.3 Effects of glibenclamide

The same experimental approach was used to study the stimulation of ENaC by glibenclamide (Renauld and Chraibi, 2009). Glibenclamide, a high affinity-blocker of the K_{ATP} channel, has been shown to stimulate *Xenopus* ENaC (but not rat ENaC) expressed in *Xenopus* oocytes. The α subunit has been shown to be critical for this activation (Chraibi and Horisberger, 1999). As described with cpt-cAMP, patch clamp recordings in the outside-out configuration showed an increase of N.Po when *Xenopus* ENaC was exposed to glibenclamide. Another study has demonstrated that the αgp subunit, but not the αrat subunit, conferred sensitivity of ENaC to glibenclamide (Schnizler et al., 2003). Using mutagenesis, these authors were able to produce other chimeric rat/gp α subunits; and they suggested that the extracellular loop or the transmembrane domain of the αgp subunit is involved in the activation of the ENaC channel by glibenclamide. Thus, similarly to cpt-cAMP activation, channels composed of the αgp subunit and the β and γ subunits from rat are sensitive to glibenclamide, while channels composed of the α, β, and γ subunits from rat are resistant. We used the chimeras of the α subunit previously generated and found that construction 4 was also important for glibenclamide stimulation of the channel. Unlike cpt-cAMP, glibenclamide had no effect on the other constructions expressed with the β and γ subunits from rat. Moreover, directed mutagenesis did not reveal particular residues involved in this regulation.

α subunit	Iglib/Ictrl	SEM	unpaired t-Test VS α rat wt
α gp wt	1.63	0.02	P<0.001
α rat wt	0.96	0.01	
construction 1	1.03	0.04	NS
construction 2	1.04	0.05	NS
construction 3	1.03	0.01	NS
construction 4	1.27	0.04	P<0.001
αr L493S	0.92	0.03	NS
αr S500N	1.01	0.03	NS
αr S507N	0.98	0.02	NS
αr I509T	0.99	0.02	NS
αr N510I	1.00	0.01	NS
αr K524T	0.97	0.02	NS
αr E531Q	0.99	0.05	NS
αr N542S	0.98	0.02	NS
αr K550N	0.91	0.01	NS
αr F554Y	0.99	0.02	NS
αr K561R	0.96	0.02	NS

Table 2. Effect of glibenclamide on different constructions and mutants of αENaC expressed in *Xenopus* oocytes together with the β and γ subunits from rat. Results are presented as a ratio of amiloride-sensitive current measured after and before glibenclamide perfusion (Iglib/Ictl). gp, guinea pig; r, rat; wt, wild type; NS, not significant relative to αrat wt

4.2.4 Effects of other molecules

Capsazepine has been described as the first selective active activator for the δENaC subunit (Yamamura et al., 2004). Indeed capsazepine specifically stimulates human-made δ subunit, but not the α subunit expressed in *Xenopus* oocytes. Moreover, this molecule can stimulate the δENaC monomer, whereas no other vanilloid compound can produce changes in sodium amiloride-sensitive current. However, the authors did not determine any amino acids involved in this activation. Directed mutagenesis could be a powerful tool to understand differences between the α and δ subunits and resolve the structure-function relationships of both proteins.

S3969 is a small molecule described as a reversible activator of human, but not mouse, αβγ ENaC through direct interaction with the extracellular side of the channel by increasing N.Po (Lu et al., 2008). Interestingly, S3969 stimulates amiloride-sensitive current in oocytes expressing the δ subunit instead of α. The authors showed that βENaC was critical for this activation. Mouse-human chimeras of the β subunit confirmed the implication of the extracellular domain. More specifically, deletion of Val348 in βENaC completely abolished S3969 activation of ENaC. Maturation and optimal transport of ENaC to the plasma membrane requires furin cleavage of the β and γ subunits at a specific Arg. Mutations of the furin cleavage site in which Arg was replaced by Ala did not prevent ENaC activation by S3969, suggesting that proteolytic activation prior to S3969 stimulation is not necessary. Mutations producing pseudohypoaldosteronism type 1 (PHA1), resulting in salt-wasting, a genetic disease, have been generated in the α (*R508STOP*) and β (*G37S*) subunits. These mutants decreased amiloride-sensitive current, but the S3969 compound was still able to stimulate ENaC activity.

5. Conclusion

The epithelial sodium channel (ENaC) has been used for decades as a therapeutic target against type 1 hypertension and Liddle syndrome. More recently, several studies pointed to ENaC as a potential target for cystic fibrosis (Zhou et al., 2011), a pathology characterized by an impaired Cl- secretion through the cystic fibrosis transmembrane conductance regulator (CFTR) and an increase of Na+ reabsorption through ENaC. The studies of mutations involved in these diseases have been extremely helpful in determining the molecular mechanisms by which they lead to a dysfunction of ENaC. Furthermore, the experiments carried out on this topic have shown the contribution of the PCR-directed mutagenesis technique in the determination of the structure-function relationships of ENaC. These studies have led to a better understanding of the domains involved in ion selectivity, gating and expression of the channel at the cell membrane. Additional studies are needed to define other key domains of ENaC. They may provide a new strategy for the treatment of pathologies linked to dysfunction of this channel.

6. References

Abriel, H., Horisberger, J.D. 1999. Feedback inhibition of rat amiloride-sensitive epithelial sodium channels expressed in Xenopus laevis oocytes. *J Physiol* 516:31-43.

Anantharam A, P.L. 2007. Determination of epithelial na+ channel subunit stoichiometry from single-channel conductances. *J Gen Physiol*. 130:55-70

Awayda, M.S., Ismailov, II, Berdiev, B.K., Fuller, C.M., Benos, D.J. 1996. Protein kinase regulation of a cloned epithelial Na+ channel. *J Gen Physiol* 108:49-65

Brouard, M., Casado, M., Djelidi, S., Barrandon, Y., Farman, N. 1999. Epithelial sodium channel in human epidermal keratinocytes: expression of its subunits and relation to sodium transport and differentiation. *J Cell Sci* 112 (Pt 19):3343-52

Bruns, J.B., Carattino, M.D., Sheng, S., Maarouf, A.B., Weisz, O.A., Pilewski, J.M., Hughey, R.P., Kleyman, T.R. 2007. Epithelial Na+ channels are fully activated by furin- and prostasin-dependent release of an inhibitory peptide from the gamma-subunit. *J Biol Chem* 282:6153-60

Caldwell, R.A., Boucher, R.C., Stutts, M.J. 2005. Neutrophil elastase activates near-silent epithelial Na+ channels and increases airway epithelial Na+ transport. *Am J Physiol Lung Cell Mol Physiol* 288:L813-9

Canessa, C.M., Horisberger, J.D., Rossier, B.C. 1993. Epithelial sodium channel related to proteins involved in neurodegeneration. *Nature* 361:467-70.

Canessa, C.M., Schild, L., Buell, G., Thorens, B., Gautschi, I., Horisberger, J.D., Rossier, B.C. 1994. Amiloride-sensitive epithelial Na+ channel is made of three homologous subunits. *Nature* 367:463-7.

Chang, S.S., Grunder, S., Hanukoglu, A., Rosler, A., Mathew, P.M., Hanukoglu, I., Schild, L., Lu, Y., Shimkets, R.A., Nelson-Williams, C., Rossier, B.C., Lifton, R.P. 1996. Mutations in subunits of the epithelial sodium channel cause salt wasting with hyperkalaemic acidosis, pseudohypoaldosteronism type 1. *Nat Genet* 12:248-53.

Chraibi, A., Horisberger, J.D. 1999. Stimulation of epithelial sodium channel activity by the sulfonylurea glibenclamide. *J Pharmacol Exp Ther* 290:341-7.

Chraibi, A., Horisberger, J.D. 2002. Na self inhibition of human epithelial Na channel: temperature dependence and effect of extracellular proteases. *J Gen Physiol* 120:133-45.

Chraibi, A., Horisberger, J.D. 2003. Dual effect of temperature on the human epithelial Na+ channel. *Pflugers Arch* 447:316-320

Chraibi, A., Schnizler, M., Clauss, W., Horisberger, J.D. 2001. Effects of 8-cpt-cAMP on the epithelial sodium channel expressed in Xenopus oocytes. *J Membr Biol* 183:15-23.

Debonneville, C., Flores, S.Y., Kamynina, E., Plant, P.J., Tauxe, C., Thomas, M.A., Munster, C., Chraibi, A., Pratt, J.H., Horisberger, J.D., Pearce, D., Loffing, J., Staub, O. 2001. Phosphorylation of Nedd4-2 by Sgk1 regulates epithelial Na(+) channel cell surface expression. *Embo J* 20:7052-9.

Diakov, A., Bera, K., Mokrushina, M., Krueger, B., Korbmacher, C. 2008. Cleavage in the {gamma}-subunit of the epithelial sodium channel (ENaC) plays an important role in the proteolytic activation of near-silent channels. *J Physiol* 586:4587-608

Dijkink, L., Hartog, A., van Os, C.H., Bindels, R.J. 2002. The epithelial sodium channel (ENaC) is intracellularly located as a tetramer. *Pflugers Arch* 444:549-55

Duc, C., Farman, N., Canessa, C.M., Bonvalet, J.P., Rossier, B.C. 1994. Cell-specific expression of epithelial sodium channel alpha, beta, and gamma subunits in aldosterone-responsive epithelia from the rat: localization by in situ hybridization and immunocytochemistry. *J Cell Biol* 127:1907-21

Eskandari, S., Snyder, P.M., Kreman, M., Zampighi, G.A., Welsh, M.J., Wright, E.M. 1999. Number of subunits comprising the epithelial sodium channel. *J Biol Chem* 274:27281-6

Firsov, D., Gautschi, I., Merillat, A.M., Rossier, B.C., Schild, L. 1998. The heterotetrameric architecture of the epithelial sodium channel (ENaC). *Embo J* 17:344-52.

Firsov, D., Schild, L., Gautschi, I., Merillat, A.M., Schneeberger, E., Rossier, B.C. 1996. Cell surface expression of the epithelial Na channel and a mutant causing Liddle syndrome: a quantitative approach. *Proc Natl Acad Sci U S A* 93:15370-5.

Garcia-Caballero, A., Dang, Y., He, H., Stutts, M.J. 2008. ENaC proteolytic regulation by channel-activating protease 2. *J Gen Physiol* 132:521-35

Grunder, S., Firsov, D., Chang, S.S., Jaeger, N.F., Gautschi, I., Schild, L., Lifton, R.P., Rossier, B.C. 1997. A mutation causing pseudohypoaldosteronism type 1 identifies a conserved glycine that is involved in the gating of the epithelial sodium channel. *Embo J* 16:899-907.

Han, D.Y., Nie, H.G., Su, X.F., Shi, X.M., Bhattarai, D., Zhao, M., Zhao, R.Z., Landers, K., Tang, H., Zhang, L., Ji, H.L. 2011. CPT-cGMP Stimulates Human Alveolar Fluid Clearance by Releasing External Na+ Self-Inhibition of ENaC. *Am J Respir Cell Mol Biol*

Hansson, J.H., Nelson-Williams, C., Suzuki, H., Schild, L., Shimkets, R., Lu, Y., Canessa, C., Iwasaki, T., Rossier, B., Lifton, R.P. 1995. Hypertension caused by a truncated epithelial sodium channel gamma subunit: genetic heterogeneity of Liddle syndrome. *Nat Genet* 11:76-82.

Hansson, J.H., Schild, L., Lu, Y., Wilson, T.A., Gautschi, I., Shimkets, R., Nelson-Williams, C., Rossier, B.C., Lifton, R.P. 1995. A de novo missense mutation of the beta subunit of the epithelial sodium channel causes hypertension and Liddle syndrome, identifying a proline-rich segment critical for regulation of channel activity. *Proc Natl Acad Sci U S A* 92:11495-9.

Horisberger, J.D., Chraibi, A. 2004. Epithelial Sodium Channel: A Ligand-Gated Channel? *Nephron Physiology* 96:37-41

Hughey, R., Bruns, J., Kinlough, C., Harkleroad, K., Tong, Q., Carattino, M., Johnson, J., Stockand, J., Kleyman, T. 2004. Epithelial sodium channels are activated by furin-dependent proteolysis. *J. Biol. Chem.* 30:18111-4

Hummler, E., Barker, P., Gatzy, J., Beermann, F., Verdumo, C., Schmidt, A., Boucher, R., Rossier, B.C. 1996. Early death due to defective neonatal lung liquid clearance in alpha- ENaC-deficient mice. *Nat Genet* 12:325-8.

Jasti, J., Furukawa, H., Gonzales, E.B., Gouaux, E. 2007. Structure of acid-sensing ion channel 1 at 1.9A° resolution and low pH. *nature* 449:316-324

Kamynina, E., Staub, O. 2002. Concerted action of ENaC, Nedd4-2, and Sgk1 in transepithelial Na(+) transport. *Am J Physiol Renal Physiol* 283:F377-87

Kellenberger, S., Gautschi, I., Rossier, B.C., Schild, L. 1998. Mutations causing Liddle syndrome reduce sodium-dependent downregulation of the epithelial sodium channel in the Xenopus oocyte expression system. *J Clin Invest* 101:2741-50.

Kellenberger, S., Schild. L. 2002. Epithelial sodium channel/degenerin family of ion channels: a variety of functions for a shared structure. *Physiol Rev.* 82:735-67

Kunzelmann, K., Mall, M. 2002. Electrolyte transport in the mammalian colon: mechanisms and implications for disease. *Physiol Rev* 82:245-89

Liebold, K.M., Reifarth, F.W., Clauss, W., Weber, W. 1996. cAMP-activation of amiloride-sensitive Na+ channels from guinea-pig colon expressed in Xenopus oocytes. *Pflugers Arch* 431:913-22

Lingueglia, E., Voilley, N., Waldmann, R., Lazdunski, M., Barbry, P. 1993. Expression cloning of an epithelial amiloride-sensitive Na+ channel. A new channel type with homologies to Caenorhabditis elegans degenerins. *FEBS Lett* 318:95-9

Lu M, Echeverri F, Kalabat D, Laita B, Dahan DS, Smith RD, Xu H, Staszewski L, Yamamoto J, Ling J, Hwang N, Kimmich R, Li P, Patron E, Keung W, Patron A, Moyer BD. 2008. Small molecule activator of the human epithelial sodium channel. *J Biol Chem* 283: 11981–11994

Masilamani, S., Kim, G.H., Mitchell, C., Wade, J.B., Knepper, M.A. 1999. Aldosterone-mediated regulation of ENaC alpha, beta, and gamma subunit proteins in rat kidney. *J Clin Invest* 104:R19-23.

Mullis, K.B., Faloona, F.A. 1987. Specific synthesis of DNA in vitro via a polymerase-catalyzed chain reaction. *Methods Enzymol* 155:335-50

Myerburg, M.M., Butterworth, M.B., McKenna, E.E., Peters, K.W., Frizzell, R.A., Kleyman, T.R., Pilewski, J.M. 2006. Airway surface liquid volume regulates ENaC by altering the serine protease-protease inhibitor balance: a mechanism for sodium hyperabsorption in cystic fibrosis. *J Biol Chem* 281:27942-9

Nie, H.G., Chen, L., Han, D.Y., Li, J., Song, W.F., Wei, S.P., Fang, X.H., Gu, X., Matalon, S., Ji, H.L. 2009. Regulation of epithelial sodium channels by cGMP/PKGII. *J Physiol* 587:2663-76

Passero, C.J., Mueller, G.M., Rondon-Berrios, H., Tofovic, S.P., Hughey, R.P., Kleyman, T.R. 2008. Plasmin activates epithelial Na+ channels by cleaving the gamma subunit. *J Biol Chem* 283:36586-91

Perucca, J., Bichet, D.G., Bardoux, P., Bouby, N., Bankir, L. 2008. Sodium excretion in response to vasopressin and selective vasopressin receptor antagonists. *J Am Soc Nephrol* 19:1721-31

Randell, S.H., Boucher, R.C. 2006. Effective mucus clearance is essential for respiratory health. *Am J Respir Cell Mol Biol* 35:20-8

Renauld S, Allache R, Chraibi A. 2008. Ile481 from the guinea pig alpha-subunit plays a major role in the activation of ENaC by cpt-cAMP. *Cell Physiol Biochem.* 22(1-4):101-8

Renauld, S., Chraibi, A. 2009. Role of the C-Terminal Part of the Extracellular Domain of the alpha-ENaC in Activation by Sulfonylurea Glibenclamide. *J Membr Biol*

Renauld, S., Tremblay, K., Ait-Benichou, S., Simoneau-Roy, M., Garneau, H., Staub, O., Chraibi, A. 2010. Stimulation of ENaC activity by rosiglitazone is PPARgamma-dependent and correlates with SGK1 expression increase. *J Membr Biol* 236:259-70

Rossier, B. 2004. The epithelial sodium channel: activation by membrane-bound serine proteases. *Proc Am Thorac Soc:* 1:4-9

Roudier-Pujol, C., Rochat, A., Escoubet, B., Eugene, E., Barrandon, Y., Bonvalet, J.P., Farman, N. 1996. Differential expression of epithelial sodium channel subunit mRNAs in rat skin. *J Cell Sci* 109 (Pt 2):379-85

Schild, L., Canessa, C.M., Shimkets, R.A., Gautschi, I., Lifton, R.P., Rossier, B.C. 1995. A mutation in the epithelial sodium channel causing Liddle disease increases channel activity in the Xenopus laevis oocyte expression system. *Proc Natl Acad Sci U S A* 92:5699-703

Schild, L., Lu, Y., Gautschi, I., Schneeberger, E., Lifton, R.P., Rossier, B.C. 1996. Identification of a PY motif in the epithelial Na channel subunits as a target sequence for mutations causing channel activation found in Liddle syndrome. *Embo J* 15:2381-7

Schnizler, M., Berk, A., Clauss, W. 2003. Sensitivity of oocyte-expressed epithelial Na+ channel to glibenclamide. *Biochim Biophys Acta* 1609:170-6

Schnizler, M., Mastroberardino, L., Reifarth, F., Weber, W.M., Verrey, F., Clauss, W. 2000. cAMP sensitivity conferred to the epithelial Na+ channel by alpha- subunit cloned from guinea-pig colon. *Pflugers Arch* 439:579-87.

Sheng, S., Bruns, J., Kleyman, T. 2004. Extracellular histidine residues crucial for Na+ self-inhibition of epithelial Na+ channels. *JBC* 279:9743-9

Sheng, S., Perry, C., Kleyman, T. 2002. External nickel inhibits epithelial sodium channel by binding to histidine residues within the extracellular domains of alpha and gamma subunits and reducing channel open probability. *J Biol. Chem.* 277:50098-111

Shimkets, R.A., Lifton, R.P., Canessa, C.M. 1997. The activity of the epithelial sodium channel is regulated by clathrin- mediated endocytosis. *J Biol Chem* 272:25537-41.

Shimkets, R.A., Warnock, D.G., Bositis, C.M., Nelson-Williams, C., Hansson, J.H., Schambelan, M., Gill, J.R., Jr., Ulick, S., Milora, R.V., Findling, J.W., et al. 1994. Liddle's syndrome: heritable human hypertension caused by mutations in the beta subunit of the epithelial sodium channel. *Cell* 79:407-14

Snyder, P.M., Cheng, C., Prince, L.S., Rogers, J.C., Welsh, M.J. 1998. Electrophysiological and biochemical evidence that DEG/ENaC cation channels are composed of nine subunits. *J Biol Chem* 273:681-4

Staub, O., Gautschi, I., Ishikawa, T., Breitschopf, K., Ciechanover, A., Schild, L., Rotin, D. 1997. Regulation of stability and function of the epithelial Na+ channel (ENaC) by ubiquitination. *EMBO J* 16:6325-6336

Stockand, J.D., Staruschenko, A., Pochynyuk, O., Booth, R.E., Silverthorn, D.U. 2008. Insight Toward Epithelial Na1 Channel Mechanism Revealed by the Acid-sensing Ion Channel 1 Structure. *IUBMB Life* 60: 620-628

Tamura, H., Schild, L., Enomoto, N., Matsui, N., Marumo, F., Rossier, B.C. 1996. Liddle disease caused by a missense mutation of beta subunit of the epithelial sodium channel gene. *J Clin Invest* 97:1780-4

Vallet, V., Chraibi, A., Gaeggeler, H.P., Horisberger, J.D., Rossier, B.C. 1997. An epithelial serine protease activates the amiloride-sensitive sodium channel. *Nature* 389:607-10.

Vallet, V., Pfister, C., Loffing, J., Rossier, B.C. 2002. Cell-surface expression of the channel activating protease xCAP-1 is required for activation of ENaC in the Xenopus oocyte. *J Am Soc Nephrol* 13:588-94

Vuagniaux, G., Vallet, V., Jaeger, N.F., Hummler, E., Rossier, B.C. 2002. Synergistic Activation of ENaC by Three Membrane-bound Channel- activating Serine Proteases (mCAP1, mCAP2, and mCAP3) and Serum- and Glucocorticoid-regulated Kinase (Sgk1) in Xenopus Oocytes. *J Gen Physiol* 120:191-201.

Vuagniaux, G., Vallet, V., Jaeger, N.F., Pfister, C., Bens, M., Farman, N., Courtois-Coutry, N., Vandewalle, A., Rossier, B.C., Hummler, E. 2000. Activation of the amiloride-sensitive epithelial sodium channel by the serine protease mCAP1 expressed in a mouse cortical collecting duct cell line. *J Am Soc Nephrol* 11:828-34.

Yamamura, H., Ugawa, S., Ueda, T., Nagao, M., Shimada, S. 2004. Protons activate the delta -subunit of epithelial Na+ channel in humans. *JBC* in preess (published on january 15, manuscript M400274200)

Zhou, Z., Duerr, J., Johannesson, B., Schubert, S.C., Treis, D., Harm, M., Graeber, S.Y., Dalpke, A., Schultz, C., Mall, M.A. 2011. The ENaC-overexpressing mouse as a model of cystic fibrosis lung disease. *J Cyst Fibros* 10 Suppl 2:S172-82

A Mutagenesis Approach for the Study of the Structure-Function Relationship of Human Immunodeficiency Virus Type 1 (HIV-1) Vpr

Kevin Hadi[1], Oznur Tastan[2],
Alagarsamy Srinivasan[3] and Velpandi Ayyavoo[1,*]
[1]University of Pittsburgh, Pittsburgh, PA,
[2]Bilkent University, Ankara,
[3]NanoBio Diagnostics, West Chester, PA,
[1]USA
[2]Turkey

1. Introduction

Before the era of molecular biology, the methods available for an understanding of gene function were limited. Such studies typically relied on the ability to identify and isolate naturally occurring variants exhibiting a defect in function. Hence, the progress was slow in this regard. The discoveries in the field of microbiology, combined with advances in technology in the later part of 20th century, dramatically changed this scenario. The current molecular biological techniques enable site-directed mutagenesis approaches for generating a gene with a specific amino acid substitution, mutation, or a deletion or for truncating a gene anywhere in a matter of days.

The present review highlights the studies conducted on human immunodeficiency virus type 1 (HIV-1) Vpr, an auxiliary protein associated with virus particles. Vpr contains 96 amino acids, and it is a multifunctional protein. To analyze the contribution of specific residues to protein function and to the cytopathic effects in HIV-1-infected individuals, investigators from several laboratories, including ours, took advantage of a novel approach. Specifically, this approach involved exchange of residues through mutagenesis. The choice of residue was based on the information available regarding the naturally occurring polymorphisms at the level of individual amino acids in Vpr. The results from these studies support a link between polymorphisms in these genes and the disease status of infected individuals, who are known as progressors or non-progressors. In addition the studies have shed light on the structure-function relationship of Vpr.

The Joint United Nations Program on HIV/AIDS (UNAIDS) (2010) reports that the worldwide prevalence of those living with human immunodeficiency virus (HIV) is between 31-35 million as of the end of 2009. Roughly 2.6 million new cases of HIV infection occurred in the same year. That number has likely remained approximately constant.

---

* Corresponding Author

Although the most common route of HIV infection is via sexual contact, the use of contaminated drug paraphernalia, mother-child transmission via pregnancy or breastfeeding, and tainted blood transfusions comprise other means of infection.

The symptomatic outcome of infection is AIDS, usually occurring ~10 years after initial infection. CD4+ T-cell counts drop below 200, and subsequent severe immune dysfunction results. This eventually leads to fatal coinfection by opportunistic pathogens. The advent of highly active retroviral therapy (HAART) in the 1990s led to a drastically improved prognosis for AIDS patients (Peters and Conway, 2011). This triple-drug cocktail controls viremia and allows immune function to recover to nearly uninfected levels, with the caveat of near-perfect patient adherence to a difficult combination of drug regimens. The extremely rigid treatment schedules have resulted in low compliance, leading to the emergence of viruses that exhibit resistance to drugs.

2. Genetic organization of HIV

The causative agents of AIDS have been identified as human immunodeficiency virus types 1 (HIV-1) and 2 (HIV-2). Both HIV-1 and HIV-2 are members of the lentivirus family of retroviruses. HIV-1 is the predominant virus responsible for AIDS throughout the world. The schematic representation of the genome organization of HIV-1 is shown in **Figure 1.** The genome of HIV-1 codes for two regulatory proteins (Tat and Rev) and four auxiliary proteins (Vif, Vpr Vpu and Nef), in addition to structural proteins Gag, Gag-Pol and Env. The genome organization of HIV-2 is similar to HIV-1. The unique genes are *vpu* and *vpx* for HIV-1 and HIV-2, respectively. With respect to viral morphogenetic events, the lentiviruses are similar to alpharetroviruses, gammaretroviruses and deltaretroviruses. During virus infection, Gag and Gag-Pol proteins are synthesized in the cytoplasm and transported to the cell membrane, where virus assembly occurs. In the case of HIV-1, the non-structural proteins Vpr, Vif, and Nef are also packaged into the virus particles.

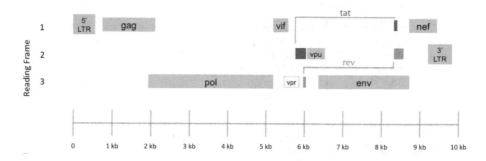

Fig. 1. Organization of the HIV-1 Genome.

3. Heterogeneity in the human immunodeficiency virus

Within HIV-1-positive patients, the high error rate of reverse transcription can produce many variants, or quasispecies. After seroconversion to HIV-positive status, the viral loads measure at 10000-50000 viral RNA copies per ml of patient sera during the asymptomatic stage; in the later stages of disease, viral RNA genomes can increase to several million copies per ml (Poropatich and Sullivan, 2011; Tungaturthi et al., 2004). Thus, a high replication rate produces tremendous variation in the viral population within a single patient. The pressures of the host immune response drive the selection of variants from early stages of infection, until the host immune response cannot cope with the viral diversity (Fischer et al., 2010; Wolinsky et al., 1996).

Studies of polymorphisms in a number of the genes of HIV-1 confirm the propensity of the virus to escape host immunity (Fischer et al., 2010; Fischer et al., 2007; Gaschen et al., 2002; Korber et al., 2001). As expected, the antigenic Env protein exhibits the highest mutability. Comparisons of current strains of HIV show a staggering 20% variability within a subtype and up to 35% variability between subtypes (Gaschen et al., 2002). To highlight the problem of this variability in vaccine development, the evolutionary dynamics of influenza virus provides a revealing picture. The influenza genome varies by 1-2% per year, enabling influenza virus escape from polyclonal vaccine responses and necessitating annual vaccine changes (Fischer et al., 2007; Gaschen et al., 2002). Along with intra- and inter-subtype variation in HIV, *env* shows on average a 10% change in genetic diversity over the course of an infection in a single patient (Korber et al., 2001). In addition to the *env* gene, Korber et al. (2001) estimated the percent variation in sequences culled from an HIV-1 database compared to the HXB2 strain consensus in *tat* and *gag*. They found 9% and 5% variation, respectively. From another study, *gag* and *pol* remained relatively conserved; but the rest of the genes exhibited high variability comparable to that of *env* (Gaschen et al., 2002) **(Figure 1)**. The vast genetic diversity seen in these studies thus far exemplifies the major obstacle to vaccine development.

The variation in the viral genome throughout the population potentially correlates with the variation among HIV-infected individuals. In the progressor group, in which AIDS develops 10 years after initial infection, CD4+ T-cell counts (the primary marker of AIDS progression) generally fall below 200 cells per µl and causing loss of cell-mediated immunity (Levy, 2009). Along with the normal progressors, several other categories of disease progression exist. Long term non-progressors (LTNPs) and elite controllers (EC) do not receive HAART and do not show the aforementioned clinical signs of AIDS for up to 20 years after infection. CD4+ T-cell counts are maintained to levels above 350 cells per µl and viral loads to under 2000 copies per ml. Viral loads in these groups can range as low as 50-2000 copies per ml (Poropatich and Sullivan, 2011). In rapid progressors (RP), CD4+ T-cell counts generally decline 3-8 years postinfection. Although individual host genetics likely play a role in differential disease progression (Goulder et al., 1997), the presence of multiple quasispecies in patients potentially explains the differences in disease outcome.

4. Polymorphisms in specific HIV-1 genes

Several genes of HIV have a functional role in disease symptoms (Caly et al., 2008; Casartelli et al., 2003; Tolstrup et al., 2006). As Nef downregulates MHC class I antigen presentation involved in generating a cytotoxic T-lymphocyte (CTL) response, polymorphisms in this gene correlate with the LTNP group (Caly et al., 2008; Casartelli et al., 2003; Tolstrup et al., 2006). Similarly the role of Tat as a transactivator of the HIV-1 promoter in the long-terminal-repeat (LTR) region and its genetic heterogeneity suggest a functional role for Tat in disease progression and outcome (Bratanich et al., 1998; Korber et al., 2001). Tat can upregulate LTR-transcriptional activity and thus replication. Cellular and viral transcription increases several hundredfold (Irish et al., 2009) in the presence of Tat. Various Tat proteins isolated from different subtypes show differences in the ability to induce viral and cellular gene transcription (Roof et al., 2002).

The multifunctional Viral Protein R (Vpr), although labeled as an accessory protein, has a vital role in efficient replication of the virus in the non-dividing macrophages and monocytes (Balliet et al., 1994; Connor et al., 1995). Vpr interacts with the Gag protein for packaging into the virion. It remains bound to the preintegration complex (PIC) upon entry of the viral contents into the target cell during infection, most likely aiding PIC entry into the nucleus. It has a highly functional role in replication as a transactivator of the LTR, binding to a number of host transcription factors, such as Sp1 (Sawaya et al., 1998) and TFIIB (Agostini et al., 1999). Extensive studies have shown that coactivation of glucocorticoid receptor by Vpr, resulting in transcription of elements in the promoter regions of the HIV LTR and host genes, enhances viral replication and downregulates host immune responses (Ayyavoo et al., 1997; Hapgood and Tomasicchio, 2010). Vpr also drives apoptosis and G2 cell cycle arrest (Morellet et al., 2009; Pandey et al., 2009; Tungaturthi et al., 2004). The processes enhance immune escape and HIV-1 replication, respectively. As Vpr seems to be essential in the viral life cycle, it is likely that *vpr* gene sequences will exhibit signature changes among the categories of HIV-1 disease progression. Due to the importance of Vpr, this review focuses mainly on the polymorphisms in *vpr*, their effects on Vpr function, and their potential consequences on disease outcome. The review will also show how using an *in vitro* model to study polymorphisms may help us better understand the mechanisms and develop therapeutics for treatment.

5. Methods used to generate mutations to study gene regulation

Identifying genes responsible for different phenotypes previously relied on isolating naturally occurring polymorphic genes within the organisms. This process is time-consuming and requires sequencing several isolates to identify the functionally relevant genes. Today there are many different means of following the general path of altering the genotype to observe the phenotype.

5.1 Site-directed mutagenesis

The use of a point mutation to observe the change of function that occurs as a result of a change in structure underscores the majority of studies on the role of polymorphisms in viral and host gene function. One method of generating a targeted mutation in a plasmid construct carrying the gene of interest uses a polymerase chain reaction (PCR) technique. This particular PCR-based method has the advantage of using a high-fidelity polymerase and methylation of DNA by common strains of *Escherichia coli*. The protocol is described as part of the commercially available QuikChange II™ Site-Directed Mutagenesis Kit (Agilent, 2010). PCR results in the amplification of daughter DNA strands that carry the mutation of interest (Agilent, 2010). In the initial PCR amplification step, two primers containing the mutation of interest bind to the complementary strands of the plasmid. This requires a mismatch of at least one nucleotide on the primers and, furthermore, optimization of the length of the primer for a suitable T^m for the annealing step of the PCR. The mismatch limits the number of mutations that can be made per reaction to about four bases, depending on the length and T^m of the primer.

PCR will copy the entire parental template. The high-fidelity *Pfu Ultra* polymerase has an error rate of one per 2.5×10^6 nucleotides (Agilent, 2011), making the errors introduced by PCR a non-issue. Once the amplification of the template is complete, the PCR product contains the original parental DNA strands and the mutated and amplified daughter DNA strands. To select against the parental strands, the protocol takes advantage of the fact that the mutated and amplified daughter strands are unmethylated. In contrast, the parental (template) strands are methylated, as they originated from methylation-capable *E. coli*. Digesting the PCR product with *Dpn*I endonuclease, which digests methylated and hemi-methylated DNA, will cleave the parental strands and leave the daughter strands intact. Since the amplified DNA has the plasmid sequence of the parent and since there is an overlap region to allow circularization by homologous recombination, plasmids with the mutation are then isolated by transformation. Any candidate must then be screened and sequenced to confirm the mutagenesis.

6. Structure and function relationship of HIV-1 Vpr

To arrive at the structure of Vpr, several NMR studies have used fragments representing different segments of the Vpr protein. Alternatively the entire Vpr molecule has been analyzed in an appropriate solvent (Morellet et al., 2003; Wecker et al., 2002). The analyses showed a flexible N-terminus with a turn, the first alpha helix, turn, the second alpha helix, turn, the third alpha helix, then a flexible C-terminal domain (Morellet et al., 2003; Wecker et al., 2002). Morellet et al. (2003) used a different solvent, allowing them to "see" the tertiary structure of Vpr. The solvent was better at revealing the hydrophobic parts that are implicated in dimerization and interaction with other proteins (Morellet et al., 2003). Ideally NMR studies of protein structure employ solvents and conditions that approximate physiological conditions. However, the oligomerization property of Vpr makes proper solvation extremely difficult. Several studies have used trifluoroethanol (TFE), a hydrophobic solvent, in different proportions to counteract the interaction of the hydrophobic domains of Vpr, some of which cause Vpr to oligomerize (Engler et al., 2001; Wecker et al., 2002). Across these studies the length of the alpha helices differed due to the proportion of TFE in the solvents used.

Morellet et al. (2003) used CD3CN, a solvent with little hydrophobicity that approaches physiological conditions. This solvent allowed the following structural analysis: alpha helices 1, 2, and 3 (17-33, 38-50, 54-77), a flexible N-terminus (1-16) and a basic C-terminus (78-96). Each helix is amphipathic, containing hydrophobic and hydrophilic residues. The hydrophobic residues can allow for interactions with other proteins, most likely other cellular factors. In the third helix, L60, L67, I74, and I81 can form a leucine zipper (Morellet et al., 2003). The N-terminus consists largely of acidic residues, while the C-terminus consists largely of basic residues, arginine being the most prominent. The 3-D structure established by Morellet et al. (2003; 2009) shows a globular structure, in which hydrophobic interactions between the helices form a lipophilic core. Each of the helices in its amphipathic portions has acidic/basic residues, which are on the external face of the modeled protein. They possibly provide contacts for interactions with other proteins. The folding of the alpha helices entails that hydrophobic portions of the protein face outwards in contact with the solvent. These unfavorable conditions could be ameliorated by binding to cellular partners. Mutational analyses of the protein have found many residues throughout the length of Vpr that maintain the structural integrity (Morellet et al., 2009). Single deletions of Y15 (see Table 1 in the chapter by Figurski et al. for the amino acid codes), K27, and Q44 result in disruption of structure. The helical domains contain residues crucial to structure, especially those that form the hydrophobic core. The basic C-terminal residues also affect structure and stability of Vpr.

6.1 Oligomerization

The property of forming Vpr oligomers has recently come under scrutiny. Zhao et al. (1994) incorporated mutagenesis in their study to elucidate a rough map of residues in Vpr necessary for this function. Deletion of individual residues in the 36-76 region diminishes formation of oligomers. Mutagenesis of individual resides in the leucine-isoleucine motif (^{60}LIRILQQLLFIHFR) in the third helix (amino acids 55-77) reduced the self-associative capacity of Vpr monomers. The residues at position 60, 61, 63, 64, 67, 68, 70, and 74 mutated from leucine or isoleucine to alanine or histidine disrupted binding between monomers (Zhao et al., 1994). The Q44 residue also plays an important role in Vpr oligomerization. Structural analysis of the second alpha helix (amino acids 38-50) reveals a hydrophilic glutamine at the 44th position (Morellet et al., 2009). Deletion of this residue by site-directed mutagenesis disrupted the secondary structure and abolished Vpr – Vpr interaction (Fritz et al., 2008). Fritz et al. (2008) revealed via 3-D modeling that the ΔQ44 mutation destabilizes the formation of the hydrophobic core and the self-interaction of the helices, thus providing an explanation for the inability of such mutants to oligomerize. Although Vpr oligomerization plays a role in other functions, this group did not find a relation between oligomerization and the ability of Vpr to induce apoptosis.

Oligomerization seems necessary for other functions, such as nuclear localization and virion incorporation (Fritz et al., 2008; Fritz et al., 2010; Venkatachari et al., 2010). Venkatachari et al. (2010) hypothesized from a structural model of the oligomerization of Vpr that residues at the predicted helical interfaces contribute to dimerization. Substitution mutagenesis of A30 to leucine abolished dimerization. The authors attributed this to the position of A30 on the external face of the tertiary structure of Vpr, where it likely affects protein-binding. Furthermore, elimination of dimerization of Vpr abolished the ability of the protein to be incorporated into HIV virions. It also diminished its nuclear localization. These results implicate a necessary role of Vpr oligomerization in its incorporation into virions and, by extension, possibly playing a role in the infection of non-replicating immune cells, important targets of HIV-1.

6.2 Nuclear import

The nuclear import of the preintegration complex (PIC) upon entry of the virus enables productive infection to occur in targeted host cells, particularly non-dividing immune cells (*e.g.*, macrophages/monocytes) (Bukrinsky et al., 1992). As Vpr is bound to the PIC, Vpr shuttles the viral contents into the nucleus, where integration occurs (Popov et al., 1998). An earlier study established the necessity of Vpr in productive viral replication in macrophages (Connor et al., 1995), eventually leading to the identification of importin-α as an essential host factor in this process (Nitahara-Kasahara et al., 2007). Sherman et al. (2001) mapped

A Mutagenesis Approach for the Study of the Structure-Function Relationship of Human Immunodeficiency
Virus Type 1 (HIV-1) Vpr

21

non-canonical nuclear localization signals (NLS) throughout the helical domains and the C-terminus of Vpr. This group used site-directed mutagenesis of its various domains in order to identify particular residues that function in nuclear import. In the C-terminus of Vpr, mutagenesis to change several arginine residues (those at positions 73, 76, 77, 85, 87, 88, and 90) to alanine resulted in the distribution of Vpr throughout the cell or cytoplasm, suggesting the importance of this segment of Vpr. In the N-terminus, leucine motifs with the consensus sequence LXXLL in the first and third helices enable nuclear localization. The sequences for each motif are as follows: in the first helix, [22]LLEEL[26]; and in the third helix, [64]LQQLL[68]. The following mutations disrupted nuclear localization most drastically: L22A, L23A, L26A, L64A, and L68A. Via site-directed mutagenesis, the authors established the importance of residues in both helices that are needed for Vpr translocation.

7. G2 cell cycle arrest

The G2 phase of the cellular life cycle serves as a checkpoint for the cell. Factors can halt the cell cycle progression into mitosis in the presence of excessive DNA damage. If these factors do not detect damage, the cell divides; but detection of chromatin disruption will activate factors, such as ATM (ataxia telangiectasia mutated) and ATR (ataxia telangiectasia mutated and Rad-3 related). These factors have downstream effects. Ultimately they hyper-phosphorylate the Cdc2-Cyclin B1 complex, the major controller of cell progression into mitosis. Hyper-phosphorylation inactivates Cdc2-Cyclin B1 to prevent cell division. This process arrests the cell in the G2 phase (Morellet et al., 2009; Pandey et al., 2009; Sherman et al., 2002). Vpr expression in various cell types leads to G2 cell cycle arrest. It does so through increased expression of p21 through the p53 pathway, a major regulator of progression through the G2 and M phases (Chowdhury et al., 2003). Vpr seems to act synergistically with p53, perhaps inducing transcription of p21 through its own transactivation mechanisms. Although the purpose of the G2 arrest function of the virus is debatable, evidence suggests that it enhances transcription from the viral promoter, the long terminal repeat (LTR) of the HIV genome (Goh et al., 1998). One mechanism by which this occurs is through enhancing transactivation of the LTR in CD4+ T-cells via Vpr itself. Vpr binds to other host transcription factors, which bind to sites in the LTR, allowing viral transcription to occur. This effect is enhanced during G2 arrest (Gummuluru and Emerman, 1999).

The C-terminal portion of the protein seems to be essential for induction of cell cycle arrest. Zhou and Ratner (2000) showed that the phosphorylated S79 residue is necessary for this function. Their mutagenesis study of substituting alanine for serine eliminated phosphorylation at this residue and abolished arrest in the G2 phase, correlating the two functions with each other. Furthermore, the mutation of G75A impaired G2 cell cycle arrest, as shown by Mahalingam et al. (1997). DeHart et al. (2007) proposed that Vpr hijacks the ubiquitin/proteasome pathway. One of the functions of this pathway is to target proteins for degradation and to cause G2 cell cycle arrest when needed. In their model, Vpr binds DCAF1, a subunit of the Cullin-4 E3 ubiquitin ligase, to lead to ubiquitination of an unknown host factor involved in halting the cell cycle progression. Change of Q65 to arginine eliminated this binding and impaired G2 cell cycle arrest. However, the group also generated an R80A substitution that also disrupted G2 cell cycle arrest; but this mutant Vpr maintained the ability to bind DCAF1. The authors interpreted these results to indicate that binding DCAF1 is necessary, but not sufficient, to cause G2 cell cycle arrest. A recent study revealed a highly conserved motif in the C-terminus ([79]SRIG[82]), in which mutations that change each amino acid prohibit G2 cell cycle arrest. This study corroborates previous evidence of the functional role of the R80 residue (DeHart et al., 2007; Maudet et al., 2011). Mutagenesis studies using the substitutions R73A and R80A eliminated the induction of p21 transcription, which is necessary for induction of G2 cell cycle arrest.

While the C-terminal portion of Vpr seems necessary for G2 cell cycle arrest, several of the previously cited studies and others revealed that certain N-terminal residues may be indispensable for this function. At least two studies indicated that mutagenesis of A30 to leucine also abolished cell cycle arrest and reduced the transcription of p21 (Chowdhury et al., 2003; Mahalingam et al., 1997). As noted above, A30L eliminated oligomerization, which suggests a correlation between the functions of oligomerization and G2 cell cycle arrest. Also K27M (methionine for lysine 27) another substitution in the first helix, disables the induction of G2 cell cycle arrest. Overall the evidence indicates that N-terminal and C-terminal moieties influence the conformational binding determinants of Vpr involved in G2 cell cycle arrest.

8. Apoptosis

Apoptosis (programmed cell death) functions as a primary means of maintaining homeostasis among cells. Apoptosis can occur as a result of irreparable DNA damage or by disruption of essential cellular processes, such as transcription or translation. Infection by HIV disrupts the normal induction of apoptosis. Despite disrupting apoptosis, infected CD4+ T-cells still die (Groux et al., 1992; Jamieson et al., 1997). However, bystander cells not directly infected by HIV-1, such as CD8+ T-cells, neurons and other cell types, undergo apoptosis (Finkel et al., 1995; Zhang et al., 2003). The Vpr protein was shown to play a key role in the induction of apoptosis in several *in vitro* studies. Vpr was able to permeabilize the mitochondrial membrane and activate caspase 9 via cytochrome C release (Chen et al., 1999; Jacotot et al., 2000; Macreadie et al., 1995; Stewart et al., 1997). Jacotot et al. (2000) established the necessity of several arginine residues inside and between a repeated H(S/F)RIG motif in the Vpr amino acid sequence, specifically [71]HFRIG[75] and [78]HSRIG[82]. Site-directed mutagenesis to generate the substitutions R73A, R77A, and R80A abolished the apoptotic effect of Vpr on Jurkat cells through the pathway involving the mitochondria.

Correlation of the function of Vpr that influences apoptosis with the function that promotes G2 cell cycle arrest remains a complicated affair. Jacquot et al. (2007) reported experiments that indicated a correlation. The group generated several Vpr substitution mutants that abolished both cell cycle arrest and apoptosis. Their results were consistent with the model of G2 cell cycle arrest leading to the induction of apoptosis. Mutants K27M and A30L in the N-terminus of Vpr and R80A and R90K in the C-terminus disabled cytostatic capacity and reduced apoptosis in a T-cell line. However, Bolton and Lenardo (2007) reported that the Vpr with the R80A substitution attenuated apoptotic effects in Jurkat cells, although it remained G2 arrest-capable. Also Maudet et al. (2011) showed that the apoptotic function of alanine mutants in the [79]SRIG[82] motif, including R80, and the K27M substitution mutant remained intact without G2 cell cycle arrest. The evidence from such mutagenesis studies may indicate that, although G2 cell cycle arrest leads to apoptosis, a G2 cell cycle arrest-independent apoptotic pathway exists.

Interestingly, Maudet et al. (2011) showed that the ability of Vpr to bind DCAF1 is not only necessary for induction of G2 cell cycle arrest, but for apoptosis as well. The substitution mutant Q65R eliminated DCAF1 binding, abolishing cell death. The S79A and K27M mutants retained their ability to cause apoptosis while losing G2 arrest function. Mutagenesis was done to produce the double mutants K27M/Q65R and S79A/Q65R. These mutants eliminated the G2 cell cycle arrest-independent induction of apoptosis. As DCAF1 is essential for the targeting of proteins to the proteasome, the authors propose a model in which Vpr binds DCAF1 at a region containing the Q65 residue and functions as an adaptor to the ubiquitin/proteasome complex. Their mutagenesis studies suggest that Vpr contains two different binding domains that interact with two separate, and as yet unidentified, host targets. This ostensibly leads to ubiquitination and subsequent degradation of these proteins. Degradation of target 1 leads to G2 cell cycle arrest, and ubiquitination of target 2 leads to apoptosis. However, these targets have yet to be identified. The model implies complex pathways.

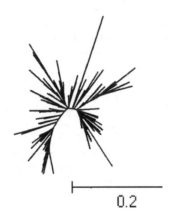

Fig. 2. Phylogeny of *vpr* Sequences Across Clades

9. Vpr polymorphisms across subtypes

A phylogenetic analysis of *vpr* sequences from clades A-D (including the highly prevalent subtype B shows a large diversity of sequences across the genetic lineages of HIV-1 subtypes **(Figure 2)**. This indicates the existence of quasispecies of Vpr existing in the population and selective pressures acting on the *vpr* gene. Upon closer analyses of the sequences within the tree, several of the species between subtypes of Vpr show closer genetic relationships than other species within subtype. The variations in Vpr may correlate more to disease progression than to the clades. A comparative analysis of the frequencies of *vpr* alleles from long-term non-progressors (LTNP) and rapid progressors (RP) indicates the presence of mutations that are of interest **(Table 1)**.

Position	Consensus Sequence of LTNP alignment Residue	Residue conserved In the LTNP alignment (n=177)	Residue In Consensus Seq PR alignment	Residue conserved In the PR alignment (n=102)
2	E	174	E	101
3	Q	168	Q	96
4	A	175	A	96
5	P	176	P	101
6	E	166	E	100
7	D	170	D	91
8	Q	175	Q	101
9	G	165	G	102
10	P	176	P	102
11	Q	174	Q	98
12	R	167	R	102
13	E	168	E	100
14	P	177	P	100
15	Y	172	Y	96
16	N	158	N	96
17	E	169	E	96
18	W	175	W	102
19	T	153	T	93
20	L	173	L	102
21	E	174	E	102
22	L	170	L	95
23	L	170	L	102
24	E	169	E	100
25	E	172	E	102
26	L	173	L	101
27	K	172	K	102
28	N	74	S	40
29	E	171	E	101
30	A	170	A	102
31	V	171	V	101
32	R	167	R	100
33	H	174	H	101
34	F	174	F	101
35	P	175	P	101
36	R	171	R	88
37	I	50	V	51
38	W	174	W	101
39	L	175	L	99
40	H	175	H	88
41	S	89	G	71
42	L	171	L	101
43	G	171	G	101

A Mutagenesis Approach for the Study of the Structure-Function Relationship of Human Immunodeficiency
Virus Type 1 (HIV-1) Vpr

25

Position	Consensus Sequence of LTNP alignment Residue	Residue conserved In the LTNP alignment (n=177)	Residue In Consensus Seq PR alignment	Residue conserved In the PR alignment (n=102)
44	Q	175	Q	101
45	H	118	H	72
46	I	174	I	101
47	Y	173	Y	101
48	E	146	E	92
49	T	174	T	98
50	Y	175	Y	101
51	G	169	G	101
52	D	175	D	100
53	T	175	T	100
54	W	174	W	98
55	A	123	A	58
56	G	171	G	100
57	V	174	V	100
58	E	164	E	97
59	A	166	A	96
60	I	149	I	62
61	I	167	I	96
62	R	170	R	96
63	I	136	I	53
64	L	176	L	99
65	Q	173	Q	97
66	Q	174	Q	96
67	L	174	L	98
68	L	174	L	82
69	F	176	F	100
70	I	145	I	93
71	H	173	H	100
72	F	174	F	99
73	R	174	R	100
74	I	172	I	97
75	G	158	G	95
76	C	171	C	99
77	R	106	R	52
78	H	174	H	100
79	S	175	S	98
80	R	172	R	99
81	I	171	I	97
82	G	174	G	100
83	I	171	I	99
84	T	99	T	66

Position	Consensus Sequence of LTNP alignment Residue	Residue conserved In the LTNP alignment (n=177)	Residue In Consensus Seq PR alignment	Residue conserved In the PR alignment (n=102)
85	R	111	Q	40
86	Q	143	Q	83
87	R	168	R	94
88	R	168	R	97
89	A	140	A	83
90	R	169	R	95
91	N	170	N	56
92	G	171	G	57
93	A	153	A	55
94	S	162	S	51
95	R	173	R	58†

Table 1†. Frequency Analysis of Amino Acids Resulting from *vpr* Alleles Found in the LTNP and RP Groups

An approach using site-directed mutagenesis to study these substitutive polymorphisms will link these variants to possible effects on Vpr function and to the long-term non-progressor and rapid progressor statuses.

10. Mutagenesis studies in the context of disease progression and pathogenesis

Wang et al. (1996) analyzed sequences of the *vpr* genes from an HIV-infected mother-child pair who showed no sign of AIDS from initial infection in 1983 to the time of the study in 1995. These investigators found that samples from the mother and child had homogeneous and similar length polymorphisms in the C-terminal region of Vpr. However these polymorphisms were not present in samples from 30 patients who developed AIDS. Several other studies showed marked heterogeneity in *vpr* sequences derived from multiple patient samples (Ge et al., 1996; Kuiken et al., 1996). In addition to length polymorphisms, numerous studies have demonstrated an association between substitution mutants in the amino acid sequence of Vpr and disease (Caly et al., 2008; Lum et al., 2003; Tungaturthi et al., 2004).

Caly et al. (2008) found a mutation, F72L, in several *vpr* sequences derived from a single patient with LTNP status. This mutation seems to be correlated with the disruption of nuclear import. However only a limited conclusion about prognosis can be drawn from this result because of the small sample size. Yedavalli and Ahmad (2001) reported several mutations from *vpr* sequences extracted from HIV-infected mothers with LTNP status who did not transmit the infection to the child during labor. They found polymorphisms from two separate patients that led to the A30S and G75R substitutions in Vpr. They also found

---

† Amino acid abbreviations follow standard conventions from the International Union of Pure and Applied Chemistry (IUPAC) (Nomenclature, 1968).

C-terminal deletions at an abundant frequency in another sample. This finding implies the possibility of structural changes to Vpr that might alter function in the context of disease progression. The involvement of Vpr at all stages of the HIV-1 life cycle suggests that this protein influences the development of disease and the severity of the outcome. Our hypothesis is that with a sufficient sample size, assessment of *vpr* sequences derived from the LTNP and RP groups will reveal signature polymorphisms that may be linked to progression of AIDS *in vivo*. This may occur through the disruption or enhancement of hallmark Vpr functions. Already several groups have suggested that unique, signature polymorphisms in Vpr culled from LTNP patient samples are associated with the reduction of host cell apoptosis (Lum et al., 2003; Somasundaran et al., 2002).

However, given the complexities of the interrelated pathways by which Vpr induces apoptosis, as revealed by the mutagenesis studies detailed above, it is likely that the two published studies explain only a fraction of the causative factors behind the progression of HIV disease. Many questions need to be answered before the role of Vpr in disease induction is clear.

11. References

Agilent (2010). QuikChange II Site-Directed Mutagenesis Kit: Instruction Manual, Revision C.

Agilent (2011). *PfuUltra* High-Fidelity DNA Polymerase AD, Revision B.

Agostini, I., J. M. Navarro, M. Bouhamdan, K. Willetts, F. Rey, B. Spire, R. Vigne, R. Pomerantz and J. Sire (1999). "The HIV-1 Vpr Co-activator Induces a Conformational Change in TFIIB." *FEBS Letters* 450(3): 235-239.

Ayyavoo, V., A. Mahboubi, S. Mahalingam, R. Ramalingam, S. Kudchodkar, W. Williams, D. Green and D. Weiner (1997). "HIV-1 Vpr Suppresses Immune Activation and Apoptosis through Regulation of Nuclear Factor Kappa B." *Nature Medicine* 3(10): 1117-23.

Balliet, J. W., D. L. Kolson, G. Eiger, F. M. Kim, K. A. McGann, A. Srinivasan and R. Collman (1994). "Distinct Effects in Primary Macrophages and Lymphocytes of the Human Immunodeficiency Virus Type 1 Accessory Genes Vpr, Vpy, and Nef: Mutational Analysis of a Primary HIV-1 Isolate." *Virology* 200(2): 623-631.

Bolton, D. L. and M. J. Lenardo (2007). "Vpr Cytopathicity Independent of G2/M Cell Cycle Arrest in Human Immunodeficiency Virus Type 1-Infeceted CD4+ T Cells." *Journal of Virology* 81(17): 8878-8890.

Bratanich, A. C., C. Liu, J. C. McArthur, T. Fudyk, J. D. Glass, S. Mittoo, G. A. Klassen and C. Power (1998). "Brain-Derived HIV-1 *Tat* Sequences from AIDS patients with Dementia Show Increased Molecular Heterogeneity." *Journal of Neurovirology* 4(4): 387-393.

Bukrinsky, M. I., N. Sharova, M. P. Dempsey, T. L. Stanwick, A. G. Bukrinskaya, S. Haggerty and M. Stevenson (1992). "Active Nuclear Import of Human Immunodeficiency virus type 1 preintegration complexes." *Proceedings of the National Academy of Science* 89(14): 6580-6584.

Caly, L., N. Saksena, S. Piller and D. Jans (2008). "Impaired nuclear import and viral incorporation of Vpr derived from a HIV long-term non-progressor." *Retrovirology* 5(1): 67.

Casartelli, N., G. Di Matteo, M. Potesta, P. Rossi and M. Doria (2003). "CD4 and Major Histocompatibility Complex Class I Downregulation by the Human Immunodeficiency Virus Type 1 Nef Protein in Pediatric AIDS Progression." *J. Virol.* 77(21): 11536-11545.

Chen, M., R. T. Elder, M. G. O'Gorman, L. Selig, R. Benarous, A. Yamamoto and Y. Zhao (1999). "Mutational analysis of Vpr-induced G2 arrest, nuclear localization, and cell death in fission yeast." *Journal of Virology* 73(4): 3236-3245.

Chowdhury, I. H., X.-F. Wang, N. R. Landau, M. L. Robb, V. R. Polonis, D. L. Birx and J. H. Kim (2003). "HIV-1 Vpr Activates Cell Cycle Inhibitor p21/Waf1/Cip1: A Potential Mechanism of G2/M Cell Cycle Arrest." *Virology* 305(2): 371-377.

Cohen, E. A., G. Dehni, J. G. Sodroski and W. A. Haseltine (1990). "Human immunodeficiency virus vpr product is a virion-associated regulatory protein." *J. Virol.* 64(6): 3097-3099.

Connor, R. I., B. K. Chen, S. Choe and N. R. Landau (1995). "Vpr Is Required for Efficient Replication of Human Immunodeficiency Virus Type-1 in Mononuclear Phagocytes." *Virology* 206(2): 935-944.

DeHart, J., E. Zimmerman, O. Ardon, C. Monteiro-Filho, E. Arganaraz and V. Planelles (2007). "HIV-1 Vpr activates the G2 checkpoint through manipulation of the ubiquitin proteasome system." *Virology Journal* 4(1): 57.

Engler, A., T. Stangler and D. Willbold (2001). "Solution structure of human immunodeficiency virus type 1 Vpr(13-33) peptide in micelles." *European Journal of Biochemistry* 268(2): 389-395.

Finkel, T. H., G. Tudor-Williams, N. K. Banda, M. F. Cotton, T. Curiel, C. Monks, T. W. Baba, R. M. Ruprecht and A. Kupfer (1995). "Apoptosis occurs predominantly in bystander cells and not in productively infected cells of HIV- and SIV-infected lymph nodes." *Nature Medicine* 1(2): 129-134.

Fischer, W., V. V. Ganusov, E. E. Giorgi, P. T. Hraber, B. F. Keele, T. Leitner, C. S. Han, C. D. Gleasner, L. Green, C.-C. Lo, A. Nag, T. C. Wallstrom, S. Wang, A. J. McMichael, B. F. Haynes, B. H. Hahn, A. S. Perelson, P. Borrow, G. M. Shaw, T. Bhattacharya and B. T. Korber (2010). "Transmission of Single HIV-1 Genomes and Dynamics of Early Immune Escape Revealed by Ultra-Deep Sequencing." *PLoS ONE* 5(8): e12303.

Fischer, W., S. Perkins, J. Theiler, T. Bhattacharya, K. Yusim, R. Funkhouser, C. Kuiken, B. Haynes, N. L. Letvin, B. D. Walker, B. H. Hahn and B. T. Korber (2007). "Polyvalent Vaccines for Optimal Coverage of Potential T-Cell Epitopes in Global HIV-1 Variants." *Nature Medicine* 13(1): 100-106.

Fritz, J. V., P. Didier, J.-P. Clamme, E. Schaub, D. Muriaux, C. Cabanne, N. Morellet, S. Bouaziz, J.-L. Darlix, Y. Mely and H. d. Rocquigny (2008). "Direct Vpr-Vpr Interaction in Cells Monitored by two Photon Fluorescence Correlation Spectroscopy and Fluroescence Lifetime Imaging." *Retrovirology* 5(87).

Fritz, J. V., D. Dujardin, J. Godet, P. Didier, J. D. Mey, J.-L. Darlix, Y. Mely and H. d. Rocquigny (2010). "HIV-1 Vpr Oligomerization but Not That of Gag Directs the Interaction between Vpr and Gag." *Journal of Virology* 84(3): 1585-1596.

Gaschen, B., J. Taylor, K. Yusim, B. Foley, F. Gao, D. Lang, V. Novitsky, B. Haynes, B. H. Hahn, T. Bhattacharya and B. Korber (2002). "Diversity Considerations in HIV-1 Vaccine Selection." *Science* 296(5577): 2354-2360.

Ge, Y. C., B. Wang, D. E. Dwyer, S.-H. Xiang, A. L. Cunningham and N. Saksena (1996). "Sequence Note: Length Polymorphism of the Viral Protein R of Human Immunodeficiency Virus Type 1 Strains." *AIDS Research and Human Retroviruses* 12(4): 351-354.

Goh, W. C., M. E. Rogel, C. M. Kinsey, S. F. Michael, P. N. Fultz, M. A. Nowak, B. H. Hahn and M. Emerman (1998). "HIV-1 Vpr increases viral expression by manipulation of the cell cycle: A mechanism for selection of Vpr *in vivo*." *Nature Medicine* 4(1): 65-71.

Goulder, P. J. R., R. E. Phillips, R. A. Colbert, S. McAdam, G. Ogg, M. A. Nowak, P. Giangrande, G. Luzzi, B. Morgana, A. Edwards, A. J. McMichael and S. Rowland-Jones (1997). "Late Escape from an Immunodominant Cytotoxic T-Lymphocyte Response Associated with Progression to AIDS." *Nature Medicine* 3(2): 212-217.

Groux, H., G. Torpier, D. Monte, Y. Mouton, A. Capron and J. C. Ameisen (1992). "Activation-induced death by apoptosis in CD4+ T cells from human immunodeficiency virus-infected asymptomatic individuals." *Journal of Experimental Medicine* 175(2): 331-340.

Gummuluru, S. and M. Emerman (1999). "Cell Cycle- and Vpr-Mediated Regulation of Human Immunodeficiency Virus Type 1 Expression in Primary and Transformed T-Cell Lines." *J. Virol.* 73(7): 5422-5430.

Hapgood, J. P. and M. Tomasicchio (2010). "Modulation of HIV-1 Virulence Via the Host Glucocorticoid Receptor: Towards Further Understanding the Molecular Mechanism of HIV-1 Pathogenesis." *Archives of Virology* 155(7): 1009-1019.

HIV/AIDS, T. J. U. N. P. o. (2010). Global Report: UNAIDS Report on the Global AIDS Epidemic 2010, Joint United Nations Programme on HIV/AIDS (UNAIDS).

Irish, B. P., Z. K. Khan, P. Jain, M. R. Nonnemacher, V. Pirrone, S. Rahman, N. Rajagopalan, J. B. Suchitra, K. Mostoller and B. Wigdahl (2009). "Molecular Mechanisms of Neurodegenerative Diseases Induced by Human Retroviruses: A review." *American Journal of Infectious Diseases* 5(3): 231-258.

Jacotot, E., L. Ravagnan, M. Loeffler, K. F. Ferri, H. L. A. Vieira, N. Zamzami, P. Costantini, S. Druillennec, J. Hoebeke, J. P. Briand, T. Irinopoulou, E. Daugas, S. A. Susin, D. Cointe, Z. H. Xie, J. C. Reed, B. P. Roques and G. Kroemer (2000). "The HIV-1 Viral Protein R Induces Apoptosis via a Direct Effect on the Mitochondrial Permeability Transition Pore." *The Journal of Experimental Medicine* 191(1): 33-46.

Jacquot, G., E. Le Rouzic, A. David, J. Mazzolini, J. Bouchet, S. Bouaziz, F. Niedergang, G. Pancino and S. Benichou (2007). "Localization of HIV-1 Vpr to the nuclear envelope: Impact on Vpr functions and virus replication in macrophages." *Retrovirology* 4(1): 84.

Jamieson, B. D., C. H. Uittenbogaart, I. Schmid and J. A. Zack (1997). "High viral burden and rapid CD4+ cell depletion in human immunodeficiency virus type 1-infected SCID-hu mice suggest direct viral killing of thymocytes in vivo." *Journal of Virology* 71(11): 8245-8253.

Korber, B., B. Gaschen, K. Yusim, R. Thakallapally, C. Kesmir and V. Detours (2001). "Evolutionary and immunological implications of contemporary HIV-1 variation." *British Medical Bulletin* 58(1): 19-42.

Kuiken, C. L., M. T. E. Cornelissen, F. Zorgdrager, S. Hartman, A. J. Gibbs and J. Goudsmit (1996). "Consistent risk group-associated differences in human immunodeficiency virus type 1 vpr, vpu and V3 sequences despite independent evolution." *Journal of General Virology* 77(4): 783-792.

Levy, J. A. (2009). "HIV Pathogenesis: 25 Years of Progress and Persistent Challenges." *AIDS* 23(2): 147-160.

Lum, J. J., O. J. Cohen, Z. Nie, J. G. Weaver, T. S. Gomez, X.-J. Yao, D. Lynch, A. A. Pilon, N. Hawley, J. E. Kim, Z. Chen, M. Montpetit, J. Sanchez-Dardon, E. A. Cohen and A. D. Badley (2003). "Vpr R77Q is associated with long-term nonprogressive HIV infection and impaired induction of apoptosis." *The Journal of Clinical Investigation* 111(10): 1547-1554.

Macreadie, I. G., L. A. Castelli, D. R. Hewish, A. Kirkpatrick, A. C. Ward and A. A. Azad (1995). "A domain of human immunodeficiency virus type 1 Vpr containing repeated H(S/F)RIG amino acid motifs causes cell growth arrest and structural defects." *Proceedings of the National Academy of Sciences* 92(7): 2770-2774.

Mahalingam, S., V. Ayyavoo, M. Patel, T. Kieber-Emmons and D. B. Weinter (1997). "Nuclear Import, Virion Incorporation, and Cell Cycle Arrest/Differentiation Are Mediated by Distinct Functional Domains of Human Immunodeficiency Virus Type 1 Vpr." *Journal of Virology* 71(9): 6339-6347.

Maudet, C., M. Bertrand, E. Le Rouzic, H. Lahouassa, D. Ayinde, S. Nisole, C. Goujon, A. Cimarelli, F. Margottin-Goguet and C. Transy (2011). "Molecular Insight into How HIV-1 Vpr Protein Impairs Cell Growth through Two Genetically Distinct Pathways." *Journal of Biological Chemistry* 286(27): 23742-23752.

Morellet, N., S. Bouaziz, P. Petitjean and B. P. Rogues (2003). "NMR Structure of the HIV-1 Regulatory Protein VPR." *Journal of Molecular Biology* 327(1): 217-227.

Morellet, N., B. P. Roques and S. Bouaziz (2009). "Structure-Function Relationship of Vpr: Biological Implications." *Current HIV Research* 7(2): 184-210.

Nitahara-Kasahara, Y., M. Kamata, T. Yamamoto, X. Zhang, Y. Miyamoto, K. Muneta, S. Iijima, Y. Yoneda, Y. Tsunetsugu-Yokota and Y. Aida (2007). "Novel Nuclear Import of Vpr Promoted by Importin {alpha} Is Crucial for Human Immunodeficiency Virus Type 1 Replication in Macrophages." *J. Virol.* 81(10): 5284-5293.

Nomenclature, I.-I. C. o. B. (1968). "A one-letter notation for amino acid sequences. Tentative rules." *Biochemistry* 7(8): 2703-2705.

Pandey, R. C., D. Datta, R. Mukerjee, A. Srinivasan, S. Mahalingam and B. E. Sawaya (2009). "HIV-1 Vpr: A Closer Look at the Multifunctional Protein from the Structural Perspective." *Current HIV Research* 7(2): 114-128.

Pandori, M., N. Fitch, H. Craig, D. Richman, C. Spina and J. Guatelli (1996). "Producer-cell modification of human immunodeficiency virus type 1: Nef is a virion protein." *J. Virol.* 70(7): 4283-4290.

Peters, B. S. and K. Conway (2011). "Therapy for HIV: Past, Present, and Future." *Advances in Dental Research* 23(1): 23-27.

Popov, S., M. Rexach, L. Ratner, G. Blobel and M. Bukrinsky (1998). "Viral Protein R Regulates Docking of the HIV-1 Preintegration Complex to the Nuclear Pore Complex." *Journal of Biological Chemistry* 273(21): 13347-13352.

Poropatich, K. and D. J. Sullivan (2011). "Human immunodeficiency virus type 1 long-term non-progressors: the viral, genetic and immunological basis for disease non-progression." *Journal of General Virology* 92(2): 247-268.

Roof, P., M. Ricci, P. Genin, M. A. Montano, M. Essex, M. A. Wainberg, A. Gatignol and J. Hiscott (2002). "Differential Regulation of HIV-1 Clade-Specific B, C, and E Long Terminal Repeats by NF-κB and the Tat Transactivator." *Virology* 296(1): 77-83.

Sawaya, B. E., K. Khalili, W. E. Mercer, L. Denisova and S. Amini (1998). "Cooperative Actions of HIV-1 Vpr and p53 Modulate Viral Gene Transcription." *Journal of Biological Chemistry* 273(32): 20052-20057.

Sherman, M. P., C. M. C. de Noronha, M. I. Heusch, S. Greene and W. C. Greene (2001). "Nucleocytoplasmic Shuttling by Human Immunodeficiency Virus Type 1 Vpr." *J. Virol.* 75(3): 1522-1532.

Sherman, M. P., C. M. C. D. Noronha, S. A. Williams and W. C. Greene (2002). "Insights into the Biology of HIV-1 Viral Protein R." *DNA and Cell Biology* 21(9): 679-688.

Somasundaran, M., M. Sharkey, B. Brichacek, K. Luzuriaga, M. Emerman, J. L. Sullivan and M. Stevenson (2002). "Evidence for a cytopathogenicity determinant in HIV-1 Vpr." *Proceedings of the National Academy of Sciences* 99(14): 9503-9508.

Stewart, S. A., B. Poon, J. B. Jowett and I. S. Chen (1997). "Human immunodeficiency virus type 1 Vpr induces apoptosis following cell cycle arrest." *Journal of Virology* 71(7): 5579-5592.

Strebel, K. (2003). "Virus-host interactions: role of HIV proteins Vif, Tat, and Rev." *AIDS* 17(4): S25-S34.

Tolstrup, M., A. L. Laursen, J. Gerstoft, F. S. Pedersen, L. Ostergaard and M. Duch (2006). "Cysteine 138 Mutation in HIV-1 Nef from Patients with Delayed Disease Progression." *Sex Health* 3(4): 281-286.

Tungaturthi, P. K., B. E. Sawaya, V. Ayyavoo, R. Murali and A. Srinivasan (2004). "HIV-1 Vpr: Genetic Diversity and Functional Features from the Perspective of Structure." *DNA and Cell Biology* 23(4): 207-222.

Venkatachari, N. J., L. A. Walker, O. Tastan, T. Le, T. M. Dempsey, Y. Li, N. Yanamala, A. Srinivasan, J. Klein-Seetharaman, R. C. Montelaro and V. Ayyavoo (2010). "Human immunodeficiency virus type 1 Vpr: oligomerization is an essential feature for its incorporation into virus particles." *Virology Journal* 7(119).

Wang, B., Y. C. Ge, P. Palasanthiran, S.-H. Xiang, J. Ziegler, D. E. Dwyer, C. Randle, D. Dowton, A. Cunningham and N. K. Saksena (1996). "Gene Defects Clustered at the C-Terminus of the vpr Gene of HIV-1 in Long-Term Nonprogressing Mother and Child Pair:In VivoEvolution of vpr Quasispecies in Blood and Plasma." *Virology* 223(1): 224-232.

Wecker, K., N. Morellet, S. Bouaziz and B. P. Rogues (2002). "NMR Structure of the HIV-1 Regulatory Protein Vpr in H2O/trifluoroethanol. Comparison with the Vpr N-terminal (1-51) and C-terminal (52-96) domains." *European Journal of Biochemistry* 269(15): 3779-3788.

Wolinsky, S. M., B. T. M. Korber, A. U. Neumann, M. Daniels, K. J. Kunstman, A. J. Whetsell, M. R. Furtado, Y. Cao, D. D. Ho, J. T. Safrit and R. A. Koup (1996). "Adaptive Evolution of Human Immunodeficiency Virus-Type 1 During the Natural Course of Infection." *Science* 272(5261): 537-542.

Yedavalli, V. R. K. and N. Ahmad (2001). "Low Conservation of Functional Domains of HIV Type 1 vif and vpr Genes in Infected Mothers Correlates with Lack of Vertical Transmission." *AIDS Research and Human Retroviruses* 17(10): 911-923.

Zhang, K., F. Rana, C. Silva, J. Ethier, K. Wehrly, B. Chesebro and C. Power (2003). "Human immunodeficiency virus type 1 envelope-mediated neuronal death: uncoupling of viral replication and neurotoxicity." *Journal of Virology* 77(12): 6899-6912.

Zhao, L.-J., L. Wang, S. Mukherjee and O. Narayan (1994). "Biochemical Mechanism of HIV-1 Vpr Function: Oligomerization Mediated by the N-Terminal Domain." *The Journal of Biological Chemistry* 269(51): 32131-32137.

Zhou, Y. and L. Ratner (2000). "Phosphorylation of Human Immunodeficiency Virus Type 1 Vpr Regulates Cell Cycle Arrest." *J. Virol.* 74(14): 6520-6527.

3

Use of Site-Directed Mutagenesis in the Diagnosis, Prognosis and Treatment of Galactosemia

M. Tang[1], K.J. Wierenga[2] and K. Lai[1]
[1]University of Utah School of Medicine,
[2]University of Oklahoma Health Sciences Center,
USA

1. Introduction

Site-directed mutagenesis (**SDM**) is undoubtedly one of the most powerful techniques in molecular biology. In this chapter, we will describe the use of SDM in the study of the human inherited metabolic disorder, Galactosemia (Type I, II, and III) and the development of novel therapies for the disease. This powerful technique not only helped confirm suspected *GAL* gene mutations in Galactosemia, but also played a significant role in unraveling the catalytic mechanisms of the GAL enzymes in the conserved Leloir pathway of galactose metabolism. To date, more than thirty disease-causing mutations in the human *GAL* genes have been characterized in great detail; and these findings have paved the way for innovative, state-of-the-art therapies, such as chaperone therapy. Recently, in order to optimize small molecule GALK inhibitors for the treatment of Type I Galactosemia, we have employed SDM to identify amino acids of the GALK enzyme that interact with its selective inhibitors. These studies exemplified the expanding roles of SDM in innovative drug design and in kinase inhibitor selectivity.

2. Background

2.1 What is galactosemia?

Galactose is a hexose that differs from glucose only by the configuration of the hydroxyl group at the carbon-4 position. Often present as an anomeric mixture of α-D-galactose and β-D-galactose, this monosaccharide exists abundantly in milk, dairy products and many other food types such as fruits and vegetables (Acosta and Gross, 1995; Berry et al., 1993). However, galactose can also be produced endogenously in human cells, mainly as products of glycoprotein and glycolipid turnover.(Berry et al., 1995, 2004). Once freely present inside the cells, β-D-galactose is epimerized to α-D-galactose through the action of a mutarotase (Beebe and Frey, 1998; Thoden and Holden, 2002a). α-D-galactose is then metabolized by the Leloir pathway (Leloir 1951), an evolutionarily conserved biochemical pathway which begins with the phosphorylation of galactose by the enzyme galactokinase (GALK) to form galactose-1 phosphate (gal-1P) (Cardini and Leloir, 1953). Gal-1P is subsequently, together with the substrate UDP-glucose, converted by galactose-1-phosphate uridylyltransferase

(GALT) to form UDP-galactose and glucose-1 phosphate (glu-1P) (Kalckar et al., 1953). The Leloir pathway is completed by reversibly forming UDP-glucose from UDP-galactose *via* UDP-galactose-4-epimerase (GALE) (Leloir 1953; Darrow and Rodstorm, 1968). Inherited deficiencies of GALK, GALT, and GALE activities in humans have all been observed, studied, and reviewed extensively (Bosch et al., 2002; Elsas 1993; Fridovich-Keil et al., 1993a). The clinical manifestations of each enzyme deficiency, however, differ markedly (Berry et al., 1995; Berry and Elsas, 2011; Fridovich-Keil et al., 1993a; Lai et al., 2009;). For instance, patients with GALK deficiency (MIM 230200) (Type II Galactosemia) have the mildest clinical consequences, as they may present only with cataracts (Bosch et al., 2002). On the other hand, GALT-deficiency (MIM 230400) (Type I or Classic Galactosemia) is potentially lethal in infancy, if undiagnosed and untreated, and is also associated with long-term, organ-specific complications (Berry et al., 1995). GALE-deficiency (MIM 230350) (Type III Galactosemia) has been somewhat controversial with regards to clinical manifestations, as this disorder is rare; and information is mostly derived from case reports (Fridovich-Keil et al., 1993a). Until newborn screening for GALE deficiency is available, the natural history will likely remain unknown. The differences in clinical outcome between GALT and GALK deficiencies reflect the differences in tissue response to the characteristic changes in the levels of galactose metabolites as a result of the respective enzyme deficiencies.

2.2 How are the different types of galactosemia detected and diagnosed?

Newborn screening programs worldwide have greatly facilitated the early detection of Galactosemia (Kaye et al., 2006; Levy 2010). The screening tests often involve the detection of elevated level of blood galactose and/or specific GAL enzyme in the dried blood spots on filter paper. Elevated galactose will detect GALK deficiency and GALT deficiency, but it may not detect GALE deficiency. Other states screen for GALT activity, and may therefore diagnose Type I Galactosemia. However, this screen will miss GALK and GALE deficiency. The final diagnosis is secured once the specific enzyme deficiency is confirmed by enzymatic assays or by DNA genotyping; these tests are available commercially in the USA (http://www.ncbi.nlm.nih.gov/sites/ GeneTests/, Tests #3437, #2229 and #53782).

2.3 What are the current treatments for galactosemia, and what is the outlook for patients?

The main aspect of management for all forms of Galactosemia is withdrawal of lactose/galactose from the diet as soon as the diagnosis is made, or even considered (Segal 1995). In infants, this means the replacement of breast/cow milk with soy-based formula. However, it has become clear that, despite early detection and (early) dietary intervention, there still is a significant burden of the disease, particularly for Classic Galactosemia where chronic problems persist through adulthood. The most common medical complications of Type I Galactosemia are speech dyspraxia, ataxia, premature ovarian insufficiency, and intellectual deficits, which are rarely seen in other forms of galactosemia (Waggoner et al., 1990; Waisbren et al., 2011). GALK deficiency (Type II Galactosemia) is managed also with lactose/galactose restriction, though the complications are mainly confined to the eye (cataracts) (Bosch et al., 2002). GALE deficiency is treated similarly, though complications of this deficiency may not be preventable with such restriction, as is GALT deficiency (Fridovich-Keil et al., 1993a).

3. Use of SDM to confirm disease-causing mutations in human GALT, GALK, and GALE genes identified clinically

3.1 The issues

Advances in federal and state newborn screening programs worldwide have resulted in the inclusion of the potentially lethal disorder, Galactosemia, in the list of diseases for which newborns are screened. Very often, once an affected newborn is identified by the biochemical assays, it is helpful to know the genotype of *GAL* gene involved because there appears to be a genotype-phenotype correlation for a few selected *GAL* gene mutations. The confirmation of the *GAL* genotypes in the affected patients will provide better prognosis. Additionally, a few well-characterized GAL enzyme variants have been shown to retain significant residual enzyme activities. Consequently, patients with selected mutations might benefit from novel therapies, such as chaperone therapies.

Unfortunately many patients with Galactosemia identified to-date have novel (private) nucleotide changes in their *GAL* genes. For instance, the *GALT* gene database set up by the ARUP Laboratories (Salt Lake City, USA) has recorded over 200 nucleotide changes of the *GALT* gene identified in patients with Type I Galactosemia (*www.arup.utah.edu/database/ GALT/GALT_welcome.php*). Without clinical correlation, it is impossible to tell if any of these novel changes actually results in impaired GALT enzyme activities seen in the patients. Moreover, many patients are compound heterozygous for the *GAL* gene mutations. In other words, a single patient may have a unique nucleotide change in each of the two *GAL* alleles; and it is difficult to conclude which one is responsible for the reduction in total enzyme activity. Thus there is a real need to perform *in vitro* expression studies of the identified "variant" *GAL* genes.

3.2 Research design

Our laboratory and others have largely used similar strategies in confirming the suspected human *GAL* gene mutations in patients with Galactosemia (Fridovich-Keil et al., 1995a; Lai et al., 1999; Reichardt et al., 1992;). In almost all cases, we sub-cloned the cDNA of the respective *GAL* gene into expression plasmid vectors, before we performed SDM of the sub-cloned fragments to obtain "mutant" cDNAs with the same sequence changes observed in patients. We then expressed the wild-type and mutated cDNAs in heterologous expression systems, such as *Escherichia coli, Saccharomyces cerevisiae* or even mammalian expression systems. Subsequently, we tested for differences in kinetic parameters of the GAL enzymes, such as K_M and V_{max}, and the expression efficiencies, such as protein and/or mRNA abundances, between mutant and control cDNAs.

3.3 The results

The primary goal for expression analysis of the suspected disease-causing mutations in the *GAL* genes is to show that the nucleotide changes observed are causing impaired GAL enzyme activities and could therefore be the causes for the diseases. In addition, in the course of the analysis, kinetic parameters of the variant enzymes are often determined, which are expected to help advance the structural knowledge of the GAL enzymes.

3.3.1 Type I (GALT-deficiency) galactosemia

As mentioned above, more than 200 nucleotide changes in the *GALT* gene have been identified so far, mostly single nucleotide substitutions. The most common human *GALT* mutation, *Q188R*, is detected in over 70% of galactosemic patients in Europe and North America. The *Q188R* mutation is associated with a poor clinical outcome, even with a galactose-restricted diet (Guerrero et al., 2000; Murphy et al., 1999; Webb et al., 2003). *K285N* is the second most common mutation found in patients in Europe, especially in the countries of central and Eastern Europe, where it can account for up to 34% of *GALT* alleles (Greber-Platzer et al., 1997; Kozak et al., 2000). In the African-American population, the *S135L* mutation is predominant. The corresponding enzyme leads to a relatively benign outcome, if the mutation is identified and the patient is treated with a galactose-restricted diet in the newborn period (Lai et al., 1996, 2001; Landt et al., 1997). A more common mutation, *N314D*, occurs in all populations mentioned above and can lead to two different phenotypes, depending on the presence or absence of a 4-bp deletion in the coding region for the carbohydrate response element. When *N314D* is associated with a four-nucleotide deletion in the promoter region (the Duarte type 2), homozygosity for N314D and this altered promoter region causes a 50% decrease of GALT activity, with a mild or even undetected phenotype (Elsas et al., 1994). In the absence of this deletion in the promoter, homozygosity for the N314D missense mutation (the Los Angeles variant) results in normal GALT in erythrocytes (Shin et al., 1998). A 5-kb deletion is found so far exclusively in Ashkenazi Jewish patients (Coffee et al., 2006).

Due to its frequency among GALT-deficiency galactosemic patients and its association with a poor clinical outcome, the *Q188R* mutation has been extensively studied. The initial study using the COS cell expression system surprisingly showed that this mutation had about 10% of normal enzymatic activity (Reichardt et al., 1991). This result was not consistent with the clinical finding that patients homozygous for *Q188R* have no detectable enzyme activity in their red blood cells. Another study, carried out in a yeast model that was completely devoid of GALT activity, used a PCR-mediated SDM technique and clarified that the *Q188R* mutation did cause loss of function of both human and yeast GALT (Fridovich-Keil et al.,1993b). Interestingly, this study also showed that the mutant yeast, with its loss of GALT activity, could not survive in galactose media if the *Q188R* missense mutation was introduced, while reconstitution of wild-type GALT resulted in normal growth (Fridovich-Keil and Jinks-Roberson, 1993b). The confounding result of the first study is likely to be explained by the presence of endogenous GALT activity in the COS cells, highlighting the importance of studying mutations in a null background system, such as the *gal7*-deleted yeast model used in the second study. Alternatively, one should use purified mutant proteins in the analysis of the enzymatic activities. Subsequent studies further confirmed that the *Q188R* mutation not only totally abolishes GALT enzyme activity, but also acts as a partial dominant-negative mutation, as the heterodimer of *Q188R*/wild type has only 15% of wild-type activity (Fridovich-Keil et al., 1995a; Elsevier and Fridovich-Keil,1996). Kinetic analysis showed this mutation mainly causes impaired specific activity of the heterodimer without altering the K_M for both substrates. In order to further understand how mutation at this site could affect the enzyme, Lai and coworkers mutated glutamine-188 (Gln[188]) to arginine and asparagine, respectively, through SDM (Lai et al., 1999). More detailed kinetic measurement showed that mutating glutamine to arginine or asparagine did not affect the first step of the double-displacement action (UDP-Glu to glu-1p). In fact, Q188R-GALT even had a better V_{max} as compared with the wild-type GALT. However, the *Q188R* mutation severely impaired the second step of the reaction. The crystal structures of *E. coli* GALT revealed that Gln[168] (equivalent to Gln[188] in human GALT) could stabilize the GALT-UMP intermediate through two hydrogen bonds formed between the amide side chain of Gln[188] and the phosphoryl oxygen of the UMP moiety (Wedekind et al., 1996). Through molecular modeling studies (or "virtual SDM"), Lai and coworkers changed glutamine to arginine and asparagine, respectively, and found that the number of hydrogen bonds formed between new amino acid residues and UMP moiety decreased to one, which could have destabilized the GALT-UMP intermediate required for the second displacement reaction (Lai et al., 1999). This destabilization was well manifested in the increased V_{max} in the Q188R mutant in the first displacement reaction, as the destabilization speeded up the recycling of the enzyme for the first reaction (Lai et al., 1999). To complete the double-displacement reaction, a stable GALT-UMP intermediate was required to bind gal-1P, which was better accomplished by the two hydrogen bonds from glutamine than by the single hydrogen bond from arginine or asparagine.

The *S135L* mutation was identified initially as a polymorphism with near normal enzymatic activity in the COS cell expression system (Reichardt et al., 1992). However, subsequent SDM studies in the yeast-expression system, defined this as a missense mutation that significantly impaired enzyme activity; but, unlike the *Q188R* mutation, it still had minor residual activity (Fridovich-Keil et al., 1995a). Later on, more detailed SDM and expression studies in yeast and *E. coli* heterologous expression systems revealed this mutation decreased the abundance of mutant protein about 2-fold compared with the wild type, as well as caused 10-fold decrease of specific activity with less than 2-fold of differences of K_M values for both substrates (Lai and Elsas, 2001; Wells and Fridovich-Keil, 1997). There was no apparent difference in releasing glu-1P between the wild type and this mutant (Lai and Elsas, 2001). Mutating this serine to alanine, cysteine, histidine, threonine or tyrosine by SDM confirmed that a hydroxyl group is required on the side chain of amino acid 135, since only the threonine substitution resulted in active enzyme (Lai and Elsas, 2001).

The *K285N* mutation compromises the activity of the enzyme, as well as its abundance, in the yeast expression system (Riehman et al., 2001). As for the *N314D* mutation, it was regarded as the reason of reduced enzymatic activity in Duarte 2 patients; but detailed enzymatic studies facilitated by SDM revealed that the mutation itself only causes isoelectric point shifting, without affecting protein abundance, subunit dimerization or activity (Fridovich-Keil et al., 1995b). The decrease in GALT activity observed in the Duarte type 2 patients is likely caused by the 4-bp deletion at the promoter region associated with the *N314D* mutation, which abolishes the binding sites of two transcription factors to the GALT gene promoter (Carney, et al., 2009). The fact that the Los Angeles variant has normal activity in the erythrocytes supports this conclusion (Carney et al., 2009).

3.3.2 Type II (GALK-deficiency) galactosemia

More than 20 mutations associated with GALK deficiency have been reported to date. Through SDM studies, the majority of the mutations have been characterized. By expressing 10 variant GALK enzymes in GALK-less *E. coli*, Timson and Reece showed that five of mutant GALK enzymes (P28T, V32M, G36R, T288M and A384P), which are associated with more severe clinical phenotypes and near-zero blood galactokinase levels, are insoluble

(Timson and Reece, 2003). Further studies showed that these mutations disrupted the secondary structure of the enzymes, which could result in misfolding of the protein (Thoden et al., 2005). Four of the five soluble mutants (H44Y, R68C, G346S, and G349S, but not A198V) have impaired enzymatic properties, such as increased K_M for one or both substrates and decreased k_{cat}. All five are associated with low blood enzyme levels and milder symptoms. From the crystal structure of human GALK, it is clear that His[44], Gly[346] and Gly[349] are close to the active site. Additionally, these residues reside in the signature motif III of the GHMP kinase superfamily (Bork et al., 1993; Thoden et al., 2005). Therefore, it is not surprising that any changes in these resides would alter the kinetic parameters of the enzyme. As for A198V, its kinetic parameters are essentially indistinguishable from the wild-type enzyme. Compared to other mutations, from which patients will develop cataracts with high incidence within the first few years (without treatment), the A198V enzyme causes only a moderate incidence of cataracts in later life.

Similarly, Park and colleagues characterized another four missense mutations and one insertion (*G137R, R256W, R277Q, V281M* and *850_851insG*) by expressing the corresponding mutated genes in COS7 cells (Park et al., 2007). The steady-state expression level of R256W was lower than that of wild type. The stability of the mutant enzyme was significantly reduced, and it had no detectable activity. No protein was detected for the insertion variant. The other three mutations manifested enzymes with similar expression levels in the soluble fraction, as compared to the wild-type level. However, the *G137R* and *R277Q* enzymes had approximately 10%-15% of wild-type activity, and no activity was detected for the *V281M* enzyme.

3.3.3 Type III (GALE-deficiency) galactosemia

GALE deficiency exists in a continuum, from generalized to peripheral *via* intermediate (Openo et al., 2006). If GALE is deficient in all tissues, it is classified as generalized; and, if it is only deficient in red and white cells but normal in other tissues, it is known as peripheral deficiency. It is possible that the presence of bi-allelic amorphic mutations is incompatible with life (Sanders et al., 2010). Infants with generalized deficiency develop disease on a lactose-containing milk diet, while infants with peripheral disease remain well, at least in the newborn period. GALE deficiency has been extensively reviewed by Fridovich-Keil and coworkers (Fridovich-Keil et al., 1993a). Genomic *GALE* is about 5 kb in length, with multiple alternatively spliced transcripts. Some of the reported mutations are deposited in the HGMD database (http://www.hgmd.org/). Few case series have been reported, including a Korean study, reporting 37 patients with reduced GALE activity (Park et al., 2005), and two US-based studies, with one reporting 35 patients (Maceratesi et al., 1998) and the other, 10 patients (Openo et al.,2006). Others have reported a few cases (Alano et al., 1998; Wohlers et al., 1999). The *V94M* mutation has been reported in the homozygous state as being associated with generalized disease (Wohlers et al., 1999). In-depth studies of the *V94M* mutation through SDM in the yeast system showed that this mutation severely damages the specific activity of the enzyme predominantly at the level of V_M without affecting its abundance and thermal stability (Wohlers et al., 1999; Wohlers and Fridovich-Keil, 2000). In the same study, the *G90E* mutation was shown to have zero enzymatic activity, rendering the mutant enzyme to high temperature and protease (Wohlers et al., 1999). A more recent study further confirmed the impact of *V94M* and *G90E* on V_M (Timson 2005). Other missense mutations have not (yet) been reported in patients, but they have been studied *in vitro* or in model systems. They are associated with severe enzyme deficiency; these include *G90E* and *L183P* (Quimby et al., 1997; Timson, 2005; Wohlers et al., 1999). Missense mutations associated with peripheral disease include *R169W*, *R239W* and *G302A* and have been described by Park and coworkers in individuals with peripheral GALE deficiency (Park et al., 2005). The *K257R* and *G319E* mutations have been described in African-Americans with peripheral deficiency (Alano et al., 1998). The *L183P* mutation encodes an enzyme that experiences severe proteolytic degradation during expression and purification. Also the authors showed that enzymes resulting from the *N34S*, *G90E* and *D103G* mutations exhibited increased susceptibility to digestion in limited proteolysis experiments (Timson 2005). An earlier study on *L183P* and *N34S* using SDM in a yeast model revealed that the L183P-hGALE mutant demonstrated 4% wild-type activity and 6% wild-type abundance, while N34S-hGALE demonstrated approximately 70% wild-type activity and normal abundance. However, yeast cells co-expressing both L183P-hGALE and N34S-hGALE exhibited only approximately 7% wild-type levels of activity, thereby confirming the functional impact of having both substitutions and raising the intriguing possibility that some form of dominant negative interaction may exist between the mutant enzymes found in this patient (Quimby et al., 1997). Two other mutations, *D130G* and *L313M*, which are associated with intermediate epimerase deficiency, manifested enzymes with near normal GALE activity, but with compromised thermal stability and protease-sensitivity (Wohlers et al., 1999). Three other mutations associated with intermediate forms (*S81R*, *T150M* and *P293L*) were analyzed for their kinetic and structural properties *in vitro* and their effects on galactose-sensitivity of *S. cerevisiae* cells in the absence of Gal10p. All three mutations result in impairment of the kinetic parameters, principally the turnover number, k_{cat}, compared to the wild-type enzyme. However, the degree of impairment was mild compared with that seen with the mutation *V94M* (Chhay et al., 2008). Studies are limited by the fact the many patients are compound heterozygotes and by the observation that dominant negative interactions may be involved in some of these cases.

4. Use of SDM in the understanding of catalytic mechanisms of the human GAL enzymes

4.1 The issue

Although the Leloir pathway is evolutionarily conserved and is indispensable for productive galactose metabolism, the catalytic mechanisms of the GAL enzymes are largely unknown.

4.2 Research design

Several groups have attempted to combine the techniques of SDM, analytical biochemistry and X-ray crystallography to advance the understanding of the catalytic mechanisms of the different GAL enzymes.

4.3 The results

4.3.1 GALK

GALK converts galactose to gal-1P by transferring γ–phosphate group of ATP to the O1 position of galactose. It belongs to a unique kinase superfamily – the GHMP kinase family, which is named after four characteristic family members: galactokinase (GALK), homoserine kinase (HSK), mevalonate kinase (MVK) and phosphomevalonate kinase (PMVK) (Bork et al., 1993). This family of proteins was first identified by three highly conserved motifs among the four kinases mentioned above by sequence alignment and analysis. Motifs I and III are located at the N-terminal and C-terminal ends; and motif II, the most conserved, is located in the middle of the protein, with the consensus sequence of GLGSS(G/A/S) (Holden et al., 2004).

Interestingly, two different catalytic mechanisms have been proposed for this family. A common catalytic strategy to achieve nucleophilic attack is to use a negative charged residue, such as aspartate or glutamate, to act as a Brønsted base. This catalytic base can then abstract a proton from the hydroxyl group of the substrate converting the weakly nucleophilic hydroxyl group into the more strongly nucleophilic alkoxide ion, which then attacks the electron-deficient phosphorus atom in ATP (Fig. 1A). In such systems, it is common to find positively-charged lysine or arginine residues close to the catalytic site to help stabilize the negative charges on the enzyme and the substrates. Studies on MVK suggest this enzyme follows this mechanism. The crystal structure of MVK reveals an aspartate (residue 204 in the rat enzyme) positioned to act as an active site base. There is also a lysine (residue 13 in rat MVK), which is close to both the putative catalytic aspartate residue and the hydroxyl group of the substrate (Fu et al., 2002; Yang et al., 2002). Replacement of the lysine residue with a methionine by SDM resulted in a reduced, but non-zero, rate (V_{max} was reduced approximately 60-fold) (Potter et al., 1997). Similar results were observed when the equivalent lysine (residue 18) was changed to methionine in yeast mevalonate diphosphate decarboxylase (Krepkiy and Miziorko, 2004). These results are consistent with this positively-charged residue playing an assisting, but non-vital, role in catalysis. Crystal structures of GALK put it into this mechanism by revealing there are aspartate and arginine residues in the active center close to the galactose C1 hydroxyl group (Asp[186] and Arg[37] in the human structure, Asp[183] and Arg[36] in *Lactococcus lactis*) (Thoden and Holden, 2003; Thoden et al., 2005). Similarly, changing Arg[37] of human GALT to alanine resulted in a nearly inactive enzyme; and lysine resulted in compromised k_{cat} and K_M for galactose (Tang, et al., 2010).

In contrast, phosphoryl transfer in HSK has been suggested to occur by direct nucleophilic attack on the γ-phosphate group of ATP by the δ-hydroxyl of homoserine (Fig. 1B) (Krishna et al., 2001). In this mechanism, the latter is stabilized by the formation of a hydrogen bond to a neighboring asparagine residue (Asn[141]), which is not conserved in the superfamily. Catalysis is proposed to be assisted through activation of the γ-phosphate of ATP by the magnesium ion, which is coordinated by a conserved glutamate residue (Glu[130]) with the deprotonation of the δ-hydroxyl possibly involving the γ-phosphate (Krishna et al., 2001).

4.3.2 GALT

GALT catalyzes the transfer of the uridine monophosphate group (UMP) from uridine diphosphate-glucose (UDP-Glu) to gal-1p to form uridine diphosphate-galactose (UDP-Gal) and glucose-1-phosphate (glu-1P) (Kalckar et al., 1953). The reaction follows the double displacement mechanism as shown in Fig. 2 (Arabshahi et al., 1986). The most characteristic

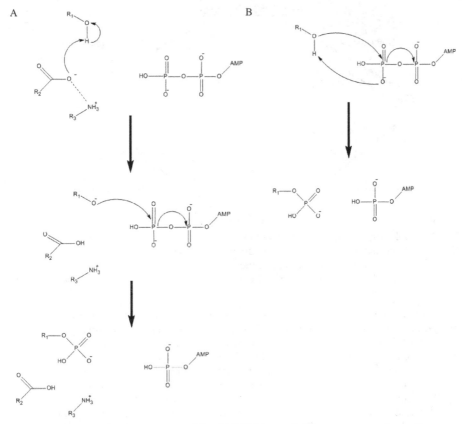

Fig. 1. Catalytic mechanisms proposed for GHMP kinase. **A.** The enzyme catalyzes the reaction through an active base residue R_1, which attracts a proton from the substrate R_3, converting the weakly nucleophilic hydroxyl to an alkoxide ion, which attacks the γ-phosphate of ATP. A positively charged residue R_2, sits close to the catalytic residue and stabilizes the alkoxide ion. **B.** There is no active base residue in the active center, the substrate directly attacks the γ-phosphate of ATP.

feature of the reaction is forming a covalent UMP-enzyme intermediate (Arabshahi et al., 1986). The intermediate was isolated by gel permeation chromatography in reaction mixtures containing the enzyme and radiolabeled UDP-Glu, and the radiolabeled intermediate could react with gal-1P or glu-1P to form the corresponding radiolabeled UDP sugar (Wong, et al., 1977a). This intermediate is very fragile in slightly acidic solutions but quite stable in strong basic solutions (Wong et al., 1977a; Yang and Frey, 1979), which indicates the intermediate is phosphoramides. Further degradation study of this intermediate confirmed that the nucleophile in GALT, to which the uridylyl group is bonded in the uridylyl-enzyme intermediate, is imidazole N3 of a histidine residue (Yang and Frey, 1979).

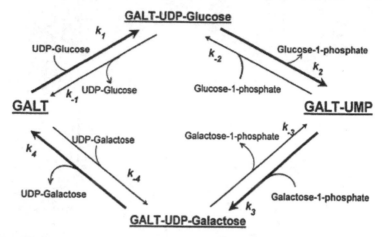

Fig. 2. Double displacement reactions of GALT. GALT binds to UDP-Glu to form a GALT-UDP-Glu intermediate. Glu-1-P is subsequently released, whereas the enzyme remains bound to UMP. Gal-1-P then reacts with the enzyme-UMP intermediate to form UDP-Gal, freeing the GALT enzyme for continued catalysis. k_n and k_{-n} denote rate constants of the forward and reverse reactions.

Substituting each of the 15 histidine residues in *E. coli* GALT with asparagines by SDM, proved that His[164] and His[166] were the only essential histidine residues in the enzyme (Field et al., 1989). In order to identify which of these two residues is the catalytic residue, two more specific mutations were introduced by SDM, *H164G* and *H166G*, which resulted in loss of function of the enzyme because of the missing imidazole ring of histidine, which might be filled and salvaged by adding exogenous imidazole ring. The experimental results showed that the activity of the H166G mutant could be recovered by adding exogenous imidazole ring, while mutant H164G could not. Therefore, His[166] provides the catalytic nucleophilic imidazole ring in the reaction (Kim et al., 1990).

Also, as mentioned earlier, by mutating Gln[188] of human GALT (equivalent to Gln[168] in *E. coli* GALT), the most common mutation found in Type I Galactosemia, to arginine and asparagine, respectively, we were able to determine that glutamine at position 188 stabilizes the UMP-GALT intermediate through hydrogen bonding and enables the double displacement of both glucose-1-phosphate (glu-1P) and UDP-galactose. The substitution of arginine or asparagine at position 188 reduces hydrogen bonding and destabilizes UMP-GALT. The unstable UMP-GALT allows single displacement of glu-1P with release of free GALT but impairs the subsequent binding of gal-1P and displacement of UDP-Gal (Lai, et al., 1999).

4.3.3 GALE

GALE catalyzes the inter-conversion of UDP-Glu and UDP-Gal to finish the Leloir pathway of galactose metabolism. There are four key steps for the reaction of GALE as shown in Fig. 3: (1) abstraction of the 4'-hydroxyl hydrogen of the sugar by an enzymatic base, (2) transfer of a hydride from C4 of the sugar to the C4 of NAD[+] leading to a 4'-ketopyranose intermediate and NADH, (3) rotation of the resulting 4'-ketopyranose intermediate in the active site, and (4) return of the hydride from NADH to the opposite face of the sugar (Maitra and Ankel, 1971). When purified, this enzyme contains tightly bound NAD+, which functions as an essential coenzyme to catalyze the reaction (Darrow and Rodstorm, 1968). The binding of the UDP group is strong, while binding with the galactosyl, glucosyl and 4-ketohexopyranosyl moieties is weak (Kang et al., 1975; Wong and Frey, 1977b). Early study on the catalytic mechanism of GALE focused on Lys[153], since it is close to the NAD[+], and the positively-charged ammonium group of Lys[153] may perturb the electron distribution in the nicotinamide ring of NAD+ through charge repulsion upon substrate binding (Swanson and Frey, 1993). Replacing this residue with alanine or methionine renders the inability of the mutant proteins to be reduced by the sugar in the presence or absence of UMP. As a result, the catalytic activities of the mutants decreased by a factor over 1000. Also the purified mutant contained much less NADH as compared with wild type (Swanson and Frey, 1993). These results indicate that Lys[153] plays an important role in the UMP-dependent reduction of GALE-NAD[+]. Further studies identified two more important residues, Tyr[149] and Ser[124], which are involved in glucose moiety binding (Thoden et al., 1996). SDM studies on the latter two residues revealed that that Tyr[149] provides the driving force for general acid-base catalysis, while Ser[124] plays an important role in mediating proton transfer (Liu et al., 1997). The crystal structure of human GALE confirmed that Tyr[149] (Tyr[157] for human GALE) sits at the proper position to interact directly with the 4′-hydroxyl group of the sugar and attracts the proton from the hydoxy group and transfers it to NAD[+] (Thoden et al., 2000).

Unlike what was observed for the E. coli enzyme, the human enzyme can also convert UDP-N-acetylglucosamine (UDP-GlcNAc) to UDP-N-acetylgalactosamine (UDP-GalNAc) (Kingsley et al., 1986; Piller et al., 1983). Through structure analysis and alignment, investigators found that, when the human enzyme equivalent of Tyr[299] in the E. coli protein is replaced with a cysteine residue (Cys[307]), the active site volume for the human protein is calculated to be approximately 15% larger than that observed for the bacterial epimerase (Thoden 2001). Substituting Tyr[299] of E. coli GALE with a cysteine residue by SDM confers UDP-GalNAc/UDP-GlcNAc converting activity to the bacterial enzyme with minimal changes in its three-dimensional structure. Specifically, although the Y299C mutation in the bacterial enzyme resulted in a loss of epimerase activity with regard to UDP-Gal by almost 5-fold, it resulted in a gain of activity against UDP-GalNAc by more than 230-fold (Thoden et al., 2002b).

5. Use of SDM in the development of novel treatment of Type I (classic or GALT-deficiency) galactosemia

5.1 The issues

Unlike Type II or the peripheral Type III Galactosemia, patients with Type I (GALT-deficiency) Galactosemia, also the most common type of Galactosemia, suffer a range of debilitating long-term complications, which include premature ovarian insufficiency, learning deficits, ataxia and speech dyspraxia (Lai et al., 2009; Berry and Elsas, 2011). The current galactose-restricted diet fails to prevent these complications, and the medical/patient communities are yearning for a more effective therapy. The causes of these organ-specific complications remain unknown, but there is a strong association with the intracellular accumulation of gal-1P. But what is the source of gal-1P in these patients with Classic Galactosemia if they limit their galactose intake? Recent studies have shown that the patients on a galactose-restricted diet are never really "galactose-free. A significant amount of galactose is found in non-dairy foodstuffs, such as vegetables and fruits (Berry et al., 1993; Acosta and Gross, 1995). More importantly, galactose is produced endogenously from the natural turnover of glycolipids and glycoproteins (Berry et al., 1995). Using isotopic labeling, Berry and coworkers demonstrated that a 50kg adult male could produce up to 2 grams of galactose per day (Berry et al., 1995, 2004). Once galactose is formed intracellularly, it is converted to gal-1P by GALK and in GALT-deficient patient cells. As a result, gal-1P is concentrated more than one order of magnitude above normal, even with strict adherence to a galactose-restricted diet. Accumulation of gal-1P is regarded as a major, if not sole, factor for the chronic complications seen in patients with Classic Galactosemia, as suggested by both clinical observation and experimental results from yeast models. Patients with inherited deficiency of GALK, who do not accumulate gal-1P, do not experience the brain and ovary complications seen in GALT-deficient patients (Gitzelmann et al., 1974; Gitzelmann 1975; Stambolian et al., 1986). While *gal7* (*i.e*, GALT-deficient) mutant yeast stops growing upon galactose challenge, a *gal7 gal1* double mutant strain (*i.e*, GALT- and GALK-deficient) is no longer sensitive to galactose (Douglas and Hawthorne, 1964, 1966). Based on these observations, in conjunction with dietary therapy, inhibiting GALK activity with a safe small-molecule inhibitor might prevent the squeals of chronic gal-1P exposure in patients with Classic Galactosemia.

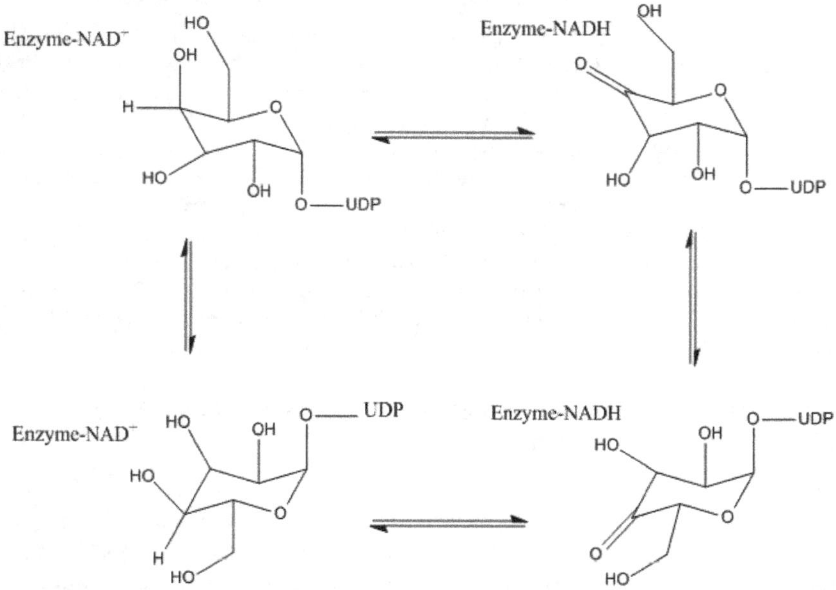

Fig. 3. Catalytic mechanism of GALE.

5.2 Research design

For the past few years, our group has conducted high-throughput screening (HTS) of small molecule compounds, which could inhibit human GALK enzyme *in vitro* (Tang et al., 2010;

Wierenga et al., 2008). To date, we have screened over 300,000 compounds of diverse chemical structures and identified a few promising hit compounds for further characterization. One of the characterization steps involved the use of SDM to change the respective amino acids of the GALK active site in order to confirm the predicted molecular interactions between the selected inhibitors and it target, GALK, through high-precision docking programs such as GLIDE (*Schrödinger*). Another characterization step that is noteworthy to mention is the assay for the kinase selectivity of the selected GALK inhibitors. As alluded above, GALK belongs to a unique small molecule kinase family, the GHMP kinase family (Bork et al., 1993). While the substrates of the GHMP kinases differ widely, the ATP-binding sites of the enzymes share a significant degree of structural homology (Tang et al., 2010). It is, therefore, important to ensure our selected GALK inhibitors did not cross-inhibit other GHMP kinases or other kinases in general.

5.3 The results

Selectivity is always one of the most important properties for developing therapeutic kinase inhibitors because of potential side-effects from unwanted inhibition of other kinases. During the characterization phase of our hit compounds, we found six compounds that selectively inhibit GALK but not any of the other GHMP kinases. These included MVK, which shares a high degree of structural similarity with GALK (Tang et al., 2010). In order to understand what structural elements conferred the specificity of these compounds, we aligned the crystal structure of human GALK and human MVK and focused on the ATP-binding site. Eight amino acid residues and the L1 loop were found to be different in these two kinases. SDM was employed to mutate each residue individually or the L1 loop, and the effects of the changes on the inhibitory capabilities of the compounds were tested. Two compounds were found to be affected by the mutation *S140G* (Table 1) (Tang et al., 2010). Ser[140] of GALK resides in the signature motif of the GHMP kinase family, Motif II; but this amino acid is not conserved among the GHMP kinases. GALK is the only member that has a serine at this site. This could explain the selectivity of these two compounds. Furthermore, computational molecular docking confirmed that these two compounds interacted with Ser[140] through hydrogen bonds; substituting serine with glycine abolished the hydrogen bonds and totally compromised the binding of the compounds to the enzymes.

Our use of SDM in the characterization of promising GALK inhibitors not only helped identify and confirm the amino acids of GALK with which these small molecules interact, but also exemplified a more rapid and cost-effective way to study the structural interactions between small molecule modifiers and their targets. This novel approach is particularly useful when large-scale co-crystallization projects are not feasible. These studies paved the way for more in-depth investigations to identify the structural determinants required for the inhibitor selectivity of GALK and GHMP kinases.

6. Concluding remarks

Using the disease Galactosemia as an example, we showed that site-directed mutagenesis (**SDM**) plays a vital role in biomedical research. As in the case of Galactosemia, in which the diagnosis begins at the bedside of the affected newborns, **SDM** can be employed in every step of basic and translational research in an attempt to improve the prognosis and treatment of the patients. Further, not only did we show that **SDM** can be applied in traditional applications, such as expression analysis, we have also expanded its use in innovative drug design and the basic understanding of kinase inhibitor selectivity.

Mutations	k_{cat} (S^{-1})	K_M of ATP (μM)	K_M of Galactose (μM)	Effects on IC$_{50}$ of compound 1	Effects on IC$_{50}$ of compound 4	Effects on IC$_{50}$ of compound 24
T77L	4.1	218.4	1305.2	None	None	None
S79N	4.8	303.4	1227.3	None	None	None
L145Y	11.6	259.7	222.7	None	None	None
L145A	6.4	379.9	356.8	None	None	None
W106A	No protein expression	-	-	-	-	-
W106T	No protein expression	-	-	-	-	-
Y109L	43.2	70.2	963.2	None	None	None
Y109A	8.7	579.3	268.7	None	None	None
GALK Loop to MVK Loop	0.1	695.4	1857.3	None	None	None
S140G	2.1	8.2	141.9	None	Increased 10-fold	Increased 20-fold
L135P	13.3	51.1	544.9	None	None	None
R37K	0.4	6.4	623.8	None	None	None
R37A	No activity	-	-	-	-	-
WT	17.5	20.9	319	-	-	-

Table 1. Effect of amino acid changes in human GALK on their enzymatic properties and the IC$_{50}$ of selected inhibitors

7. Acknowledgement

We acknowledge that we could not have completed this manuscript without the outstanding contributions made by our scientific and clinical colleagues, as well as patient volunteers. Research grant support to Kent Lai includes NIH grants 5R01 HD054744-04 and 3R01 HD054744-04S1.

8. References

Acosta, P. B. and Gross, K. C., 1995. Hidden sources of galactose in the environment. Eur. J. Pediatr. 154(7 Suppl 2): S87-92.

Alano, A., Almashanu, S., Chinsky, J. M., Costeas, P., Blitzer, M. G., Wulfsberg, E. A., and Cowan, T.M., 1998. Molecular characterization of a unique patient with epimerase-deficiency galactosaemia. J. Inherit. Metab. Dis. 21(4): 341-350.

Alano A, Almashanu, S., Maceratesi, P., Reichardt, J., Panny, S., and Cowan T.M., 1997. UDP-galactose-4-epimerase deficiency among African-Americans: evidence for multiple alleles. J. Invest. Med. 45: :191A.

Arabshahi, A., Brody, R. S., Smallwood, A., Tsai, T.C., and Frey, P.A., 1986. Galactose-1-phosphate uridylyltransferase. Purification of the enzyme and stereochemical course of each step of the double-displacement mechanism. Biochemistry. 25(19): 5583-5589.

Beebe, J. A. and Frey, P. A., 1998. Galactose mutarotase: purification, characterization, and investigations of two important histidine residues. Biochemistry. 37(42): 14989-14997.

Berry, G. T., Palmieri, M., Gross, K. C., Acosta, P. B., Henstenburg, J. A., Mazur, A., Reynolds, R., and Segal, S., 1993. The effect of dietary fruits and vegetables on urinary galactitol excretion in galactose-1-phosphate uridyltransferase deficiency. J. Inherit. Metab. Dis. 16(1): 91-100.

Berry, G. T., Nissim, I., Lin, Z., Mazur, A. T., Gibson, J. B., and Segal, S., 1995. Endogenous synthesis of galactose in normal men and patients with hereditary galactosaemia. Lancet. 346(8982): 1073-1074.

Berry, G. T., Moate, P. J., Reynolds, R.A., Yager, C. T., Ning, C., Boston, R. C., and Segal, S., 2004. The rate of de novo galactose synthesis in patients with galactose-1-phosphate uridyltransferase deficiency. Mol. Genet. Metab. 81(1): 22-30.

Berry, G. T. and Elsas, L. J., 2011. Introduction to the Maastricht workshop: lessons from the past and new directions in galactosemia. J. Inherit. Metab. Dis. 34(2): 249-255.

Bork, P., Sander, C., and Valencia, A., 1993. Convergent evolution of similar enzymatic function on different protein folds: the hexokinase, ribokinase, and galactokinase families of sugar kinases. Protein. Sci. 2(1): 31-40.

Bosch, A. M., Bakker, H. D., van Gennip, A. H., van Kempen, J. V., Wanders, R. J., and Wijburg, F. A., 2002. Clinical features of galactokinase deficiency: a review of the literature. J. Inherit. Metab. Dis. 25(8): 629-634.

Cardini, C. E. and Leloir, L. F., 1953. Enzymic phosphorylation of galactosamine and galactose. Arch. Biochem. Biophys. 45(1): 55-64.

Carney, A. E., Sanders, R. D.,, Garza, K. R., McGaha, L. A., Bean, L. J., Coffee, B. W., Thomas, J. W., Cutler, D. J., Kurtkaya, N. L., and Fridovich-Keil, J. L., 2009. Origins, distribution and expression of the Duarte-2 (D2) allele of galactose-1-phosphate uridylyltransferase. Hum. Mol. Genet. 18(9): 1624-1632.

Chhay, J. S., Vargas, C. A.,, McCorvie, T. J., Fridovich-Keil, and Timson, D. J., 2008. Analysis of UDP-galactose 4'-epimerase mutations associated with the intermediate form of type III galactosaemia. J. Inherit. Metab. Dis. 31(1): 108-116.

Coffee, B., Hjelm, L. N., DeLorenzo, A., Courtney, E. M., Yu, C., and Muralidharan, K., 2006. Characterization of an unusual deletion of the galactose-1-phosphate uridyl transferase (GALT) gene. Genet. Med. 8(10): 635-640.

Darrow, R. A. and Rodstrom, R., 1968. Purification and properties of uridine diphosphate galactose 4-epimerase from yeast. Biochemistry. 7(5): 1645-1654.

Douglas, H. C. and Hawthorne, D. C., 1964. Enzymatic Expression And Genetic Linkage Of Genes Controlling Galactose Utilization In Saccharomyces. Genetics 49: 837-844.

Douglas, H. C. and Hawthorne, D. C., 1966. Regulation of genes controlling synthesis of the galactose pathway enzymes in yeast. Genetics 54(3): 911-916.

Elsas, L. J., 1993. Galactosemia. (http://www.ncbi.nlm.nih.gov/books/NBK1518/)

Elsas, L. J., Dembure, P. P., Langley, S., Paulk, E. M., Hjelm, L. N., and Fridovich-Keil, J., 1994. A common mutation associated with the Duarte galactosemia allele. Am. J. Hum. Genet. 54(6): 1030-1036.

Elsevier, J. P. and Fridovich-Keil, J. L., 1996. The Q188R mutation in human galactose-1-phosphate uridylyltransferase acts as a partial dominant negative. J. Biol. Chem. 271(50): 32002-32007.

Field, T. L., Reznikoff, W. S., and Frey, P. A.., 1989. Galactose-1-phosphate uridylyltransferase: identification of histidine-164 and histidine-166 as critical residues by site-directed mutagenesis. Biochemistry 28(5): 2094-2099.

Fridovich-Keil, J., Bean, L., He, M., and Schroer, R., 1993a. Epimerase Deficiency Galactosemia. (http://www.ncbi.nlm.nih.gov/books/NBK51671/)

Fridovich-Keil, J. L. and Jinks-Robertson, S., 1993b. A yeast expression system for human galactose-1-phosphate uridylyltransferase. Proc. Natl. Acad. Sci. U. S. A. 90(2): 398-402.

Fridovich-Keil, J. L., Langley, S. D., Mazur, L. A., and Elsevier, J. P., 1995a. Identification and functional analysis of three distinct mutations in the human galactose-1-phosphate uridyltransferase gene associated with galactosemia in a single family. Am. J. Hum. Genet. 56(3): 640-646.

Fridovich-Keil, J. L., Quimby, B. B., Wells, L., Mazur, L. A., and Elsevier, J. P., 1995b. Characterization of the N314D allele of human galactose-1-phosphate uridylyltransferase using a yeast expression system. Biochem. Mol. Med. 56(2): 121-130.

Fu, Z., Wang, M., Potter, D., Miziorko, H. M., and Kim, J. J., 2002. The structure of a binary complex between a mammalian mevalonate kinase and ATP: insights into the reaction mechanism and human inherited disease. J. Biol. Chem. 277(20): 18134-18142.

Gitzelmann, R., Wells, H. J., and Segal, S., 1974. Galactose metabolism in a patient with hereditary galactokinase deficiency. Eur. J. Clin. Invest. 4(2): 79-84.

Gitzelmann, R., 1975. Letter: Additional findings in galactokinase deficiency. J. Pediatr. 87(6 Pt 1): 1007-1008.

Greber-Platzer, S., Guldberg, P., Scheibenreiter, S., Item., C., Schuller, E., Patel, N., and Strobl, W., 1997. Molecular heterogeneity of classical and Duarte galactosemia: mutation analysis by denaturing gradient gel electrophoresis. Hum. Mutat. 10(1): 49-57.

Guerrero, N. V., Singh, R. H., Manatunga, A., Berry, G. T., Steiner, R. D., Elsas, L. J., 2nd, 2000. Risk factors for premature ovarian failure in females with galactosemia. J. Pediatr. 137(6): 833-841.

Holden, H. M., Thoden, J. B., Timson, D. J., and Reece, R. J., 2004. Galactokinase: structure, function and role in type II galactosemia. Cell. Mol. Life. Sci. 61(19-20): 2471-2484.

Kalckar, H. M., Braganca, B., Munch-Petersen, H. M., 1953. Uridyl transferases and the formation of uridine diphosphogalactose. Nature. 172(4388): 1038.

Kang, U. G., Nolan, L. D., and Frey, P.A., 1975. Uridine diphosphate galactose-4-epimerase. Uridine monophosphate-dependent reduction by alpha- and beta-D-glucose. J. Biol. Chem. 250(18): 7099-7105.

Kaye, C. I., Accurso, F., La Franchi, S., Lane, P. A., Hope, N., Sonya, P., S, G. B., and Michele, A. L., 2006. Newborn screening fact sheets. Pediatrics. 118(3): e934-963.

Kim, J., Ruzicka, F., and Frey, P.A., 1990. Remodeling hexose-1-phosphate uridylyltransferase: mechanism-inspired mutation into a new enzyme, UDP-hexose synthase. Biochemistry. 29(47): 10590-10593.

Kingsley, D. M., Kozarsky, K. F., Hobbie, L., and Krieger, M., 1986. Reversible defects in O-linked glycosylation and LDL receptor expression in a UDP-Gal/UDP-GalNAc 4-epimerase deficient mutant. Cell. 44(5): 749-759.

Kozak, L., Francova, H., Fajkusova, L., Pijackova, A., Macku, J., Stastna, S., Peskovova, K., Martincova, O., Krijt, J., and Bzduch, V., 2000. Mutation analysis of the GALT gene in Czech and Slovak galactosemia populations: identification of six novel mutations, including a stop codon mutation (X380R). Hum. Mutat. 15(2): 206.

Krepkiy, D. and Miziorko, H. M., 2004. Identification of active site residues in mevalonate diphosphate decarboxylase: implications for a family of phosphotransferases. Protein Sci. 13(7): 1875-1881.

Krishna, S. S., Zhou, T., Daugherty, M., Osterman, A., and Zhang, H., 2001. Structural basis for the catalysis and substrate specificity of homoserine kinase. Biochemistry. 40(36): 10810-10818.

Lai, K., Langley, S. D., Singh, R. H., Dembure, P. P., Hjelm, L. N., and Elsas, L. J., 2nd, 1996. A prevalent mutation for galactosemia among black Americans. J. Pediatr. 128(1): 89-95.

Lai, K., Willis, A. C., and Elsas, L. J., 1999. The biochemical role of glutamine 188 in human galactose-1-phosphate uridyltransferase. J. Biol. Chem. 274(10): 6559-6566.

Lai, K. and Elsas, L. J., 2001. Structure-function analyses of a common mutation in blacks with transferase-deficiency galactosemia. Mol. Genet. Metab. 74(1-2): 264-272.

Lai, K., Elsas, L. J., and Wierenga, K. J., 2009. Galactose toxicity in animals. IUBMB Life. 61(11): 1063-1074.

Landt, M., Ritter, D., Lai, K., Benke, P. J., Elsas, L. J., and Steiner, R. D., 1997. Black children deficient in galactose 1-phosphate uridyltransferase: correlation of activity and immunoreactive protein in erythrocytes and leukocytes. J. Pediatr. 130(6): 972-980.

Leloir, L. F., 1951. The enzymatic transformation of uridine diphosphate glucose into a galactose derivative. Arch. Biochem. Biophys. 33(2): 186-190.

Leloir, L. F., 1953. Enzymic isomerization and related processes. Adv. Enzymol. Relat. Subj. Biochem. 14: 193-218.

Levy, H. L., 2010. Newborn screening conditions: What we know, what we do not know, and how we will know it. Genet. Med. 12(12 Suppl): S213-214.

Liu, Y., Thoden, J. B., Kim, J., Berger, E., Gulick, A. M., Ruzicka, F. J., Holden, H. M., and Frey, P. A., et al., 1997. Mechanistic roles of tyrosine 149 and serine 124 in UDP-galactose 4-epimerase from Escherichia coli. Biochemistry. 36(35): 10675-10684.

Maceratesi, P., Daude, N., Dallapiccola, B., Novelli, G., Allen, R., Okano, Y., and Reichardt, J., 1998. Human UDP-galactose 4' epimerase (GALE) gene and identification of five missense mutations in patients with epimerase-deficiency galactosemia. Mol. Genet. Metab. 63(1): 26-30.

Maitra, U. S. and Ankel, H., 1971. Uridine diphosphate-4-keto-glucose, an intermediate in the uridine diphosphate-galactose-4-epimerase reaction. Proc. Natl. Acad. Sci. U. S. A. 68(11): 2660-2663.

Murphy, M., McHugh, B., Tighe, O., Mayne, P., O'Neill, C., Naughten, E., Croke, D. T., 1999. Genetic basis of transferase-deficient galactosaemia in Ireland and the population history of the Irish Travellers. Eur. J. Hum. Genet. 7(5): 549-554.

Openo, K. K., Schulz, J. M., Vargas, C. A., Orton, C. S., Epstein, M. P., Schnur, R. E., Scaglia, F., Berry, G. T., Gottesman, G. S., Ficicioglu, C., Slonim AE, Schroer RJ, Yu C, Rangel VE, Keenan J, Lamance K, and Fridovich-Keil, J., 2006. Epimerase-deficiency galactosemia is not a binary condition. Am. J. Hum. Genet. 78(1): 89-102.

Park, H. D., Park, K. U., Kim, J. Q., Shin C. H., Yang, S. W., Lee, D. H., Song, Y. H., and Song, J , 2005. The molecular basis of UDP-galactose-4-epimerase (GALE) deficiency galactosemia in Korean patients. Genet. Med. 7(9): 646-649.

Park, H. D., Bang, Y. L., Park, K. U., Kim, J. Q., Jeong, B. H., Kim, Y.S., Song, Y. H., and Song, J., 2007. Molecular and biochemical characterization of the GALK1 gene in Korean patients with galactokinase deficiency. Mol. Genet. Metab. 91(3): 234-238.

Piller, F., Hanlon, M. H., and Hill, R. L., 1983. Co-purification and characterization of UDP-glucose 4-epimerase and UDP-N-acetylglucosamine 4-epimerase from porcine submaxillary glands. J. Biol. Chem. 258(17): 10774-10778.

Potter, D., Wojnar, J. M., Narasimhan, C., and Miziorko, H. M., 1997. Identification and functional characterization of an active-site lysine in mevalonate kinase. J. Biol. Chem. 272(9): 5741-5746.

Quimby, B. B., Alano, A., Almashanu S., DeSandro, A. M., Cowan, T. M., and Fridovich-keil, J. L., 1997. Characterization of two mutations associated with epimerase-deficiency galactosemia, by use of a yeast expression system for human UDP-galactose-4-epimerase. Am. J. Hum. Genet. 61(3): 590-598.

Reichardt, J. K., Packman, S., and Woo, S. L., 1991. Molecular characterization of two galactosemia mutations: correlation of mutations with highly conserved domains in galactose-1-phosphate uridyl transferase. Am. J. Hum. Genet. 49(4): 860-867.

Reichardt, J. K., Levy, H. L., and Woo, S. L., 1992. Molecular characterization of two galactosemia mutations and one polymorphism: implications for structure-function analysis of human galactose-1-phosphate uridyltransferase. Biochemistry. 31(24): 5430-5433.

Riehman, K., Crews, C., and Fridovich-Keil, J. L., 2001. Relationship between genotype, activity, and galactose sensitivity in yeast expressing patient alleles of human galactose-1-phosphate uridylyltransferase. J. Biol. Chem. 276(14): 10634-10640.

Sanders, R. D., Sefton, J. M., Moberg, K. H., and Fridovich-Keil, J. L., 2010. UDP-galactose 4' epimerase (GALE) is essential for development of Drosophila melanogaster. Dis. Model. Mech. 3(9-10): 628-638.

Segal, S., Berry, GT., 1995. Disorders of galactose metabolism. The Metabolic Basis of Inherited Diseases. B. A. Scriver D, Sly W, Valle D. New York, McGraw-Hill. I: 967-1000.

Shin, Y. S., Koch, H. G., Kohler, M., Hoffmann, G., Patsoura, A., Podskarbi, T., 1998. Duarte-1 (Los Angeles) and Duarte-2 (Duarte) variants in Germany: two new mutations in the GALT gene which cause a GALT activity decrease by 40-50% of normal in red cells. J. Inherit. Metab. Dis. 21(3): 232-235.

Stambolian, D., Scarpino-Myers, V., Eagle, R. C., Jr., Hodes, B., and Harris, H., 1986. Cataracts in patients heterozygous for galactokinase deficiency. Invest. Ophthalmol. Vis. Sci. 27(3): 429-433.

Swanson, B. A. and Frey, P. A., 1993. Identification of lysine 153 as a functionally important residue in UDP-galactose 4-epimerase from Escherichia coli. Biochemistry. 32(48): 13231-13236.

Tang, M., Wierenga, K., Elsas, L. J., and Lai, K., 2010. Molecular and biochemical characterization of human galactokinase and its small molecule inhibitors. Chem. Biol. Interact. 188(3): 376-385.

Thoden, J. B., Frey, P. A., and Holden, H. M., 1996. Molecular structure of the NADH/UDP-glucose abortive complex of UDP-galactose 4-epimerase from Escherichia coli: implications for the catalytic mechanism. Biochemistry. 35(16): 5137-5144.

Thoden, J. B., Wohlers, T. M., Fridovich-Keil, J. L., and Holden H. M., 2000. Crystallographic evidence for Tyr 157 functioning as the active site base in human UDP-galactose 4-epimerase. Biochemistry. 39(19): 5691-5701.

Thoden, J.B., Wholers. T., Fridovich-Keil, J.L., Holden, H.M., 2001. Human UDP-galactose 4-epimerase. Accommodation of UDP-N-acetylglucosamine within the active site. J. Biol. Chem. 4;276(18):: 15131-15136.

Thoden, J. B. and Holden, H. M., 2002a. High resolution X-ray structure of galactose mutarotase from Lactococcus lactis. J. Biol. Chem. 277(23): 20854-20861.

Thoden, J. B., Henderson, J. M., Fridovich-Keil, J. L., and Holden, H. M., 2002b. Structural analysis of the Y299C mutant of Escherichia coli UDP-galactose 4-epimerase. Teaching an old dog new tricks. J. Biol. Chem. 277(30): 27528-27534.

Thoden, J. B. and Holden, H. M., 2003. Molecular structure of galactokinase. J. Biol. Chem. 278(35): 33305-33311.

Thoden, J. B., Timson, D. J., Reece, R. J., and Holden, H. M., 2005. Molecular structure of human galactokinase: implications for type II galactosemia. J. Biol. Chem. 280(10): 9662-9670.

Timson, D. J. and Reece, R. J., 2003. Functional analysis of disease-causing mutations in human galactokinase. Eur. J. Biochem. 270(8): 1767-1774.

Timson, D. J., 2005. Functional analysis of disease-causing mutations in human UDP-galactose 4-epimerase. FEBS J 272(23): 6170-6177.

Waggoner, D. D., Buist, N. R., and Donnell, G. N., 1990. Long-term prognosis in galactosaemia: results of a survey of 350 cases."J. Inherit. Metab. Dis. 13(6): 802-818.

Waisbren, S. E., Potter, N. L., Gordon C. M., Green, R. C., Greenstein, P., Gubbels, C. S., Rubio-Gozalbo, E., Schomer, D., Welt, C., Anastasoaie, V., D'Anna, K, Gentile, J., Guo, C.Y., Hecht, L., Jackson, R., Jansma, B. M., Li, Y., Lip, V., Miller, D. T., Murray, M., Power, L., Quinn, N., Rohr, F., Shen, Y., Skinder-Meredith, A., Timmers, I., Tunick, R., Wessel, A., Wu, B. L, Levy, H., Berry, G. T., 2011. The adult galactosemic phenotype. J. Inherit. Metab. Dis.

Webb, A. L., Singh, R. H., Kennedy, M. J., and Elsas, L. J., 2003. Verbal dyspraxia and galactosemia. Pediatr. Res. 53(3): 396-402.

Wedekind, J. E., Frey, P. A., and Raymond, I., 1996. The structure of nucleotidylated histidine-166 of galactose-1-phosphate uridylyltransferase provides insight into phosphoryl group transfer. Biochemistry. 35(36): 11560-11569.

Wells, L. and Fridovich-Keil, J. L., 1997. Biochemical characterization of the S135L allele of galactose-1-phosphate uridylyltransferase associated with galactosaemia. J. Inherit. Metab. Dis. 20(5): 633-642.

Wierenga, K. J., Lai, K., Buchwald, P. and Tang, M., 2008. High-throughput screening for human galactokinase inhibitors. J. Biomol. Screen. 13(5): 415-423.

Wohlers, T. M., Christacos, N. C., Harreman, M. T., and Fridovich-Keil, J. L., 1999. Identification and characterization of a mutation, in the human UDP-galactose-4-epimerase gene, associated with generalized epimerase-deficiency galactosemia. Am. J. Hum. Genet. 64(2): 462-470.

Wohlers, T. M. and Fridovich-Keil, J. L. 2000. Studies of the V94M-substituted human UDPgalactose-4-epimerase enzyme associated with generalized epimerase-deficiency galactosaemia. J. Inherit. Metab. Dis. 23(7): 713-729.

Wong, L. J., Sheu, K. F., Lee, S. I. and Frey, P. A., 1977a. Galactose-1-phosphate uridylyltransferase: isolation and properties of a uridylyl-enzyme intermediate. Biochemistry. 16(5): 1010-1016.

Wong, S. S. and Frey, P. A., 1977b. Fluorescence and nucleotide binding properties of Escherichia coli uridine diphosphate galactose 4-epimerase: support for a model for nonstereospedific action. Biochemistry. 16(2): 298-305.

Yang, D., Shipman, L. W., Roessner, C. A., Scott, A. I., Sacchettini, J. C., 2002. Structure of the Methanococcus jannaschii mevalonate kinase, a member of the GHMP kinase superfamily. J. Biol. Chem. 277(11): 9462-9467.

Yang, S. L. and Frey, P. A., 1979. Nucleophile in the active site of Escherichia coli galactose-1-phosphate uridylyltransferase: degradation of the uridylyl-enzyme intermediate to N3-phosphohistidine. Biochemistry. 18(14): 2980-2984.

4

Inherited Connective Tissue Disorders of Collagens: Lessons from Targeted Mutagenesis

Christelle Bonod-Bidaud and Florence Ruggiero

Institut de Génomique Fonctionnelle de Lyon, ENS de Lyon, UMR CNRS 5242,
University Lyon 1,
France

1. Introduction

The extracellular matrix (ECM) is the cell structural environment in tissues and organs. The ECM is a dynamic structure that it is constantly remodelled. It contributes to tissue integrity and mechanical properties. It is also essential for maintaining tissue homeostasis, morphogenesis and differentiation, which it does, through specific interactions with cells. The ECM is composed of a mixture of water and macromolecules classified into four main categories: collagens, proteoglycans, elastic proteins, and non-collagenous glycoproteins (also called adhesive glycoproteins). The nature, concentration and ratio of the different ECM components are all important factors in the regulation of the assembly of complex tissue-specific networks tuned to meet mechanical and biological requirements of tissues.

Collagens form a superfamily of 28 trimeric proteins, distinguishable from the other ECM components by their particular abundance in tissues (collagens represent up to 80-90% of total proteins in skin, tendon and bones) and their capacity to self-assemble into supramolecular organized structures (the best known being the banded fibers). The collagen superfamily is highly complex and shows a remarkable diversity in structure, tissue distribution and function (Ricard-Blum and Ruggiero, 2005).

The importance of collagens has been illustrated by the wide range of mutations in collagen genes that result in minor and severe human diseases. Various mutations (point, null or structural mutations, insertions, exon skipping, deletions) in genes encoding collagens are known to be responsible for a large spectrum of human disorders (*e.g.*, Elhers-Danlos syndrome, epidermolysis bullosa, chondrodysplasia, osteogenesis imperfecta, Alport syndrome, Bethlem myopathy, Ulrich congenital muscular dystrophy, Fuchs' endothelial dystrophy, Knobloch syndrome) that affect different tissues and organs, such as skin, blood vessels, cartilage, bones, kidney, muscle, cornea and retina. Considering the variety of collagen-related diseases and the complexity of collagen biology, there is a clear need to understand how mutations alter collagen synthesis, cell trafficking, cell and molecular interactions to result in tissue dysfunction. In the eighties targeted mutagenesis emerged as a new approach to help establish the structure-function relationship of collagens. Along with the emergence of protein engineering and genetically modified mice, site-directed mutagenesis has become instrumental in understanding the physiopathology of diseases, as well as in developing new and specific therapies and drugs for the treatment of human diseases. To date about 20 distinct genes encoding collagen chains have been ablated (by knock-out mutations) in mice or are involved in naturally occurring mutations. Only a few knock-in modified mice has been generated, in which a single point mutation or an exon deletion, for example, has been generated in a specific gene. This is likely due to the very large size of collagen genes. Site-directed knock-in mutations in mice have often proven to be more useful than knock-out mutations (which inactivate genes) for the analysis of the genotype-phenotype relationship, since small mutations represent the primary bases of inherited diseases.

The aim of this chapter is to describe the use of targeted mutagenesis in the understanding of the physiopathology of inherited connective tissue disorders. Specifically we are concerned with mutations in collagen genes. We will focus on the use of site-directed mutagenesis to analyze the causative effects of human-identified collagen gene mutations. Recombinant molecules were used to analyze the effects of these mutations on collagen structure, biosynthesis, posttranslational modifications and interactions with binding partners and cells. This work has considerably improved our knowledge in development and in human disorders. These results will then be compared with the limited information about the introduction of subtle targeted mutations into murine collagen genes.

2. The collagen superfamily at a glance

The 28 members of the collagen superfamily exhibit considerable complexity and diversity in structure, assembly and function. However, collagens also share common features. (i) All members are modular proteins composed of collagenous (COL) domains flanked by non collagenous (NC) domains or linker regions. (ii) They are trimeric molecules formed by the association of three identical or different α-chains, which are characterized by repetitions of the G-x-y tripeptide (with the x and y positions often occupied by proline and hydroxyproline, respectively). (Abbreviations and single-letter codes for amino acids are given in Table 1 of the chapter by Figurski *et al.*) (iii) They are able to assemble into supramolecular aggregates in the extracellular space, although this property has not been proven for all recently identified collagen members. Collagens also undergo various post-translational modifications, including proteolytic processing, fibril formation, reticulation, shedding of transmembrane collagens and production of functional domains (also called matricryptins) (Ricard-Blum and Ruggiero, 2005). The mechanisms of collagen biosynthesis are far from being completely understood. Our knowledge is primarily based on the biosynthesis of fibril-forming collagens. Triple-helix formation commonly starts at the C-terminus (C-propeptide) of the α-pro-chain and proceeds toward the N-terminus (N-propeptide) in a zipper-like fashion. Prior to and simultaneously with triple-helix formation, specific prolines and lysines are chemically modified by addition of hydroxyl group. These modifications play a pivotal role in stabilization and resistance to temperature. Completed trimeric procollagens are secreted from the cells, proteolytically processed and assemble into collagen fibrils (Ricard-Blum and Ruggiero, 2005).

Based on their structure and supramolecular organization, collagens have been divided into several subfamilies (Myllyharju and Kivirikko, 2001). They are (i) the fibril-forming collagens I, II, III, V, XI, XXIV and XVII, which share the capacity to assemble into organized fibrils; (ii) the network-forming collagens IV, VIII and X and the FACIT (Fibril-Associated Collagen with Interrupted Triple-helix collagens) collagens IX, XII, XIV, XVI, XIX, XX, XI and XXII, which are known to mediate protein-protein interactions; (iii) the basement membrane multiplexin (multiple triple-helix domains and interruptions) collagens XV and XVIII; (iv) the transmembrane collagens, including the neuronal XXV collagen and types XIII, XVII, XXIII; and finally (v) other unconventional collagens, such as the anchoring fibrils collagen VII and the ubiquitous collagen VI, which assembles into characteristic beaded filaments (Table 1).

The length of the triple helical domains varies noticeably among different collagen types. Fibril-forming collagens consist of a long central COL domain with about 1000 amino acids (330 G-x-y tripeptide repeats), flanked by small terminal globular extensions (NC domains). After proteolytic processing of the N and C-terminal extensions, the mature molecules aggregate into highly ordered fibrils with a banded pattern observable by transmission electron microscopy. In other collagens, the COL domains are shorter and/or contain interruptions. The NC domains can represent the main part of the molecule, as for the FACIT collagen XII. Most, if not all, collagen types are recognized by specific cell receptors, such as the major ECM integrin receptors, collagen-specific discoidin domain receptors (DDR) and the transmembrane proteoglycan syndecans (Humphries et al., 2006; Xian et al., 2010; Leitinger et al., 2007). Through various interactions with these cell receptors, collagens can induce intracellular pathways directly or indirectly and regulate cell functions, such as migration, proliferation and differentiation. Certain collagens can also bind to growth factors and control their bioavailability by acting as reservoirs. The controlled release of growth factors by proteolytic activity or expression of a splice variant that does not contain the binding site controls morphogenesis, as described for the cartilage collagen II (Zhu et al., 1999).

3. A large spectrum of mutations in collagen genes causes inherited disorders

A myriad of mutations has been characterized in collagen genes (Table 1). The function of the gene product and its tissue localization are criteria that lead to a number of inherited connective tissues disorders (reviewed in Bruckner-Tuderman and Bruckner, 1998; Bateman et al., 2009). Typically mutations in collagen genes are null-mutations, i.e., those resulting in the translation of an α-chain that cannot assemble into a triple helix and is consequently degraded intracellularly. Null mutations reduce the overall quantity of collagen in tissue and generally cause a human disorder. Small deletions and base substitutions can lead to synthesis of a mutated α-chain that is able to form a triple helix. The molecule is secreted, but its structure is compromised for supramolecular assembly, which normally occurs in the extracellular space. In fine collagen gene mutations result in defective matrix assembly and organization that in turn can affect cell function (Figure 1). In cases of large multimeric molecules, such as collagens, dominant-negative mutations can be more deleterious than null mutations. However, a growing body of evidence shows that the synthesis of a large quantity of abnormal collagen molecules in cells during development can induce endoplasmic reticulum stress, with consequences ranging from cell recovery to death (Tsang et al., 2010). The correlation between phenotype severity and the location of a point mutation in the gene is not clear. However, a mutation located in the coding region for the amino-terminus of the fibrillar collagen triple helix generally results in a mild phenotype, whereas a mutation in the coding region for the carboxy-terminus of the molecule is often lethal. This observation may be related to the C- to N-terminus directional propagation of the triple helix and the role of the C-propeptides in α-chain registration and triple helix nucleation. The nature of glycine substitution in the G-x-y repeats and the neighboring amino-acid sequence may have different biochemical and clinical consequences. These consequences include (i) delay of the triple-helix formation and over-glycosylation (Raghunath et al., 1994); (ii) alteration of procollagen processing (Lightfoot et al., 1994); (iii) retention of unfolded abnormal proteins intracellularly, leading to ER stress; and (iv) formation of abnormal unstable trimeric molecules, leading to disrupted fibrillogenesis.

The presence of a glycine in every third position is critical for triple-helix formation, since only glycine, the smallest amino acid, fits into the center of the triple helix. The majority of dominant-negative mutations in collagen genes are due to replacements of one of the glycines in the collagenous domains of the α-chains with a larger amino acid. Glycine substitution mutations in collagen genes underlie heritable connective tissue diseases, such as osteogenesis imperfecta (OI), chondrodysplasias, certain subtypes of Ehlers-Danlos syndrome (EDS), or Alport's syndrome (reviewed in Bruckner-Tuderman and Bruckner, 1998; Bateman et al., 2009). Since a non-glycine amino acid does not easily fit into the interior space of the triple helix, helix formation is distorted, thereby affecting its structure and stability and impeding fibrillogenesis. Delay in triple-helix formation can result in over-modification and may affect collagen function.

Osteogenesis imperfecta (OI), also known as brittle bone disease, is caused by mutations in genes for collagen I, the most abundant collagen in organisms. OI is characterized by fragile bones that break easily and reduced bone mass. Most OI cases are believed to be associated with glycine substitution mutations in the COL1A1 or COL1A2 genes. Over 200 mutations have been reported for the COL1A1 (located on chromosome 17) and COL1A2 (located on chromosome 7) genes, which code for the collagen I pro-α1 and pro-α2 chains, respectively. This fact may explain the wide range of clinical characteristics and degrees of severity that are seen in the disease (Kuivaniemi et al., 1991, Byers and Steiner 1992, Dalgleish, 1998). Because collagen I is found in other tissues of the body, OI has non-skeletal manifestations as well. People with OI may also suffer from muscle weakness, hearing loss, fatigue, joint laxity, distensible skin, or dentinogenesis imperfecta. The fibril-forming collagen I is mostly synthesized as the $[\alpha1(I)]_2\alpha2(I)$ heterotrimer chain, though a minor form $[\alpha1(I)]_3$ is expressed in embryonic tissues. COL1A1 and COL2A1 are both susceptible to various mutations responsible for the production of quantitatively or qualitatively deficient fibrils. The clinical severity of OI relates to the extent of the conformational change in the collagen triple helix induced by the glycine substitution. These mutations result in altered fibrillogenesis. However, no general mechanism can be drawn from genotype/phenotype analyses.

Collagen VII, encoded by COL7A1, is the major component of the anchoring fibrils at the dermo-epidermal junction (Burgeson, 1993). COL7A1 gene mutations cause dystrophic epidermolysis bullosa (DEB), a skin-blistering disorder (Bruckner-Tuderman, 1999). Approximately 200 mutations of COL7A1 have been characterized, leading to a very high molecular heterogeneity of collagen VII defects (Dunnill et al., 1996). Almost all cases of dominant DEB are caused by a glycine substitution in the triple helical region of collagen VII, and most of the mutations are unique to individual families. Some glycine substitutions in collagen VII interfere with biosynthesis of the protein in a dominant-negative manner, whereas others may lead to collagen VII retention within the rough endoplasmic reticulum.

Mutations in the *COL5A1* and *COL5A2* genes, encoding respectively the pro-α1 and pro-α2 chains of the fibril-forming collagen V, have been identified in approximately 50% of patients with a clinical diagnosis of classic Ehlers-Danlos syndrome (EDS) (Malfait *et al.*, 2010). Collagen V contains a third chain, proα3(V); but no mutation in *COL5A3* has been reported so far. Classic EDS is a heritable disorder of connective tissues characterized by skin hyperextensibility, fragile and soft skin, delayed wound healing with formation of atrophic scars, easy bruising, and generalized joint hypermobility. The majority of mutations lead to a non-functional *COL5A1* allele. One mutant *COL5A1* transcript showed a premature stop codon. A minority of mutations affect the structure of the central helical domain. In approximately one-third of patients, the disease is caused by a mutation leading to a non-functional *COL5A1* allele, resulting in collagen V haploinsufficiency. Structural mutations in *COL5A1* or *COL5A2*, resulting in the production of a functionally defective protein, account for a small proportion of patients.

Collagen V is a quantitatively minor fibril-forming collagen that co-polymerizes with collagen I to form heterotypic fibrils (Fichard *et al.*, 1995). Co-polymerisation has a critical role in the nucleation and growth of fibrils in tissues. A collagen V feature is to retain in the mature molecule a major part of the α1(V) N-propeptide which projects beyond the surface of collagen fibrils. This domain was proposed to limit heterotypic fibril growth by steric hindrance and electrostatic interactions (Linsenmayer *et al.*, 1993). Skin biopsies revealed abnormalities in fibril formation (altered diameter, contour, or shape of dermal fibrils). However, abnormalities of fibril structure affected less than 5% of fibrils (reviewed in Fichard *et al.*, 2003). Moreover, the clinical phenotype of classical EDS supports an important role of collagen V in the biomechanical integrity of the skin, tendon and ligaments, although collagen V is only a minor component of the affected tissues. Thus, collagen V may be involved in functions other than the control of fibril growth in classical EDS. A likely hypothesis is that collagen V might be involved in the physiopathology of EDS through interactions with other fibril-associated components and/or with cell receptors. Along this line, it has been shown that mutations in the genes for the collagen V-binding partners, tenascin-X (*TNXB* gene) and collagen I (*COL1A1* gene), resulted in EDS (Lindor and Bristow, 2005).

Although mutant gene products are thought to impair matrix structure and assembly that eventually alters tissue function, growing evidence links ER stress and the unfolded protein response (UPR) to the initiation and progression of a broad repertoire of connective tissue disorders, including those caused by collagen gene mutations. Some mutant chains cannot be incorporated into procollagen molecules, consequently causing protein degradation with important downstream effects. Misfolded or slowly folding collagens are retained within the endoplasmic reticulum (ER) and ultimately targeted for degradation by a mechanism initially called "protein suicide." Because connective-tissue cells typically produce large quantities of collagens, the contribution of ER stress induced by misfolded collagens in disease pathogenesis has certainly been underrated. The current knowledge on the implications of unfolded protein response and ER stress in connective tissue diseases has been recently reviewed, and readers are referred to these reviews for further reading (Boot-Handford and Briggs, 2010; Tsang *et al.*, 2010). Notably, mutations in genes encoding collagen I (*COL1A1* and *COL1A2*) (osteogenesis imperfecta), collagen II (*COL2A1*) (spondyloepiphyseal dysplasia), and collagen X (*COL10A1*) (metaphyseal

Collagen subfamily	Collagen type	Associated diseases	Gene	Modified mice	References (mouse models)
Fibril forming collagens	I	osteogenesis imperfecta	COL1A1	knock-out	Löhler et al., 1984
			COL1A1	single mutation	Forlino et al., 1999
			COL1A2	single mutation	Daley et al., 2010
	II	chondrodysplasias	COL2A1	knock-out	Aszodi et al., 1998
	III	vascular Ehlers-Danlos syndrome	COL3A1	knock-out	Liu et al., 1997
	V	classic Ehlers-Danlos syndrome	COL5A1	knock-out	Wenstrup et al., 2004
			COL5A2	knock-in	Andrikopoulos et al., 1995
		nd	COL5A3	knock-out	Huang et al., 2011
	XI	chondrodysplasias	COL11A1	knock-out	Wenstrup et al., 2011
			COL11A2	knock-out	Mc Guirt et al., 1999
FACIT	IX	Stickler syndrome	COL9A1	knock-out	Fässler et al., 1994
			COL9A2	functional knock-out	Hagg et al., 1997
			COL9A3		
	XII	nd	COL12A1	knock-out	Izu et al, 2011
	XIV	nd	COL14A1	knock-out	Ansorge et al., 2009
	XIX	nd	COL19A1	knock-out	Sumiyoshi et al., 2004
Multiplexins	XV	nd	COL15A1	knock-out	Eklund et al., 2001
	XVIII	Knobloch syndrome	COL18A1	knock-out	Seppinen et al., 2008
Network forming collagens	IV	nd	COL4A1	nd	
			COL4A2		
		Alport syndrome	COL4A3	knock-out	Cosgrove et al., 1996
			COL4A4	knock-out	Lu et al., 1999
			COL4A5	knock-out	Rheault et al., 2004
			COL4A6	nd	
	VIII	corneal dystrophia	COL8A1	knock-out	Hopfer et al., 2005
			COL8A2	knock-out	Hopfer et al., 2005
	X	chondrodysplasias	COL10A1	knock-out	Kwan et al., 1997
Beaded-filament forming collagen	VI	Bethlem myopathy and Ullrich congenital muscular dystrophy	COL6A1	knock-out	Bonaldo et al., 1998
			COL6A2	nd	
			COL6A3		
		nd	COL6A4	nd	
			COL6A5		
			COL6A6		
Anchoring collagens	VII	epidermolysis bullosa	COL7A1	knock-out	Heinonen et al., 1999
Transmembrane collagens	XVII	epidermolysis bullosa	COL17A1	knock-out	Tanimura et al., 2011

nd, not determined

Table 1. Collagen types, associated-diseases and mouse models.

chondrodysplasia) have been shown to induce ER dilatation in patient cells. Mutations that affect the triple helix, the C-propeptide for the fibril-forming collagens, and splice donor sites, as well as single amino-acid substitutions, were shown to cause ER stress. Recently, mutations that affected the signal peptide domain of the proα1(V)-collagen chain were shown to cause classic EDS. The signal peptides are the addresses of proteins destined for secretion. The mutant procollagen V is retained within the cell, leading to a collagen V haploinsufficiency and altered collagen fibril formation. It is probable that the signal peptide mutation also causes accumulation of the mutated protein within the ER and eventually to ER stress, as described for other collagen-related disorders (Symoens *et al.*, 2009).

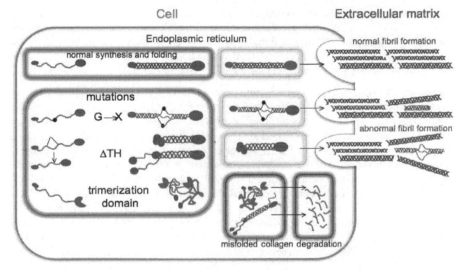

Fig. 1. Schematic diagram illustrating the biological consequences of point mutations or small deletions in collagen genes on chain synthesis, protein folding and subsequent fibril assembly in the extracellular matrix.

Mutations in the three major collagen VI genes (*COL6A1, COL6A2* and *COL6A3*) cause multiple muscle disorders, including the severe Ullrich congenital muscular dystrophy (UCMD) and the mild Bethlem myopathy, which is characterized by muscle weakness with striking joint laxity and progressive contractures. Three genetically distinct novel chains α4(VI), α5(VI), and α6(VI) have recently been identified; but very little is known about their molecular assembly and biosynthesis and their possible involvement in human diseases (Gara *et al.*, 2011). Collagen VI biosynthesis is a complex multistep process. Monomer formation results from the heterotrimeric association of the three chains [α1(VI), α2(VI), and α3(VI)] encoded by the *COL6A1, COL6A2* and *COL6A3* genes. Monomers first assemble into antiparallel dimers that associate laterally to form tetramers stabilized by disulphide bonds. The tetramers associate linearly to form the unique beaded filaments, the ultimate step of collagen VI biosynthesis. Dominant and recessive autosomal mutations in *COL6A1, COL6A2,* and *COL6A3* primarily result in dysfunctional microfibrillar collagen VI in muscle extracellular matrix. However they also affect other connective tissues, such as skin and tendons. Different mutations have been shown to have variable effects on protein assembly, secretion, and its ability to form a functioning extracellular network. As observed in other collagen-related diseases, glycine-substitution mutations in *COL6A1*, *COL6A2*, or *COL6A3* that disrupt the triple-helix motif constitute a frequent pathogenic mechanism. Triple-helix distortion may exert a dominant-negative effect by reducing the ability of mutated monomers to form beaded filaments. Interestingly, mitochondrial dysfunction was implicated in the pathogenesis of a myopathic phenotype. Muscles lacking collagen VI are characterized by the presence of a dilated sarcoplasmic reticulum and dysfunctional mitochondria. This condition triggers apoptosis and leads to myofiber degeneration. Recently, it was shown that the persistence of abnormal organelles and apoptosis observed in some congenital muscular dystrophies are caused by defective activation of the autophagic machinery. Autophagy has a key role in the clearance of damaged organelles and in the turnover of cell components and is thus essential for tissue homeostasis. Recently, 56 novel mutations have been described, allowing a clinical classification and revealing the complexity of genotype-phenotype relationships (Briñas *et al.*, 2010).

The paucity of evidence-based data regarding correlations of genotype and phenotype is in part due to the large spectrum of mutations reported for the collagen genes [*e.g.*, about 200 mutations for the collagen I genes responsible for OI (Dalgleish, 1998); 160 mutations in the *COL4A5* gene encoding collagen IV α5 chain responsible for Alport syndrome; 200 mutations in *COL7A1* responsible for EDB]. Things are not as simple as one gene-numerous mutations-one phenotype. Sometimes a combination of a mutation for a connective tissue disorder and a specific collagen gene mutation will result in another disease. Some patients with UCMD show clinical characteristics typical of classical disorders of connective tissue, such as EDS. Ultrastructure of skin biopsy samples from patients with UCMD showed alterations of collagen fibril morphology in skin that resemble those described in patients with EDS (Kirschner *et al.*, 2005). Recently, using the yeast two-hybrid approach, we showed a direct interaction between collagen V and collagen VI that may nicely explain the overlap of UCMD and classic EDS (Symoens *et al.*, 2011). Unexpectedly an arginine-->cysteine substitution localized at position 134 of the α1(I) collagen chain resulted in classical EDS (Nuyntick *et al.*, 2000). This finding is indicative of genetic heterogeneity in collagen-related disorders.

A powerful approach to study the biochemical consequences of mutation and the protein structure/function relationship is to engineer a specific mutation into a functional domain of the molecule. Targeted mutagenesis approaches, including the use of alanine-scanning mutagenesis techniques, have led to important insights into the effects of collagen mutations on protein structure and function. A major limitation of mutagenesis strategies to investigate collagens is the large number of collagen gene mutations to be investigated in order to have a better understanding of the molecular mechanisms of "collagenopathies." Knowledge about the impact of collagen mutations has also been hampered by the technical difficulty of introducing targeted mutations of very large collagen genes into mice.

4. Lessons from site-directed mutagenesis of recombinant collagen genes and derived fragments

Production of a recombinant collagen gene represents a powerful technique to introduce a human mutation into the gene of interest by site-directed mutagenesis. It allows one to analyze the impact of the mutation on collagen assembly and secretion. Collagen biosynthesis is a complex multistep process that takes place in the intracellular and extracellular space and includes various post-translationnal modifications, such as prolyl- and lysyl-hydroxylation, glycosylation, trimerization, proteolytic processing, polymerization and cross-links. Because of recombinant technology, these large multimeric proteins have been produced in large amounts in almost all existing expression systems (Ruggiero and Koch, 2008). This technological breakthrough enabled researchers to analyze in detail the effects of collagen mutations on biosynthesis, molecular and cell interactions, processing and, in some cases, self-assembly. Researchers can also address the question of the correlation of genotype, protein structure and function.

Mutations occurring in collagen I genes are the most extensively studied mutations among all collagen types. A first set of experiments substituted glycine 859 of the proα1(I) chain with cysteine or arginine by site-directed mutagenesis to reproduce two mutations identified in OI patients. In order to study the expression of the mutant molecule in the presence or absence of the wild-type proα1(I) chain, the mutated constructs were transfected into normal fibroblasts to look for a dominant-negative effect in the presence of the wild-type gene or in fibroblasts isolated from Mov13 homozygous mice (referred to as Mov13 fibroblasts hereafter), whose cells carry a provirus that prevents transcription initiation of the natural proα1(I) gene (Schnieke et al., 1987). In agreement with observations of collagen I in OI patients, the mutated collagens were poorly secreted from the cells and exhibited reduced thermal stability and increased sensitivity to degradation. This supported the idea that the strict preservation of the G-x-y triplets is absolutely required for proper formation of the triple helix.

The integrity of the C-propeptide is pivotal for the trimerization of all fibril-forming collagens. The C-propeptides of the proα1(I) and proα2(I) chains contain an Asn-Ile-Thr sequence. That sequence fits a consensus sequence for the addition of N-linked oligosaccharides. To analyze the role of this post-translational modification, the asparagine residue of the proα1(I) chain was changed to glycine by site-directed mutagenesis. The expression of the corresponding molecule was analyzed in transfected normal and Mov13 fibroblasts (Lamandé and Bateman, 1995). The mutation did not impair heterotrimeric assembly and secretion of hybrid procollagen I into the extracelllular space. Only a slight effect on C-proteinase cleavage efficiency was observed with the unglycosylated molecule. To circumvent the difficulty of producing a large repertoire of full-length mutated collagens I in order to undertake a genotype/phenotype analysis, a recombinant trimeric mini-collagen I was recently expressed in an *Escherichia coli* system. Recombinant mini-collagens can be obtained by fusing the sequence encoding a fragment of the proα1(I) chain triple-helix to the sequence encoding the C-terminal domain (called "foldon") of the bacteriophage T4 fibritin, which is capable of trimerization (Xu et al., 2008). Two mutations (G901S and G913S), corresponding to mild and severe types of OI, respectively, were introduced into the recombinant mini-collagen I. Biophysical measurements and protease cleavage analysis revealed that the G913S mutant chain resulted in the formation of an unstable collagen I triple helix by disrupting salt bridges important for maintaining the chains in a triple-helix conformation (Yang et al., 1997; Xu et al., 2008). A very recent study utilized a recombinant bacterial collagen to develop a mutagenesis scheme in which a glycine residue within the triple-helix sequence is substituted with arginine or serine. The purpose was to analyze the positional effect of glycine mutations on triple-helix formation and stability (Cheng *et al.*, 2011). Interestingly, all glycine mutations provoked a significant delay in the triple-helix formation. However, a more severe defect was observed when the mutation was located near the trimerization domain of the triple-helix where folding is initiated.

COL7A1 mutations cause dystrophic epidermolysis bullosa (DEB), a skin blistering disorder. Woodley and collaborators (2008) have used site-directed mutagenesis to elucidate the effect of human mutations on the function of collagen VII, which is the major component of the epidermal anchoring fibrils. To undertake a comprehensive analysis of the impact of human mutations in the formation, folding and stability of collagen VII and, particularly relevant to the DEB phenotype, its effect on cell attachment and migration, four distinct substitutions occurring in collagen VII (G2049E, R2063W, G2569R, and G2575R) were introduced using *COL7A1* cDNA. The authors demonstrated that the G2049E and R2063W mutants caused local destabilization of the triple helix and reduced the capability of collagen VII to elicit cell adhesion and migration. The G2569R and G2575R mutants interfered with triple-helix formation and stability. Alterations of protein stability and/or cell attachment to collagen VII mutants help explain the fragility of the dermal-epidermal junction observed in DEB patients. Naturally occurring *COL7A1* mutations were investigated in a separate study (Hammami-Hauasli *et al.*, 1998). As commonly described for glycine-substitution mutants of collagens, the authors showed that three glycine substitutions located in the same triple-helix portion affected folding, stability and secretion of procollagen VII in a dominant-negative manner. However, the glycine substitution G1519D located in another segment of the triple helix had no effect on procollagen VII secretion or its ability to anchor fibril assembly. These data showed that the biological impact of glycine substitutions can depend on their position within the triple helix, as shown for collagen I (Cheng *et al.*, 2011).

Human collagen IV mutations, thought to affect the biosynthesis of this basement membrane collagen, were extensively investigated. These mutations were known to cause Alport syndrome, a severe renal disease leading eventually to kidney failure. Collagen IV chains, $\alpha1(IV)$-$\alpha6(IV)$, are encoded by 6 genes, *COL4A1-COL4A6*, respectively. Although mutations have also been identified in *COL4A3 and COL4A4*, about 30% of known missense mutations occur in the *COL4A5* gene, which encodes the human $\alpha5(IV)$ chain. Most of them are glycine substitutions. One glycine-substitution mutation in *COL4A5* could prevent correct α-chain folding or/and the association with other α-chains to form a stable triple helix. To address this question, the authors took advantage of the bacterial system. A DNA encoding a 22-kDa recombinant domain of the $\alpha5(IV)$ triple helix in its wild-type form or harboring the G1015V or G1030S mutations was expressed in *E. coli* (Wang *et al.*, 2004). The recombinant wild-type and mutant proteins were purified and assayed for changes in triple-helix assembly and stability by circular dichroism. The two different glycine-substitution mutants displayed different defects in the secondary structures of their protein products that matched with the severity of the patient phenotypes. However, the use of a bacterial system to analyze the effects of specific human mutations on mini-collagen assembly and stability presents several disadvantages. Because collagens are large multimeric proteins, full-length molecules cannot be produced in a bacterial host. Most importantly, the bacterial system is limited. Not all post-translational modifications needed for the triple-helix formation and stability, such as hydroxylation, glycosylation, and disulfide-bond formation, are present in bacteria. A few years later, the bacterial limitations were bypassed by the development of the production of full-length recombinant collagen molecules in mammalian cells (Fichard *et al.*, 1997; Ruggiero and Koch, 2008). No less than eighteen human mutations (11 substitutions and 7 deletions) were introduced into the sequence encoding the trimerization NC1 domain of the α5(IV) chain gene. The constructs were transfected into cells together with constructs containing the wild-type sequences of α3(IV) and α4(IV) chains to analyze the impact of the mutations in the NC1 domain on the formation of the α3α4α5 collagen IV heterotrimer. Twelve out of 15 mutant chains did loose their capacity to assemble into heterotrimeric molecules. The three remaining mutants formed heterotrimers, but the mutations prevented their secretion into the extracellular space (Kobayashi *et al.*, 2008). The authors nicely demonstrated, using site-directed mutagenesis, that amino acid substitutions in the α5(IV) NC1 trimerization domain are specifically responsible for impairment of collagen IV heterotrimer assembly. This defect may be a main molecular mechanism for the pathogenesis of Alport syndrome. Interestingly, an interactome (a map of known and predicted molecular interactions, as well as phenotypic and structural landmarks) of collagen IV was recently constructed to identify functional and disease-associated domains and genotype-phenotype relationships (Parkin *et al.*, 2011). Construction of such interactomes will greatly improve our capacity to integrate all data from different site-directed mutagenesis experiments. This advance will greatly help our understanding of the molecular mechanisms underlying "collagenopathies"; and, consequently, it may lead to the development of specific treatments.

Collagens undergo a great variety of proteolytic modifications. The fate and functions of the released fragments derived from collagens are still under intensive investigation, but the consequences of mutations in the coding regions for the cleavage sites on collagen structure, self-assembly and function have not been investigated in detail. A large repertoire of proteinases is responsible for these processing interactions. Included among such enzymes are the ADAMTS (a disintegrin and metalloprotease with thrombospondin motifs) and the BMP-1/tolloid families of metalloproteinases and more recently the furin-like proprotein convertases (Ricard-Blum and Ruggiero, 2005). To investigate collagen processing, fastidious extraction and purification steps were often necessary to obtain limited amounts of unprocessed proteins and enzymes with full activity in order to undertake *in vitro* enzymatic assays. To circumvent this problem, we recently described a new cell system allowing a rapid and straightforward analysis of processing interactions. Our system relies on the use of site-directed mutagenesis. This strategy was particularly instrumental in analyzing the complex procollagen V processing during maturation. We showed it to be unique among the fibril-forming collagens (Bonod-Bidaud *et al.*, 2007). Collagen V is a minor fibrillar collagen that can be distinguished from the others by its capacity to control fibrillogenesis (Fichard *et al.*, 1995). In addition this molecule undergoes a particular form of processing; and it is involved in fundamental processes, such as development and human connective tissues disorders. The proα1(V) N-terminus can be processed by the procollagen proteinases ADAMTS-2 and BMP-1 (Colige et *al.*, 2005; Bonod-Bidaud *et al.*, 2007), whereas the C-propeptide can be cleaved by furin and BMP-1 (Kessler *et al.*, 2001). The proα1(V) C-propeptide furin cleavage site, which occurs immediately downstream of the recognition sequence RTRR, was double-mutated to alanine residues (R1584A/R1585A) to abolish furin cleavage. All constructs were introduced into cells, along with a BMP-1-expressing construct; and the cleavage products were directly analyzed in conditioned medium of the transiently transfected cells. We were able to show that BMP-1 is capable of processing the α1(V) C-propeptide in absence of furin activity (Bonod-Bidaud et al., 2007). In the same way, the determinant for α1(V) N-propeptide processing by BMP-1 activity was identified by introducing in the coding region for the cleavage site (S254/Q255-D256) three single mutations (S254A, Q255A and D256A), two double mutations (S254A/Q255N and Q255A/D256A) and one triple mutation (S254A/Q255A/D256A). The data highlighted the unexpected importance of the aspartic acid in the P2' position of the BMP-1 cleavage site (Bonod-Bidaud et al., 2007). Processing, proteolytic release of functional domains and shedding of collagens are involved in fundamental processes. It is likely that substitutions located in the proteolytic cleavage sites may represent a molecular cause of connective tissues disorders. A reported mutation in the α1(V) N-propeptide in one patient with classic EDS resulted in a protein product missing the sequence of exon 5 that encompasses the BMP-1 cleavage site. The abnormal-sized N-propeptide present in the mutated collagen V caused dramatic alterations in fibril structure (Takahara et al., 2002).

5. Lessons from site-directed mutagenesis in mice

In vitro studies are useful and necessary approaches to understand the mechanisms of collagen biosynthesis and to establish structure-function relationship. However, they do not always reflect the normal and pathological in vivo situations. Genetically modified mice appear to be a powerful technique to better understand the physiopathology of connective tissue disorders. Several different genetically modified mice have been created during the last 10 years (reviewed in Aszódi et al., 2006). This clearly opened doors to better understand collagen function in developing tissues and provide reliable mouse models for inherited collagen diseases. Along this line, a targeted disruption of Col4a3 gene led to renal failure and eventually to the death of mice at 3-4 months of age (Cosgrove et al., 1996; Miner and Sanes, 1996). This result is consistent with defects described for Alport disease.

In most cases, the gene of interest was disrupted and knock-out mice were preferably generated. Few transgenic mice harbouring point mutations or small deletion in collagen genes have been generated (Table 1). Naturally occurring mutations in mice disrupting collagen genes have also been identified and characterized. The oim mice present a spontaneously acquired deletion in the Col1a2 gene that leads to an accumulation of $[\alpha(I)]_3$ collagen homotrimer in the extracellular matrix. These mice develop a phenotype similar to moderate OI in humans, providing a good model for this collagen disorder (Chipman et al., 1993). It was shown that homozygous Mov13 embryos harboring an inactivated proα1(I) chain (due to the insertion of the Moloney murine leukaemia virus into the first exon of the Col1a1 gene) died in utero around day 12 because of vascular failure (Löhler et al., 1984). However, in 1999 Forlino et al. developed the first knock-in mouse model for human OI by introducing a G349C mutation into the Col1a1 gene. Along this line, a knock-in mouse model for OI, harboring a point mutation (G610C) in Col1a2 was recently created (Daley et al., 2010). These mice had reduced body mass and bone strength and exhibited bone fracture susceptibility consistent with the clinical features of human OI. Thus, the G610C knock-in mouse represents a novel model for the study of OI pathogenesis and also for testing potential therapies for OI.

Another example concerns collagen V deficiency/dysfunction, which is responsible for Ehlers-Danlos syndrome (EDS). In the absence of the *Col5a1* gene, the mice died at the onset of organogenesis at approximately embryonic day 10 (Wenstrup *et al.*, 2004). Interestingly, a targeted deletion in the *Col5a2* gene, encoding the proα2(V) chain, recapitulated many of the clinical, biomechanical, morphologic, and biochemical features of the classical EDS. The deletion removes the sequence encoding the N-telopeptide (*pN*), a 20-residue region that confers flexibility to the N-terminal part of the molecule (Andrikopoulos *et al.*, 1995). A detailed study of the skin at the morphological, histological, ultrastructural and biochemical levels indicated that the *Col5a2* deletion impairs assembly and/or secretion of the $[\alpha1(V)]_2\alpha2(V)$ heterotrimer. Consequently, the $[\alpha1(V)]_3$ homotrimer, and not the $[\alpha1(V)]_2\alpha2(V)$ heterotrimer, is the predominant species deposited into the matrix, which in turn severely impaired extracellular matrix organization (Chanut-Delalande *et al.*, 2004). These data underscored the importance of the collagen V $[\alpha1(V)]_2\alpha2(V)$ heterotrimer in dermal fibrillogenesis and can explain defects observed in the dermis of EDS patients.

6. Concluding remarks

Site-directed mutagenesis has been extensively used in collagen engineering and has shed light on collagen structure, expression, folding, secretion, interactions and self-assembly in the extracellular space. It also opened the way for the analysis of specific functional domains. It allowed the study of the wide variety of collagen types, including those expressed in trace amounts in tissues but nevertheless display pivotal functions. While it is true that site-directed mutagenesis has yielded important information on the functional consequences of a range of collagen mutations responsible for human diseases, only few studies have approached the consequences of collagen gene mutations on cell adaptation to ER stress. Collagen gene mutations affect protein synthesis, folding and secretion imbalance, which eventually induces ER stress. *In vitro* studies have been done on transfected cells, in which expression and trafficking of mutant collagen can be easily manipulated and analysed at the cellular level. The effects of gene manipulation can be studied *in vivo* using mice. The effect of collagen gene mutations on induction of an ER stress response could be straightforwardly addressed in the near future. It may be a key factor in pathogenesis (Boot-Handford and Briggs, 2010).

Mouse models are particularly useful for analysing the biological significance of collagens in pathological situations. Knock-out mice often lead to embryonic lethality, which hampers in-depth analysis of the phenotype. A few knock-in mice have been created with subtle mutations or small deletions that reproduce human mutations. The major reason for the paucity of knock-in mice is certainly that collagen genes are very large. Thus, they are difficult to manipulate. The introduction of a small deletion or a single point mutation in murine collagen genes still represents a considerable challenge. Nevertheless, the few examples of knock-in mouse lines tend to prove that mouse models can bring new information about *in vivo* consequences of collagen dysfunction that cannot be predicted by *in vitro* approaches. Knock-in mice are also indispensable models for assessing the effects of subtle mutations on tissue function, development, and aging. They are also valuable for developing specific gene therapy approaches to combat collagen-related disorders. The combination of site-directed mutagenesis in transfected cells and knock-in approaches in mice to address the impact of specific mutations will enable us to identify mechanisms underlying the vast repertoire of collagen-related diseases. The implications may lead to the development of a specific therapy.

7. References

Andrikopoulos, K., Liu, X., Keene, D.R., Jaenisch, R. & Ramirez, F. (1995). Targeted mutation in the *col5a2* gene reveals a regulatory role for type V collagen during matrix assembly. *Nat Genet.* 9(1):31-36.

Aszodi, A., Legate, K.R., Nakchbandi, I. & Fässler, R. (2006). What Mouse Mutants Teach Us About Extracellular Matrix Function. *Annu. Rev. Cell Dev. Biol.* 22:591–621.

Bateman, J.F., Boot-Handford, R.P. & Lamandé, S.R. (2009). Genetic diseases of connective tissues: cellular and extracellular effects of ECM mutations. *Nat Rev Genet.* 10(3):173-83.

Bonaldo, P., Braghetta, P., Zanetti, M., Piccolo, S., Volpin, D. & Bressan, G.M. (1998). Collagen VI deficiency induces early onset myopathy in the mouse: an animal model for Bethlem myopathy. *Hum Mol Genet.* 7(13):2135-2140.

Bonod-Bidaud, C., Beraud, M., Vaganay, E., Delacoux, F., Font, B., Hulmes, D.J. & Ruggiero, F. (2007). Enzymatic cleavage specificity of the proalpha1(V) chain processing analysed by site-directed mutagenesis. *Biochem J.* 405(2):299-306.

Boot-Handford, R.P. & Briggs, M.D. (2010). The unfolded protein response and its relevance to connective tissue diseases. *Cell Tissue Res.* 339(1):197-211.

Briñas, L., Richard, P., Quijano-Roy, S., Gartioux, C., Ledeuil, C., Lacène, E., Makri, S., Ferreiro, A., Maugenre, S., Topaloglu, H., Haliloglu, G., Pénisson-Besnier, I., Jeannet, P.Y., Merlini, L., Navarro, C., Toutain, A., Chaigne, D., Desguerre, I., de Die-Smulders, C., Dunand, M., Echenne, B., Eymard, B., Kuntzer, T., Maincent, K., Mayer, M., Plessis, G., Rivier, F., Roelens, F., Stojkovic, T., Taratuto, A.L., Lubieniecki, F., Monges, S., Tranchant, C., Viollet, L., Romero, N.B., Estournet, B., Guicheney, P., Allamand, V. (2010). Early onset collagen VI myopathies: Genetic and clinical correlations. *Ann Neurol.* 68(4):511-520.

Bruckner-Tuderman, L. & Bruckner, P. (1998). Genetic diseases of the extracellular matrix: more than just connective tissue disorders. *J Mol Med.* 76(3-4):226–237.

Bruckner-Tuderman, L., Höpfner, B. & Hammami-Hauasli, N. (1999). Biology of anchoring fibrils: lessons from dystrophic epidermolysis bullosa. *Matrix Biol.* 18(1):43-54.

Burgeson, R.E. (1993). Type VII collagen, anchoring fibrils, and epidermolysis bullosa. *J Invest Dermatol.* 101(3):252-255.

Byers, P.H. & Steiner, R.D. (1992). Osteogenesis imperfecta. *Annu Rev Med.* 43:269-282.

Chanut-Delalande, H., Bonod-Bidaud, C., Cogne, C., Malbouyres, M., Ramirez, F., Fichard, A. & Ruggiero, F. (2004). Development of a functional skin matrix requires deposition of collagen V heterotrimers. *Mol. Cell. Biol.*, 24(13):6049-6057

Cheng, H., Rashid, S., Yu, Z., Yoshizumi, A., Hwang, E. & Brodsky, B. (2011). Location of glycine mutations within a bacterial collagen protein affects degree of disruption of triple-helix folding and conformation. J Biol Chem. 286(3):2041-2046.

Chipman, S.D., Sweet, H.O., McBride, D.J. Jr, Davisson, M.T., Marks, S.C. Jr, Shuldiner, A.R., Wenstrup, R.J., Rowe, D.W. & Shapiro, J.R. (1993). Defective pro alpha 2(I) collagen synthesis in a recessive mutation in mice: a model of human osteogenesis imperfecta. *Proc Natl Acad Sci U S A.* 90(5):1701-1705.

Colige, A., Ruggiero, F., Vandenberghe, I., Dubail, J., Kesteloot, F., Van Beeumen, J., Beschin, A., Brys, L., Lapière, C.M. & Nusgens B. (2005). Domains and maturation processes that regulate the activity of ADAMTS-2, a metalloproteinase cleaving the aminopropeptide of fibrillar procollagens types I-III and V. *J Biol Chem.* 280(41):34397-34408.

Cosgrove, D., Meehan, D.T., Grunkemeyer, J.A., Kornak, J.M., Sayers, R., Hunter, W.J. & Samuelson, G.C. (1996).Collagen COL4A3 knockout: a mouse model for autosomal Alport syndrome. *Genes Dev.* 10(23):2981-92.

Daley, E., Streeten, E.A., Sorkin, J.D., Kuznetsova, N., Shapses, S.A., Carleton, S.M., Shuldiner, A.R., Marini, J.C., Phillips, C.L., Goldstein, S.A., Leikin, S. & McBride D.J. Jr. (2010). Variable bone fragility associated with an Amish COL1A2 variant and a knock-in mouse model. *J Bone Miner Res.* 25(2):247-261.

Dalgleish, R. (1998). The Human Collagen Mutation Database 1998. Nucleic Acids Res. 26(1):253-255.

Dunnill, M.G., McGrath, J.A., Richards, A.J., Christiano, A.M., Uitto, J., Pope, F.M. & Eady, R.A. (1996). Clinicopathological correlations of compound heterozygous COL7A1 mutations in recessive dystrophic epidermolysis bullosa. *J Invest Dermatol.* 107(2):171-177.

Eklund, L., Piuhola, J., Komulainen, J., Sormunen, R., Ongvarrasopone, C., Fässler, R., Muona, A., Ilves, M., Ruskoaho, H., Takala, T.E. & Pihlajaniemi, T. (2001). Lack of type XV collagen causes a skeletal myopathy and cardiovascular defects in mice. *Proc Natl Acad Sci U S A.* 98(3):1194-1199.

Fässler, R., Schnegelsberg, P.N., Dausman, J., Shinya, T., Muragaki, Y., McCarthy, M.T., Olsen, B.R. & Jaenisch, R. (1994). Mice lacking alpha 1 (IX) collagen develop noninflammatory degenerative joint disease. *Proc Natl Acad Sci U S A.* 91(11):5070-5074.

Fichard, A., Kleman, J.P. & Ruggiero F. (1995). Another look at collagen V and XI molecules. *Matrix Biol.* 14(7):515-531.

Fichard, A., Tillet, E., Delacoux, F., Garrone, R. & Ruggiero, F. (1997). Human recombinant alpha1(V) collagen chain. Homotrimeric assembly and subsequent processing. *J Biol Chem.* 272(48):30083-7.

Fichard, A., Chanut-Delalande, H. & Ruggiero, F. (2003). The Ehlers-Danlos syndrome: the extracellular matrix scaffold in question. *Med/Sci* (Paris). 19(4):443-452.

Forlino, A., Porter, F.D., Lee, E.J., Westphal, H. & Marini, J.C. (1999).Use of the Cre/lox recombination system to develop a non-lethal knock-in murine model for osteogenesis imperfecta with an alpha1(I) G349C substitution. Variability in phenotype in BrtlIV mice. *J Biol Chem.* 274(53):37923-37931.

Gara, S.K., Grumati, P., Squarzoni, S., Sabatelli, P., Urciuolo, A., Bonaldo, P., Paulsson, M. & Wagener, R. (2011). Differential and restricted expression of novel collagen VI chains in mouse. *Matrix Biol.* 30(4):248-257.

Hagg, R., Hedbom, E., Möllers, U., Aszódi, A., Fässler, R. & Bruckner, P. (1997). Absence of the alpha1(IX) chain leads to a functional knock-out of the entire collagen IX protein in mice. J Biol Chem. 272(33):20650-20654.

Hammami-Hauasli, N., Schumann, H., Raghunath, M., Kilgus, O., Lüthi, U., Luger, T. & Bruckner-Tuderman, L. (1998). Some, but not all, glycine substitution mutations in

COL7A1 result in intracellular accumulation of collagen VII, loss of anchoring fibrils, and skin blistering. *J Biol Chem.* 273(30):19228-19234.

Heinonen, S., Männikkö, M., Klement, J.F., Whitaker-Menezes, D., Murphy, G.F. & Uitto, J. (1999). Targeted inactivation of the type VII collagen gene (Col7□□1) in mice results in severe blistering phenotype: a model for recessive dystrophic epidermolysis bullosa. *J Cell Sci.* 112 (Pt 21):3641-3648.

Humphries, J.D., Byron, A. & Humphries, M.J. (2006). Integrin ligands at a glance. *J Cell Sci.* 119(Pt 19):3901-3903.

Huang, G., Ge, G., Wang, D., Gopalakrishnan, B., Butz, D.H., Colman, R.J., Nagy, A. & Greenspan, D.S. (2011). α3(V) collagen is critical for glucose homeostasis in mice due to effects in pancreatic islets and peripheral tissues. *J Clin Invest.* 121(2):769-783.

Hopfer, U., Fukai, N., Hopfer, H., Wolf, G., Joyce, N., Li, E. & Olsen, B.R. (2005). Targeted disruption of Col8a1 and Col8a2 genes in mice leads to anterior segment abnormalities in the eye. *FASEB J.* 19(10):1232-1244.

Izu, Y., Sun, M., Zwolanek, D., Veit, G., Williams, V., Cha, B., Jepsen, K.J., Koch, M. & Birk D.E. (2011). Type XII collagen regulates osteoblast polarity and communication during bone formation. *J Cell Biol.* 193(6):1115-1130.

Kessler, E., Fichard, A., Chanut-Delalande, H., Brusel, M. & Ruggiero, F. (2001). Bone morphogenetic protein-1 (BMP-1) mediates C-terminal processing of procollagen V homotrimer. *J Biol Chem.* 276(29):27051-27057.

Kirschner, J., Hausser, I., Zou, Y., Schreiber, G., Christen, H.J., Brown, S.C., Anton-Lamprecht, I., Muntoni, F., Hanefeld, F. & Bönnemann, C.G. (2005). Ullrich congenital muscular dystrophy: connective tissue abnormalities in the skin support overlap with Ehlers-Danlos syndromes. *Am J Med Genet A.* 132A(3):296-301.

Kobayashi, T., Kakihara, T. & Uchiyama, M. (2008). Mutational analysis of type IV collagen alpha5 chain, with respect to heterotrimer formation. *Biochem Biophys Res Commun.* 366(1):60-65.

Kuivaniemi, H., Tromp, G. & Prockop, D.J. (1991). Mutations in collagen genes: causes of rare and some common diseases in humans. *FASEB J.* 5(7):2052-2060.

Kwan, K.M., Pang, M.K., Zhou, S., Cowan, S.K., Kong, R.Y., Pfordte, T., Olsen, B.R., Sillence, D.O., Tam, P.P. & Cheah, K.S. (1997). Abnormal compartmentalization of cartilage matrix components in mice lacking collagen X: implications for function. *J Cell Biol.* 136(2):459-471.

Lamandé, S.R & Bateman, J.F. (1995). The type I collagen pro alpha 1(I) COOH-terminal propeptide N-linked oligosaccharide. Functional analysis by site-directed mutagenesis. *J Biol Chem.* 270(30):17858-17865.

Leitinger, B. & Hohenester, E. (2007). Mammalian collagen receptors. *Matrix Biol.* 26(3):146-155.

Lightfoot, S.J., Atkinson, M.S., Murphy, G., Byers, P.H. & Kadler, K.E.(1994). Substitution of serine for glycine 883 in the triple helix of the pro alpha 1 (I) chain of type I procollagen produces osteogenesis imperfecta type IV and introduces a structural change in the triple helix that does not alter cleavage of the molecule by procollagen N-proteinase. *J Biol Chem.* 269(48):30352-30357.

Lindor, N.M. & Bristow, J. (2005). Tenascin-X deficiency in autosomal recessive Ehlers-Danlos syndrome. *Am J Med Genet A.* 135(1):75-80.

Linsenmayer, T.F., Gibney, E., Igoe, F., Gordon, M.K., Fitch, J.M., Fessler, L.I. & Birk, D.E. (1993). Type V collagen: molecular structure and fibrillar organization of the chicken alpha 1(V) NH2-terminal domain, a putative regulator of corneal fibrillogenesis. *J Cell Biol*. 121(5):1181-1189.

Liu, X., Wu, H., Byrne, M., Krane, S. & Jaenisch, R. (1997). Type III collagen is crucial for collagen I fibrillogenesis and for normal cardiovascular development. *Proc Natl Acad Sci U S A*. 94(5):1852-1856.

Löhler, J., Timpl, R. & Jaenisch, R. (1984). Embryonic lethal mutation in mouse collagen I gene causes rupture of blood vessels and is associated with erythropoietic and mesenchymal cell death. *Cell*. 38(2):597-607.

Lu, W., Phillips, C.L., Killen, P.D., Hlaing, T., Harrison, W.R., Elder, F.F., Miner, J.H., Overbeek, P.A. & Meisler, M.H. (1999). Insertional mutation of the collagen genes Col4a3 and Col4a4 in a mouse model of Alport syndrome. *Genomics*. 61(2):113-124.

Malfait, F., Wenstrup, R.J. & De Paepe, A. (2010). Clinical and genetic aspects of Ehlers-Danlos syndrome, classic type. Genet Med. 12(10):597-605.

Miner, J.H., Sanes, J.R. (1996). Molecular and functional defects in kidneys of mice lacking collagen alpha 3(IV): implications for Alport syndrome. *J Cell Biol*. 135(5):1403-1413.

Myllyharju, J. & Kivirikko, K.I. (2001). Collagens and collagen-related diseases. *Ann Med*. 33(1):7-21.

Nuytinck, L., Freund, M., Lagae, L., Pierard, G.E., Hermanns-Le, T. & De Paepe, A. (2000). Classical Ehlers-Danlos syndrome caused by a mutation in type I collagen. *Am J Hum Genet*. 66(4):1398-1402.

Parkin, J.D., San Antonio, J.D., Pedchenko, V., Hudson, B., Jensen, S.T. & Savige, J. (2011). Mapping structural landmarks, ligand binding sites, and missense mutations to the collagen IV heterotrimers predicts major functional domains, novel interactions, and variation in phenotypes in inherited diseases affecting basement membranes. *Hum Mutat*. 32(2):127-143.

Raghunath, M., Bruckner, P & Steinmann, B. (1994). Delayed triple helix formation of mutant collagen from patients with osteogenesis imperfecta. *J Mol Biol*. 236(3):940-949.

Rheault, M.N., Kren, S.M., Thielen, B.K., Mesa, H.A., Crosson, J.T., Thomas, W., Sado, Y., Kashtan, C.E. & Segal, Y. (2004). Mouse model of X-linked Alport syndrome. *J Am Soc Nephrol*. 15(6):1466-1474.

Ricard-Blum, S. & Ruggiero, F. (2005). The collagen superfamily: from the extracellular matrix to the cell membrane. *Pathol Biol*. 53(7):430-442.

Ruggiero, F. & Koch, M. (2008). Making recombinant extracellular matrix proteins. *Methods*. 45(1):75-85.

Schnieke, A., Dziadek, M., Bateman, J., Mascara, T., Harbers, K., Gelinas, R. & Jaenisch, R. (1987). Introduction of the human pro alpha 1(I) collagen gene into pro alpha 1(I)-deficient Mov-13 mouse cells leads to formation of functional mouse-human hybrid type I collagen. *Proc Natl Acad Sci U S A*. 84(3):764-768.

Seppinen, L., Sormunen, R., Soini, Y., Elamaa, H., Heljasvaara, R. & Pihlajaniemi, T. (2008). Lack of collagen XVIII accelerates cutaneous wound healing, while overexpression of its endostatin domain leads to delayed healing Matrix Biol. 27(6):535-546.

Sumiyoshi, H., Mor, N., Lee, S.Y., Doty, S., Henderson, S., Tanaka, S., Yoshioka, H., Rattan, S. & Ramirez, F. (2004). Esophageal muscle physiology and morphogenesis require assembly of a collagen XIX-rich basement membrane zone. *J Cell Biol.* 166(4):591-600.

Symoens, S., Malfait, F., Renard, M., André, J., Hausser, I., Loeys, B., Coucke, P. & De Paepe, A. (2009). COL5A1 signal peptide mutations interfere with protein secretion and cause classic Ehlers-Danlos syndrome. *Hum Mutat.* 30(2):E395-403.

Symoens, S., Renard, M., Bonod-Bidaud, C., Syx, D., Vaganay, E., Malfait, F., Ricard-Blum, S., Kessler, E., Van Laer, L., Coucke, P., Ruggiero, F. & De Paepe, A. (2011). Identification of binding partners interacting with the □1-N-propeptide of type V collagen. *Biochem J.* 433(2):371-381.

Takahara, K., Schwarze, U., Imamura, Y., Hoffman, G.G., Toriello, H., Smith, L.T., Byers, P.H. & Greenspan, D.S. (2002). Order of intron removal influences multiple splice outcomes, including a two-exon skip, in a COL5A1 acceptor-site mutation that results in abnormal pro-alpha1(V) N-propeptides and Ehlers-Danlos syndrome type I. Am J Hum Genet. 71(3):451-465.

Tanimura, S., Tadokoro, Y., Inomata, K., Binh, N.T., Nishie, W., Yamazaki, S., Nakauchi, H., Tanaka, Y., McMillan, J.R., Sawamura, D., Yancey, K., Shimizu, H. & Nishimura, E.K. (2011). Hair follicle stem cells provide a functional niche for melanocyte stem cells. *Cell Stem Cell.* 8(2):177-187.

Tsang, K.Y., Chan, D., Bateman, J.F. & Cheah, K.S. (2010). In vivo cellular adaptation to ER stress: survival strategies with double-edged consequences. *J Cell Sci.* 123(Pt 13):2145-54.

Wang, Y.F., Ding, J., Wang, F. & Bu, D.F. (2004). Effect of glycine substitutions on alpha5(IV) chain structure and structure-phenotype correlations in Alport syndrome. *Biochem Biophys Res Commun.* 316(4):1143-1149.

Wenstrup, R.J., Florer, J.B., Brunskill, E.W., Bell, S.M., Chervoneva, I. & Birk, D.E. (2004). Type V collagen controls the initiation of collagen fibril assembly. *J Biol Chem.* 279(51):53331-53337.

Wenstrup, R.J., Smith, S.M., Florer, J.B., Zhang, G., Beason, D.P., Seegmiller, R.E., Soslowsky, L.J. & Birk, D.E. (2011). Regulation of collagen fibril nucleation and initial fibril assembly involves coordinate interactions with collagens V and XI in developing tendon. *J Biol Chem.* 286(23):20455-204565.

Woodley, D.T., Hou, Y., Martin, S., Li, W. & Chen, M. (2008). Characterization of molecular mechanisms underlying mutations in dystrophic epidermolysis bullosa using site-directed mutagenesis. *J Biol Chem.* 283(26):17838-17845.

Xian, X., Gopa, S. & Couchman, J.R. (2010). Syndecans as receptors and organizers of the extracellular matrix. *Cell Tissue Res.* 339(1):31-46.

Xu, K., Nowak, I., Kirchner, M. & Xu, Y. (2008). Recombinant collagen studies link the severe conformational changes induced by osteogenesis imperfecta mutations to the disruption of a set of interchain salt bridges. *J Biol Chem.* 283(49):34337-34344.

Yang, W., Battineni, M.L. & Brodsky B. (1997). Amino acid sequence environment modulates the disruption by osteogenesis imperfecta glycine substitutions in collagen-like peptides. *Biochemistry.* 36(23):6930-6935.

Zhu, Y., Oganesian, A., Keene, D.R. & Sandell, L.J. (1999). Type IIA procollagen containing the cysteine-rich amino propeptide is deposited in the extracellular matrix of prechondrogenic tissue and binds to TGF-beta1 and BMP-2. *J Cell Biol.* 144(5):1069-1080.

Molecular Genetics in Applied Research

5

Site-Directed Mutagenesis as Applied to Biocatalysts

Juanita Yazmin Damián-Almazo and Gloria Saab-Rincón
Departamento de Ingeniería Celular y Biocatálisis,
Instituto de Biotecnología, Universidad Nacional Autónoma de México,
Cuernavaca, Morelos,
México

1. Introduction

Enzymes are biological catalysts responsible for supporting almost all of the chemical reactions in living organisms. Their *activities, specificities* and *selectivities* make them attractive as biocatalysts for a wide variety of industries. Examples are agrochemicals, detergents, starch, textiles, personal care, pulp and paper, food processing, and animal feed. The chemo-, enantio- and regioselectivities of biological catalysts are hallmarks that make them especially attractive for use in the synthesis of fine chemicals and pharmaceutical intermediates. They are a viable alternative to chemical synthesis, which is usually characterized by low yield and the accumulation of undesirable secondary products. The imminent decrease in the use of fossil fuels has turned attention to new enzyme-based developments for the production of biofuels (*e.g.*, biodiesel) that use renewable raw materials (Cherry & Fidantsef, 2003).

Nevertheless, natural enzymes are often not optimal for use in industrial conditions. It is usually necessary to change the conditions of the process or, most commonly, to alter one or more of the properties of the enzyme. Desirable changes in the enzyme are those that affect substrate specificity, expression level, solubility, stability, activity, selectivity, or thermal stability. Other desired effects could include tolerance to organic solvents or to extreme pH values (Hibbert et al., 2005; Turner, 2009).

Protein engineering usually involves the modification of amino acid sequences at the DNA sequence level by means of chemical or genetic techniques. The resultant protein is then tested for novel, optimal or improved physical and/or catalytic properties (Ulmer, 1983). There are two different basic approaches to engineer proteins, although it is common to combine both approaches for better results.

a. *Rational design.* Mutations are introduced at specific places in the protein-encoding gene. Positions to mutagenize are based on the knowledge of possible relationships of sequence, structure, function and/or the catalytic mechanism of the protein. Recently computational predictive algorithms have been developed and used to preselect promising target sites (Bolon & Mayo, 2001; Kaplan & DeGrado, 2004; Kuhlman et al., 2003; Meiler & Baker, 2006; Pavelka et al., 2009; Zanghellini et al., 2006). However, a deep knowledge of the structure and energy functions is required in order to predict the changes required to modify some parameter of the enzyme. This is especially true if one wishes to change the reaction mechanism.

b. *Directed evolution.* This approach involves repeated cycles of random mutagenesis of and/or recombination with variants of the gene to create a library of genes with slightly different sequences. The enzyme variants thus obtained are submitted to genetic selection or to high-throughput screening to identify those enzyme variants with improvements in the desired property (Stemmer, 1994a; b; Zhao et al., 1998). Directed evolution has been demonstrated to be a very powerful technique, especially for increasing stability (Giver et al., 1998; Ladenstein & Antranikian, 1998; Song & Rhee, 2000; Uchiyama et al., 2000; Zhao & Arnold, 1999) or to change the specificity of an enzyme (Castle et al., 2004; Cohen et al., 2004; Christians et al., 1999; Joerger et al., 2003; Jurgens et al., 2000; Levy & Ellington, 2001; Matsumura & Ellington, 2001; Sakamoto et al., 2001; Song et al., 2002; Zhang et al., 1997). It is a particularly useful approach, since no structural or mechanistic information is required. In many cases, changes that contribute to the improved properties are far from the active sites. They would not have been targeted by a rational strategy.

Both techniques have strengths, but they also have some limitations. For this reason, it is often common to find rational design work combined with directed evolution. It may be desirable to tune some properties of a designed protein (Savile et al., 2010; Siegel et al., 2010) or randomization may be directed to specific regions of a protein that were identified in the design process.

This review summarizes the most common strategies used to identify possible targets for site-directed mutagenesis to enhance biocatalysis. We included sequence- and structure-based strategies for generating enzymes with desired properties. To illustrate a number of the points discussed above, special attention was paid to the site-directed mutagenesis of glycosyl hydrolases. We also used the modification of alpha amylases as a case study. We have described the sequence-based mutagenesis approach that was used to change the transglycosylation/hydrolysis ratio of alpha amylases. Residues involved in the hydrophobicity and electrostatic environment of the active site were identified by sequence and structural alignments with other glycosyltransferases. As a result, certain residues were targeted for mutagenesis. We also used a multiple sequence alignment and structural information in an approach to reduce the hydrolytic activity of the alpha amylase from *Thermotoga maritima*, while increasing its alcoholytic activity. Unlike the wild-type parent, the modified enzyme was able to synthesize alkyl-glucosides.

2. Approaches for selecting targets to mutagenize

Any biochemical, structural, or protein sequence information may be useful for identifying residues that may influence a desired enzyme property. The information may indicate changes that increase or decrease the overall fitness of the enzyme.

A common approach is to focus on regions or positions that may be directly related to the catalytic property. For example, amino acid residues that alter substrate specificity or selectivity are commonly non-conserved residues. They are often in close contact with catalytic residues in or near the active site, cofactors or substrates (Morley & Kazlauskas,

2005; Paramesvaran et al., 2009; Park et al., 2005). Another approach is the identification of sequence motifs that are thought to have been conserved during evolution (Saravanan et al., 2008). In contrast, residues thought to be involved in thermostabilization are spread throughout the entire sequence. Each such residue is thought to make a small contribution to thermostability. However, the additive effect can be significant. For this reason, random mutagenesis is a powerful tool for achieving protein stabilization. However, some features that are known to contribute to protein stability can be implemented by site-directed mutagenesis strategies. These include the introduction of additional disulfide bridges (Mansfeld et al., 1997); decrease of loop entropy by replacement of some amino acid residues to P or by the shortening of loops (Nagi & Regan, 1997); change of α-helix propensity by mutations to replace G residues (low α-helix propensity) to A residues (high α–helix proepnsity); or by the introduction of salt bridges to increase electrostatic interactions in the protein (Kumar et al., 2000; Lehmann & Wyss, 2001; Spector et al., 2000).

2.1 Sequence-based mutagenesis

2.1.1 Alanine scanning

Alanine scanning is a method used to determine the contribution of the side-chains of specific residues in a protein. Substitution of residues with alanine removes all side chain atoms past the β-carbon, without introducing additional conformational changes into the protein backbone. Although mutagenesis by alanine scanning can be a laborious method (because each alanine-mutated protein must be constructed, expressed and analyzed separately), it has nevertheless been useful for the study of interactions at protein-protein interfaces or for the identification of residues involved in substrate recognition (Gibbs & Zoller, 1991), protein stability (Blaber et al., 1995), or binding (Ashkenazi et al., 1990; Cunningham & Wells, 1989).

Alternatives to conventional alanine scanning are computational methods for modeling alanine-scanning mutants. This approach has proven to be useful for predicting active-site residues important for activity (Funke et al., 2005) and to identify amino acid residues important in protein-protein interactions (Kortemme et al., 2004).

2.1.2 Protein sequence alignment

Nature has had the opportunity to explore the protein sequence space through millions of years of evolution. Genetic drift is thought to be the driving force that is responsible for the sequence diversity observed today. However, residues that are indispensable to function and/or stability have been maintained by selective pressure. Multiple sequence alignments are useful tools for identifying positions that are unchangeable in a protein. They will also identify those regions with the plasticity to allow multiple changes. Briefly, when residues with a common evolutionary origin or having structural or functional equivalence are arranged so that the highly conserved residues are aligned, their alignment serves as an anchor for the alignment of the sequences in a set. Analysis of position-specific residue usage (residue profiles) gives information about amino acid conservation or variability at each position. When a multiple sequence alignment is combined with phylogenetic information, it is possible to explore ancestral relationships among groups of homologous protein sequences. It is also useful for identifying important amino acids that probably cannot be modified.

2.1.2.1 Correlating amino acid sequence patterns to specific properties

An approach for identifying residues that may be functionally relevant is to correlate an enzyme property with the amino acid patterns observed in a multiple sequence alignment. For example, comparison of more stable proteins with less stable ones is a strategy for identifying possible thermostabilizing residues (Ditursi et al., 2006; Gromiha et al., 1999; Kumar & Nussinov, 2001; Perl et al., 2000).

Sequence patterns can also be used to identify the determinants of specificity. Good examples are the attempts to change cofactor specificity in dehydrogenases to NAD^+, since NAD^+ is considerably less energetically demanding for the cell to make than is $NADP^+$ (Flores & Ellington, 2005; Kristan et al., 2007; Rodríguez-Zavala, 2008; Rosell et al., 2003).

Even distant mutations can significantly affect the properties of an active site. They may alter slightly the geometry, electrostatic properties or dynamics of amino acids in the active site. Distant residues that are important for their interactions with the active site may be seen as conserved in a multiple sequence alignment. Multiple sequence alignments have also revealed that some residues are infrequent in a sequence; but nevertheless are frequently adjacent. Such cluster-forming residues have probably coevolved. These "protein sectors" are often critical for specific functional roles, including substrate binding, stability, allosteric regulation or catalytic activity (Halabi et al., 2009).

2.1.2.2 Consensus sequence

The method of using a consensus sequence is based on the assumption that, in an amino acid sequence alignment of homologous proteins, the consensus amino acid at a given position contributes more to the stability or the function of the protein than does a non-consensus residue. This assumption is based on the belief that a consensus sequence may closely mimic the sequence of an ancestral protein. One hypothesis posits that many proteins were originally thermophilic or hyperthermophilic (Di Giulio, 2003). Under this premise, the consensus sequence has been used to improve the thermostability of several enzymes. This was achieved by mutation of several residues towards the consensus sequence obtained from a multiple sequence alignment. There are numerous examples in which this approach has been used to increase the thermostability of proteins. Some proteins reconstruct the complete consensus sequence (Lehmann et al., 2000; Sullivan et al., 2011). In others, point mutations were used to identify the residues that increased stability. They were then combined to increase the thermostability of the protein (Maxwell & Davidson, 1998; Nikolova et al., 1998; Yamashiro et al., 2010).

Similarly, Jochens et al. (2010) showed the improvement of an enzyme property by mutagenizing the codon for a residue to codons for those amino acids that appear frequently in natural enzymes at identical positions. Evolution probably selected these residues. They are unlikely to perturb the folding or the function of the protein. In contrast, absent and rarely occurring residues are the ones that are probably not allowed. Their rareness suggests that they may be deleterious to the protein. This approach was used to improve the activity and enantioselectivity of an esterase from *Pseudomonas fluorescens* (PFE). The amino acid distribution at four positions near the active site of PFE, previously reported to influence the enantioselectivity of the enzyme, was determined by a structure-guided multiple-sequence alignment of 171 esterases generated by the 3DM database (Kuipers et al., 2010). A library was created by site-directed mutagenesis of the coding regions for the four active site positions in PFE. Substitutions were limited to frequently occurring residues. Almost all mutants in the library showed significantly improved activity towards a commonly used esterase substrate. Moreover, one mutant had its specific activity enhanced 240-fold relative to that of the wild-type enzyme. The mutant also exhibited substantially higher enantioselectivity in the hydrolysis of 3-phenyl butyric acid p-nitrophenyl ester (E=80) compared to that of the almost nonselective wild-type enzyme (E=3.2) (Jochens & Bornscheuer, 2010).

2.1.2.3 Design of ancestral proteins

As mentioned above, one hypothesis suggests that ancestral proteins were able to withstand the harsh conditions prevalent on earth at that time (Di Giulio, 2001). In addition to their thermostability, ancestral enzymes may have been promiscuous with respect to substrates. The evolution theory of proteins holds that current proteins evolved from low-specificity ancestral proteins. Because of their low specificity, the ancestral proteins evolved to become more efficient at using specific substrates. Thus, the reconstruction of ancestral sequences from multiple sequence alignments and phylogenetic trees may provide the opportunity to change enzyme specificity. Several methods based on this approach have been reported. Some of them are given below.

In the *Ancestral Library* method, all residues located close to or within the enzyme's active site are mutated to residues predicted by phylogenetic analysis and ancestral inference. The substitutions are those residues found in the hypothetical proteins at various nodes and branches of the evolutionary trajectories of a given enzyme family. They do not reflect the entire diversity seen in existing family members (Alcolombri et al., 2011).

Alcolombri and coworkers (2011) used serum paraoxonases and cytosolic sulfotransferases (SULTs) as models. In order to promote changes in substrate specificity, they constructed ancestral libraries of enzymes. Their mutagenesis was directed to residues near or within the active site of an enzyme. From a phylogenetic tree, the most probable ancestral sequences were obtained for all nodes. Using these sequences as templates and the three-dimensional structure of the enzyme, residues in and near the active site were located. The ancestral residues were identified, and the relevant altered enzymes constituted a library of mutants. After activity screening, several variants with different activities and specificities were identified. Some mutants had up to 50-fold higher activity than the activity of the starting enzyme.

REAP (Reconstructing Evolutionary Adaptive Paths) analysis uses phylogeny to identify mutations in gene sequences that are thought to have emerged from a common universal ancestor during functional divergence. The findings are used to generate focused and functionally enriched enzyme libraries (Chen et al., 2010).

REAP was implemented to identify differences in the sequence of promiscuous viral polymerases and non-viral polymerases. The differences may be responsible for the functional divergence without loss of catalytic activity. Sequence alignments and a phylogenetic tree of 719 polymerases were constructed. Ancestral proteins sequences were inferred from the collection of sequences at the nodes of the tree. REAP identified sites that may have changed during the separation of viral and non-viral polymerases. In one example, mutations of the residues identified by REAP analysis for the DNA polymerase of *Thermus aquaticus* yielded 8 mutants that showed a change in the substrate specificity for unnatural dNTP's (Chen et al., 2010).

Similarly, the *Evolutionary Trace* method correlates evolutionary variations within a gene of interest with divergence in the phylogenetic tree of that sequence family. This method has been shown to reveal the functional importance of residues (Lichtarge et al., 1996; Lichtarge et al., 2003).

2.1.2.4 SCA (Statistical Coupling Analysis)

Well-separated mutations can significantly affect the activity, specificity, or enantioselectivity of an enzyme by slightly altering the geometry, electrostatic properties or dynamics of amino acids in the active site. Moreover, it is known that physically contiguous residues form "protein sectors" that can be critical for specific functional roles, including substrate binding, stability, allosteric regulation and catalytic activity (Halabi et al., 2009).

Statistical Coupling Analysis (SCA) is a method that estimates the co-evolution between pairs of amino acids in the multiple sequence alignment of a protein family. SCA shows that proteins can be divided into "protein sectors." In several different proteins, the sectors correspond to amino acids that are physically contiguous. These amino acids often underlie various aspects of function, allosteric regulation, binding, catalytic specificity, and/or fold stability. An application of SCA revealed networks of small subsets of residues that link distant functional sites and cooperate in allosteric communication (Suel et al., 2003).

2.2 Structure-based mutagenesis

The evolution of proteins involves mutations of single residues, insertions, deletions (Pascarella & Argos, 1992), gene duplications, fusions, exon duplications and shuffling (Grishin, 2001). Such changes, which accumulate over time, make the identification of sequence similarities very difficult. However, structure is more preserved than sequence and can be used as an evidence of homology among proteins. Comparative analyses of protein sequences and structures are important approaches for the identification of structural, evolutionary and functional relationships between proteins.

The rapidly growing number of protein structures in the Protein Database (PDB) and advances in homology modeling are of great value for generating structural alignments. In general, these methods provide a measure of structural similarity between proteins. They also generate an alignment that defines the residues that have structurally equivalent positions in the proteins being compared. Homology modeling can be done even when no sequence similarity is detected. Based on structural alignments, it is possible to identify residues in direct contact with the substrate or near the active-site cavity. A more complex analysis can even look into enzyme locations that are far from the active site, but are part of a network of interactions that hold the active site together. The residues at these locations can be targeted for mutagenesis. There is a server called HotSpot Wizard that combines information from extensive sequence and structure database searches with functional data to create a mutability map for a target protein. This approach was validated by comparing "hot spot" predictions with mutations extracted from the literature (Pavelka et al., 2009).

2.2.1 Site-directed Saturation Mutagenesis (SDSM)

The Site-directed Saturation Mutagenesis (SDSM) approach consists of using all 20 amino acids at a position in a protein. Based on structural knowledge, it may be sufficient to target the active-site residues (Park et al., 2005; Schmitzer et al., 2004; Wilming et al., 2002; Woodyer et al., 2003). SDSM can be used to complement error-prone PCR. One of the limitations of using error-prone PCR to generate variants of a protein is that the sequence exploration is limited to an average of seven amino acid substitutions per residue. Once positions that seem to be important for improving a property of a protein are identified by error-prone PCR, SDSM will tune the optimization by testing all 20 amino acids in those positions. Multiple positions can be mutagenized simultaneously by SDSM if only a few positions are being explored. However, the number of variants increases exponentially with the number of positions being explored. Therefore, if more than 3 positions are being randomized, it is better to carry out successive targeted randomizations at the positions.

2.2.2 Combinatorial Active-site Saturation Test (CAST)

The Combinatorial Active-site Saturation Test (CAST) was developed to increase the enantioselectivity and/or the substrate specificity of enzymes. The basis of the method is the generation of small libraries of mutant enzymes that are easy to screen for activity. The mutants are produced by simultaneous randomization of sets of two or three spatially close amino acids, whose side chains form part of the substrate-binding pocket.

Application of this methodology allowed the expansion of the substrate specificity of *Pseudomonas aeruginosa* lipase (PAL). Based on the crystal structure, pairs of amino acids surrounding the binding pocket were defined; and the corresponding libraries were created separately by simultaneous saturation mutagenesis at each pair. The libraries were screened for activity with different substrates. The best-performing variants were selected (Reetz et al., 2005). Further optimization was achieved by iterative cycles of CASTing (Reetz et al., 2006a). Mutants that enhanced a given catalytic property were selected. The residue positions thought to be responsible were organized into groups of two or three. Each group was randomized by saturation mutagenesis to create libraries that were subsequently screened. The best hit of those libraries was used as the template for the next round of mutagenesis. Variability at the other sites was introduced by another round of saturation mutagenesis. The process was continued until the desired degree of catalyst improvement had been achieved. Iterative screening has been applied to enhance very different catalytic properties, including thermostability (Reetz et al., 2006b), substrate acceptance and enantioselectivity (Clouthier et al., 2006; Reetz et al., 2006a).

2.2.2.1 B-factor iterative test

The B-factor iterative test is used to modify enzyme thermostability by increasing rigidity at sites to help prevent unfolding. The selection of target residues is made on the basis of crystallographic B-factor data. This value reflects the degree to which the measured electron density for a particular atom spreads out. It is strongly influenced by thermal fluctuations and the mobility of the atom. Residues with the highest B factors have high flexibility. Appropriate mutations lead to enhanced rigidity and, therefore, to higher thermostability. Target sites are chosen as sites for iterative saturation mutagenesis (ISM), in which each of the target residues is mutagenized to saturation. The best mutant of the first screening is then used as the template for a second round of saturation mutagenesis at one of the other selected sites. The cycle is repeated in an iterative manner (Reetz et al., 2006a).

This method has been used to enhance the thermostability and the tolerance to organic solvents of the mesophilic lipase (LipA) from *Bacillus subtilis*. After several rounds of ISM at residues with high B-factors, a mutant was obtained with the inactivation temperature increased from 48°C to 93°C and with an improved robustness towards organic solvents without affecting the activity of the enzyme (Reetz et al., 2006b; Reetz et al., 2010). Similarly, Jochens and coworkers (Jochens et al., 2010) increased the Tm of the esterase from *Pseudomonas fluorescens* by 9 °C. Smart libraries were guided by the B-factor. By guiding ISM at residues displaying the lowest B-factors, Reetz and coworkers (Reetz et al., 2009) were able to create a lipase from *Pseudomonas aeruginosa* that has a decrease in the Tm value from 71.6 °C to 35.6 °C without affecting the catalytic profile of the enzyme.

2.3 Site-directed homologous recombination

Most proteins are only marginally stable. For this reason, the accumulation of a few mutations is sometimes sufficient to destabilize a protein. The introduction of variability by recombination with the structural gene for a homolog is often less perturbing for folding than is mutagenesis to introduce point mutations. The reason is that some amino acid changes introduced by recombination have already been selected by Nature to give a particular structure and function. Because recombination is more conservative than mutagenesis, several research groups have tried to introduce variability in a sequence by constructing chimeras with genes for homologous proteins. This can be done either randomly (Crameri et al., 1998; Minshull & Stemmer, 1999) or in a site-directed fashion (Landwehr et al., 2007; Li et al., 2007; Pantazes et al., 2007). By substituting a homologous segment, some interactions may be perturbed; and the protein might not be functional. The less perturbing sites for chimeragenesis are thus identified, and a library of recombinants is then constructed by recombining DNA fragments from different homologous genes. The resulting library is screened for a specific property. The properties have included thermostability, activity towards different natural and non-natural substrates, and/or specificity. The power of this technique, compared to random recombination strategies, is that the libraries constructed have a high percentage of folded proteins, thus making it easier to find interesting variants.

2.4 Site-directed loop exchange in proteins

With the same idea as the site-directed chimeragenesis described above, site-directed loop exchange is based on introducing variability only in the binding and/or catalytic loops; the rest of the structure is perturbed very little. The basis of this strategy is that a good portion of catalytic and molecular recognition sites are in loops, while the rest of the structure is the scaffold that maintains the residues and the network of interactions in place to create an environment suitable for catalysis. By exchanging loops, the sequence can be different from that of the parental protein; but the stability and the folding of the protein is maintained. The technique has been widely used, especially for antibodies. It has been recognized that the binding specificity of antibodies relies on the loops of the variable regions (Clark et al., 2009). More recently, import of loops from natural enzymes has been carried out to explore novel activities in a given scaffold (Park et al., 2006). In our laboratory, we have developed a strategy that allows systematic loop exchange from eight different proteins with a TIM barrel fold into a TIM barrel scaffold. (A TIM fold is characterized by a barrel formed by eight parallel β-strands surrounded by seven or eight α-helices. The loops that join the β-strands to the α-helices at the top of the barrel conform the active site in proteins sharing this fold.) We demonstrated that the libraries generated had a high percentage of properly folded proteins (Ochoa-Leyva et al., 2009).

3. Glycosyl hydrolases

Glycosyl hydrolases (also called glycosidases) constitute a widespread group of enzymes that catalyze the hydrolysis of the glycosidic bond. Glycosyl hydrolases can hydrolyze the glycosidic linkage between two or more carbohydrates or between a carbohydrate and a non-carbohydrate moiety to release smaller sugars. They are classified as exo- and endoglycosidases, depending on their ability to cleave a substrate at the end (the non-reducing end) or in the middle of an oligosaccharide or polysaccharide chain, respectively. Glycossyl hydrolases have been classified into more than 100 families based on amino acid sequence similarities (Davies & Henrissat, 1995; Henrissat, 1991; Henrissat & Bairoch, 1993; 1996; Henrissat & Davies, 1997). This classification system, available on the CAZy (CArbohydrate-Active EnZymes) web site (Cantarel et al., 2009), allows reliable prediction of evolutionary relationships, mechanism (retaining/inverting), active site residues and possible substrates. It is even a reasonable tool for newly sequenced enzymes for which function has not yet been biochemically demonstrated.

3.1 Reactions catalyzed by glycosyl hydrolases

In most cases, the hydrolysis of the glycosidic bond is catalyzed by two amino acid residues - a general acid (proton donor) and a nucleophile/base. Depending on the spatial positions of these catalytic residues, hydrolysis occurs via overall retention or overall inversion of the anomeric configuration (Davies & Henrissat, 1995; McCarter & Withers, 1994; Sinnott, 1990).

3.1.1 Inverting glycosyl hydrolases

Inverting enzymes act by a single-step, acid/base-catalyzed mechanism. Two residues, typically glutamic or aspartic acids located 6-11 Å apart, act as acid and base. The leaving group is directly displaced by the nucleophilic water with a single inversion at the anomeric centre (Fig. 1).

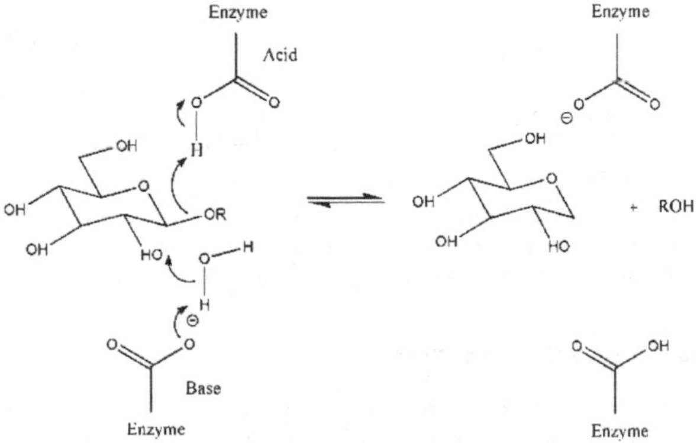

Fig. 1. Inversion hydrolysis mechanism of glycosyl hydrolases.

3.1.2 Retaining glycosyl hydrolases

Retaining glycosidases act through a double-displacement mechanism (each step resulting in inversion at the anomeric centre) involving a covalent glycosyl-enzyme intermediate (Fig. 2). The reaction is catalyzed with acid/base and nucleophilic assistance provided by two amino acid side chains, typically glutamate or aspartate, located 5.5 Å apart. In the first step (glycosylation), one residue plays the role of a nucleophile. It attacks the anomeric centre to displace the aglycon and form a glycosyl enzyme intermediate. At the same time, the other residue functions as an acid catalyst and protonates the glycosidic oxygen as the bond cleaves. In the second step (deglycosylation), the now deprotoned acid-base carboxylate functions as a base to activate the incoming nucleophile (water, saccharide or alcohol) to which the glycosyl is transferred from the enzyme intermediate to give the product.

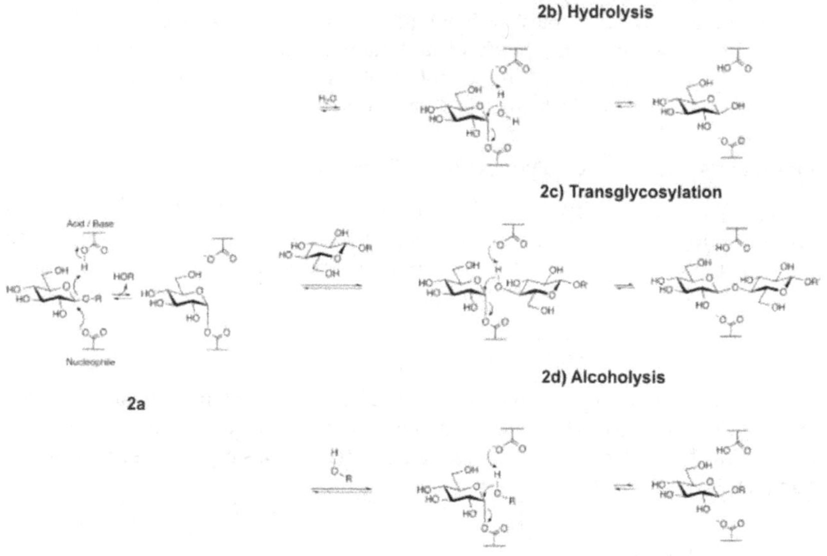

Fig. 2. Retaining mechanism of glycosyl hydrolases.
In a first reaction (a), glycosidic bond breakage of the donor saccharide is carried out and a glycosyl enzyme complex is formed. In (b), the incoming acceptor molecule (water) is activated to promote the release of sugar. If the incoming nucleophile is different from water, the enzyme carries out a transfer reaction, called transglycosylation if the incoming molecule is an oligosaccharide (c), or alcoholysis if the incoming nucleophile is an alcohol (d).

3.2 Industrial uses of glycosyl hydrolases

In nature, glycosyl hydrolases catalyze the degradation of diverse glycosylated polymers, like starch, glucogen, cellulose and hemicellulose. They also participate in anti-bacterial defense strategies (*e.g.*, lysozyme), in pathogenesis mechanisms (*e.g.*, viral neuraminidases) and in normal cellular functions. Glycosyl hydrolases are also of great importance to industry. For example, in the food industry, enzymes like invertase and amylase are employed for the manufacture of invert sugar or maltodextrins; in the paper and pulp industry, xylanases are used to remove hemicelluloses from paper pulp; cellulases are widely used in the textile industry and in laundry detergents; and recently, cellulases and xylanases have been used in the conversion of lignocellulosic biomass into forms suitable for biofuel production.

However, in most cases, glycosyl hydrolases are not optimal in industrial conditions. It is often necessary to alter their stabilities, catalytic activities and/or substrate specificities by protein engineering methods. One of the most frequently altered enzyme properties is thermostability. It can be a limiting factor in the selection of enzymes for industrial applications due to the elevated temperatures or the extreme pH of many biotechnological processes. The stability of an enzyme can be improved by site-directed mutagenesis (Ben Mabrouk et al., 2011; Ghollasi et al., 2010; Leemhuis et al., 2004; Liu et al., 2008; Yin et al., 2011). One of the most exhaustive efforts was done by Palackal and coworkers (Palackal et al., 2004), who used saturation mutagenesis for each of the 189 amino acid residues of a xylanase. They generated a library of modified enzymes, each altered at single position. This library was then screened for variants with increased thermostabiltity, and nine single amino acid changes that contribute to increased stability were identified. These nine single substitutions were then combinatorially assembled to generate all 512 possible variants. Another round of screening identified eleven enzymes with melting temperatures up to 35°C higher than that of the wild-type enzyme.

Another enzyme property that is desirable to modify is the optimum pH. For example, in soybean β-amylase, the hydrogen bond networks around the catalytic base residue (E380) of the enzyme were removed by point mutations, raising the optimal pH from 5.4 to the more neutral pH range of between 6 and 6.6 (Hirata et al., 2004a; Hirata et al., 2004b).

In vivo, glycosidases catalyze the hydrolysis of glycosidic linkages. However, *in vitro*, they can be used as synthetic catalysts to form glycosidic bonds. This process is called the kinetic approach, and it can be accomplished by reverse hydrolysis or by transglycosylation (Fig. 2 c and d). The utility of glycosidases in the synthesis of glucosides through transglycosylation reactions has been employed to synthesize unusual products that are difficult to obtain by other methods. Several site-directed mutagenesis strategies have been used to increase the translycosylation activity of glycosidases or to change substrate specificity. For example, rational modification of the β-glycosidase from *Sulfolobus sulfataricus* to accept a wider range of substrates in transglycosylation reactions has been done. Site-directed mutagenesis was used to alter two key residues involved in substrate recognition to provide access to many different glycoside linkages, including the especially problematic β-mannosyl and β-xylosyl linkages (Hancock et al., 2005). We will focus our discussion on the protein engineering work on α-amylases carried out by us and others.

3.3 alpha-amylases

α-Amylases (EC number 3.2.1.1) are part of the family 13 of glycosyl hydrolases. They catalyze the hydrolysis of internal α-1,4-glycosidic linkages of starch, liberating poly- and oligosaccharides chains of varying lengths. They are found in both eubacteria and eukaryotes. They have a large number of different substrate specificities, as well as huge variations in both temperature and pH optima (Vihinen & Mantsala, 1989).

α-Amylases are the starting enzymes in the industry of modification and conversion of starch. This is because of their capacity to catalyze reactions under environmentally friendly conditions and without the addition of expensive activated sugars (Buchholz & Seibel, 2008). In the sugar-producing industry (Nielsen & Borchert, 2000), bacterial and fungal α-amylases of family GH13, particularly those of the *Bacillus species*, play a vital role in the starch liquefaction process. Starch from wheat, maize and tapioca is hydrolyzed to produce oligosaccharides by the thermostable α-amylases from *Bacillus licheniformis*. The oligosaccharides are then saccharified to glucose by glucoamylase (Crabb & Shetty, 1999). According to the degree of hydrolysis of starch, α-amylases are divided into two categories: (1) saccharifying α-amylases, which hydrolyze 50 to 60% of the saccharide bonds and (2) liquefying enzymes, which process about 30 to 40% of starch hydrolysis (Fukumoto & Okada, 1963). The enzyme commonly used in the industrial process is the α-amylase from *Bacillus licheniformis*. It has the great advantage of being thermostable. This enzyme thus allows the fast hydrolysis of starch at the high temperatures required to dissolve it, with the consequent decrease in viscosity, before decreasing the temperature for the addition of the next enzyme in the process. Some of the disadvantages of using different enzymes are that the pH or temperature conditions may need to be adjusted during the process, with consequent increase in time, costs, and the introduction of salts (buffers) that will have to be removed from the final product. Thus, several research groups, including ours, have tried to engineer α-amylases to change their product profiles (*i.e.*, to make them more saccharifying) to increase their optimal temperatures and to widen their pH spectra.

All α-amylases consists of three domains, called A, B and C. Domain A contains the catalytic residues and has four conserved sequence regions (numbered I-IV) (Mackay et al., 1985; Nakajima et al., 1986; Rogers, 1985), which have been postulated to be essential for the function of α-amylase. Among α-amylase sequences, the four regions align and are spaced at similar intervals along the proteins. These regions presumably form the active site cleft, the substrate-binding site, and the site for binding the stabilizing calcium ion. Domain B forms a large part of the substrate binding cleft, and it is presumed to be important for the substrate specificity differences observed among α-amylases (MacGregor, 1988). It is the least conserved domain among α-amylases (Guzman-Maldonado & Paredes-Lopez, 1995). Finally, domain C constitutes the C-terminal part of the sequence and seems to be involved in substrate binding. All known α-amylases contain a conserved calcium ion located at the interface between domains A and B (Boel et al., 1990; Kadziola et al., 1998; Machius et al., 1998; Machius et al., 1995). The calcium ion is known to be essential for enzyme stability (Vallee et al., 1959).

Depending on the enzyme, the active site cleft can accommodate from four to ten glucose units, each one bound by amino acid residues that constitute the binding subsite for that glucose unit. Subsites are numbered according to the location of the scissile bond. In α-amylases there are two or three subsites on the reducing end of the scissile bond (subsites +1, +2 and +3). The number of subsites on the non-reducing side of scissile bond varies between two and eleven (subsites -1, -2, ... -11) (Brzozowski et al., 2000; Davies et al., 1997; MacGregor, 1988). The number of subsites and their affinities are some of the determinant factors of the final product profiles of α-amylases (Gyemant et al., 2002; Kandra et al., 2002; Matsui et al., 1992a; b).

3.3.1 Transglycosylation reactions in alpha-amylases

As other retaining glycosidases, α-amylases, particularly saccharifying amylases, can also catalyze transfer reactions, which are the result of employing molecules other than water (*e.g.*, carbohydrates or alcohols) as glucosyl acceptors (Fig. 2 c and b, respectively). When a high molecular-weight alcohol is used as an acceptor, the products are alkyl-glucosides. These molecules have a high surface tension activity that has important applications in several industries. Although various retaining glucosidases, like β-galactosidase (Moreno-Beltran et al., 1999; Svensson, 1994), β-xylosidase (Shinoyama et al., 1988), β-fructofuranosidase (Rodríguez et al., 1996; Straathof et al., 1988), and β-glucosidase (Chahid et al., 1992; Vulfson et al., 1990), have been used in alcoholysis reactions, the use of a readily available substrate, like starch, gives α-amylases great potential in the catalysis of this type of reaction.

We found a correlation between the efficiency of hydrolysis and the capacity of the enzymes to carry out transglycosylation reactions. A plausible hypothesis is that those enzymes that are able to transglycosylate can recycle intermediate size oligosaccharides produced during hydrolysis to generate longer ones that are better substrates. This would result in a more saccharifying pattern at equilibrium. Transglycosylation activity is not reported in the bacillary α-amylases used in the starch process industry. We decided to introduce this activity by engineering liquefying α-amylases from *Bacillus stearothermophilus* (Saab-Rincon et al., 1999) and *Bacillus licheniformis* (Rivera et al., 2003). We tried to identify residues that could be responsible for transferase activity. Kuriki and coworkers (Kuriki et al., 1996) suggested three residues that are likely to be responsible for controlling the water activity in the active site of the neopullulanase, a natural transferase from *Bacillus stearothermophilus*. When one of these residues (Y377) was mutated to a non-polar residue, the transglycosylation reaction was favored due to a change in the transglycosylation/ hydrolysis ratio. We carried out a multiple sequence alignment of α-amylases and cyclodextrin glycosyltransferases (CGTases) and identified a residue (A289 in the *Bacillus stearothermophilus* α-amylase) that is analogous to Y377 in the *Bacillus stearothermophilus* neopullulanase. The *Bacillus stearothermophilus* α-amylase is a liquefying enzyme unable to carry out transglycosylation reactions (Fig. 3). We used site-directed mutagenesis to change the A at residue 289 to Y and F, which are present in natural transferases, like neopollullanases and CGTases. The two mutants that were generated were able to carry out the transfer reaction not only to other saccharides (Fig. 4) but also to alcohols, like methanol, to produce methyl-glucosides. The A289Y mutant was more efficient at catalyzing transfer reactions than was A289F (Fig. 5) (Saab-Rincon et al., 1999). Apparently the hydrophobic nature of the mutant residues and the electrostatic interactions that may affect the geometry of the side chains in the active site are important for the transglycosylation reaction. In contrast, when the same mutations were introduced at the equivalent position (V286) in the α-amylase from *Bacillus licheniformis*, the V286Y mutant showed an increase of hydrolytic activity, whereas the V286F mutant had a higher translgycosylation/hydrolysis ratio (Rivera et al., 2003).

In contrast to bacterial liquefying α-amylases from *B. licheniformis* and *B. stearothermophilus*, several fungal amylases like those from *Aspegillus niger* and *Aspergillus oryzae* have the ability to carry out alcoholysis reactions. These two fungi amylases are responsible for saccharifying enzymes that produce maltose, maltotriose and some glucose. Santamaria and coworkers (Santamaria et al., 1999) demonstrated that these enzymes were able to carry out alcoholysis reactions in the presence of methanol and starch as substrate, even at high methanol (20%) and starch (15%) concentrations. Although the alcoholysis reaction was reported in α-amylase from *A. oryzae* using aryl-maltoside and either, methanol, ethanol or butanol as substrates (Matsubara, 1961), the alcoholysis reaction with starch as substrate is less efficient (Santamaria et al., 1999).

Enzyme	Region I		Region II		Region III		Region IV		
		*		*			*		
T. maritima	KVIMDLVINHT	129	DGFRIDAAKHI	223	VGEVHSGN	263	FNFALM	280	FLENHDL	311	
A. oryzae	YLMVDVVANHM	123	DGLRIDTVKHV	211	IGEVLDGD	235	LNYPIY	255	FVENHDN	298	
A. niger	YLMVDVVPDHM	123	DGLRIDSVLEV	211	VGEIDNGN	235	LNYPIY	255	FIENHDN	298	
H. sapiens sal	RIYVDAVINHM	102	AGFRIDASKHM	202	YQEVIDLG	238	TEFKYG	259	FVDNHDM	301	
H. sapiens pan	RIYVDAVINHM	102	AGFRLDASKHM	202	YQEVIDLG	238	TEFKYG	259	FVDNHDN	301	
Pig Panc	RIYVDAVINHM	102	AGFRIDASKHM	202	FQEVIDLG	238	TEFKYG	259	FVDNHDN	301	
Tenebrio	RIYVDAVINHM	100	AGFRVDAAKHM	190	YQEVIDLG	227	LEFQFG	248	FVDNHDN	288	α-Amylases
Alteromonas	DIYVDTLINHM	90	KGFRFDASKHV	179	FQEVIDQG	205	TEFKYS	226	FVDNHDN	265	
Barley	KAIADIVINHR	93	DGWRFDFAKGY	184	VAEIWTSL	209	FDFTTK	249	FVDNHDT	286	
B. lichen	NVYGDVVINHK	106	DGFRLDAVRHI	236	VAEYWQND	266	FDNPLH	289	FVDNHDT	329	
B. amylo	QVYGDVVLNHK	104	DGFRIDAAKHI	236	VAEYWQNN	266	FDVPLH	289	FVDNHDT	329	
B. stearo	QVYADVVFDHK	107	DGFRLDAVKHI	239	VGEYWSYD	269	FDNPLH	292	FVDNHDT	332	
B. sub	KVIVDAVINHT	103	DGFRFDAAKHI	180	YGEILQDS	213	TASNYG	232	WVESHDT	270	
B. circl	KVIIDFAPNHT	141	DGIRMDAVKHM	234	FGEWFLGV	262	LDFRFA	286	FIDNHDM	329	
B. circ2	KIVIDFAPNHT	141	DGIRVDAVKHM	234	FGEWFLGS	262	LDFRFN	286	FIDNHDM	329	
B. specie	KVIIDFAPNHT	141	DGIRVDAVKNM	234	FGEWFLGV	262	LDFRFA	286	FIDNHDM	329	CGTases
B. stearo	KVIIDFAPNHT	137	DGIRMDAVKHM	230	FGEWFLSE	257	LDFRFG	281	FIDNHDM	325	
Termo sulfu	KVIIDFAPNHT	142	DGIRLDAVKHM	235	FGEWFLGT	263	LDFRFS	283	FIDNHDM	330	
Thermus sp	RVMLDAVFNHC	248	DGWRLDVANEI	233	LGEIWHDA	362	MNYPLA	380	LLGSHDT	422	
B. lichen	KIMLDAVFNHI	252	DGWRLDVANEV	334	LGEIWHQA	363	MNYPFT	381	LLDSHDT	426	Maltogenic.
B. stearo	RVMLDAVFNHS	248	DGWRLDVANEV	333	LGEIWHDA	361	MNYPFT	380	LLGSHDT	425	

Fig. 3. Multiple sequence alignment around the four characteristic regions observed in members of the glycoside hydrolase family 13.

The catalytic residues, conserved in all the sequences are marked with asterisks. Aromatic residues involved in the transglycosylation activity of the CGTases are indicated in grey. Residues mutagenized are shown in red. Sequences from α-amylases, GCTases and maltogenic amylases are taken from Damian-Almazo et al. (2008). The abbreviated terms follow: T. maritima, *Thermotoga maritima* α-amylase; A. oryzae, *Aspergillus oryzae* α-amylase; A. niger, *Aspergillus niger* α-amylase; H. sapiens sal, *Homo sapiens* salivary α-amylase; H. sapiens pan, *Homo sapiens* pancreatic α-amylase; Pig panc, *Sus scrofa* pancreatic α-amylase; Tenebrio, *Tenebrio molitor* (mealworm) α-amylase; Alteromonas, *Pseudoalteromonas haloplanktis* α-amylase; Barley, *Hordeun vulgare* α-amylase; B. lichen, *Bacillus licheniformis* α-amylase; B. amylo, *Bacillus amyloliquefaciens* chimera α-amylase; B. stearo, *Bacillus stearothermophilus* α-amylase; B. sub, *Bacillus subtilis* 2633 α-amylase; B. circl, *Bacillus circulans* 251 cyclodextrin glycosyltransferase (CGTase); B. circ2, *Bacillus circulans* 8 CGTase; B. specie, Bacillus sp. 1011 CGTase; B. stearo, *B. stearothermophilus* CGTase; Termo sulfu, *Thermoanaerobacter thermosulfurogenes* CGTase; Thermus sp., Thermus sp.; B. lichen, *B. licheniformis* maltogenic amylase; B. stearo, *B. stearothermopilus* maltogenic amylase.

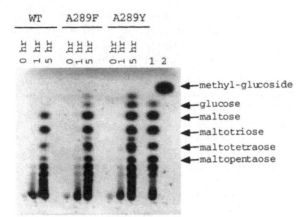

Fig. 4. Product profiles of the alcoholysis reaction with wild-type *Bacillus stearothermophilus* α-amylase and the A289F and A289Y variants at different times of reaction using starch and methanol as substrates.

The product profiles obtained with wild-type (WT) and mutant (A289F and A289Y) *Bacillus stearothermophilus* α-amylases are compared at 0, 1, 5 hours of reaction. We used as standards a mixture of oligosaccharides (1) and methyl-glucoside (2).

Although the wild-type enzyme and the A289F and A289Y mutants showed similar hydrolysis and transglycosylation patterns, the mutants showed products between the glucose and methyl-glucoside standards that could be attributed to alcoholysis reactions. Presumably, those spots for which there are no molecular weight markers correspond to alkyl-oligosaccharides.

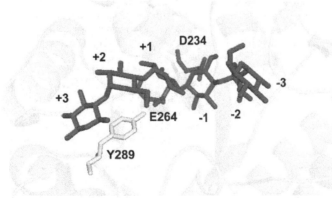

Fig. 5. Structural model of the A289Y mutant of *Bacillus stearothermophilus* α-amylase (PDB code 1HVX) crystallized with the substrate analogue acarbose. (The structure is based on the structure of the α-amylase from *Aspergillus oryzae*, PDB code 7TAA.) Catalytic residues are represented in blue sticks and the mutated residue (Y289) is represented in yellow.

However, the direct use of these enzymes for the production of alkyl-glucosides is precluded by the high temperature required for starch solubilization. The use of a thermophilic saccharifying α-amylase would be attractive, not only in the development of alcoholysis reactions, but also in the starch-processing industry.

Liebl et al. (1997) described an extracellular α-amylase (AmyA) produced by the hyperthermophillic bacterium *Thermotoga maritima* MSB8. The enzyme is a saccharifying amylase with an optimum temperature of 85°C. It can hydrolyze internal α-1-4-glycosidic bonds in various α-glucans, such as starch, amylose, amylopectin and glycogen, to yield mainly glucose and maltose as final products. Because AmyA has the advantage of being a saccharifying enzyme in a stable scaffold, we explored its properties in the transglycosylation and alcoholysis reactions (Damián-Almazo et al., 2008; Damian-Almazo et al., 2008; Moreno et al., 2010). In addition to the characterization reported by Lieb et al., we found that AmyA is capable of using small oligosaccharides (G2 to G7) as substrates for the transglycoslation reactions at short reaction times. This was followed by hydrolysis to yield glucose and maltose as final products. The ability of AmyA to use maltose as a substrate is unusual, as most α-amylases are not capable of using maltose to transfer glucosyl units to other oligosaccharides. Moreover, in the presence of various substrates, AmyA is able to form neotrehalose, a non-reducing disaccharide composed of two glucose molecules joined by α-1, β-1 linkage. It uses 6% maltose as a substrate. Like other saccharifying enzymes, AmyA is capable of transferring glycosyl units to methanol and butanol to produce alkyl-glucosides. When compared to other saccharifiyng α-amylases, AmyA has a high transfer capacity. The enzyme generates 7.5 mg/ml of methyl-glucoside (Moreno et al., 2010), almost three times the maximum amount found for the *A. niger* α-amylase and almost eight times the maximum amount found for the *A. oryzae* α-amylase (Santamaria et al., 1999).

In order to increase the alcoholytic activity present in AmyA, we constructed a structural homology model based on the structure of the α-amylase from *A. oryzae*. The low sequence identity between these enzymes precluded the use of the automatic modeler function in the Swiss Prot server (Sali et al., 1995; Sanchez & Sali, 1997). Therefore, the sequence alignment of the proteins had to be manually adjusted using as anchors the four highly conserved regions of the α-amylases, as shown in Fig. 3. Once a model was generated, the inhibitor molecule acarbose was placed in the active site using the coordinates of the *A. oryzae* α-amylase (PDB code 7TAA). A close-up of the active site model (Fig. 6) supported our hypothesis of the relationship between the presence of an aromatic residue at the position equivalent to Y377 in neopullulanase and the transglycosylation activity of the enzyme and a saccharifying profile. We identified other residues in subsite +1 that are involved in the transglycosylation activity of other glycosyl hydrolases (Kim et al., 2000; Leemhuis et al., 2004; van der Veen et al., 2001). One of these (H222) is part of the second highly conserved region among glycosyl hydrolases and has also been implicated in calcium ion coordination. In the AmyA model, this residue points toward the sugar moiety at subsite +1. Mutagenesis of the equivalent residue in other amylases has been shown to change transferase activity. In the case of the *B. stearothermophilus* α-amylase, the replacement of the equivalent H238 with aspartic acid generated an enzyme with a reduced hydrolysis rate and a modified final product profile (Vihinen & Mantsala, 1990). We constructed a small library by site-directed mutagenesis to explore the effects of replacing H222 with D, Q and E. All mutants showed a greater amount of methyl-glucoside than did the wild-type enzyme, as a result of a change in the alcoholysis/hydrolysis ratio. Mutant H222Q showed an increase in the alcoholysis events as a consequence of an increase in alcoholysis and a reduction in hydrolytic activity of almost 30%. The same change was observed in mutants H222D and H222E. The instability of these mutants toward alcohols decreased the final yield of alkyl-glucoside, as shown in Fig. 7 (Damian-Almazo et al., 2008).

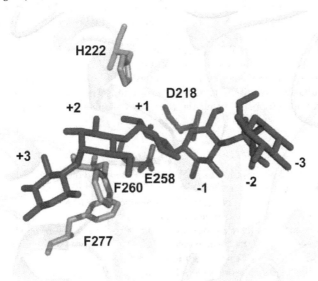

Fig. 6. Homology model obtained for AmyA active site.
The inhibitor acarbose (red) is surrounded by catalytic residues D218 and E258 (blue) and various mutated residues (green). The F277 residue, equivalent to Y377 of neopullulanase, is shown in orange.

(A)

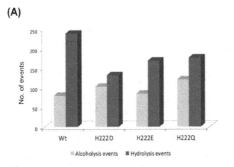

(B)

Enzyme	Alcoholysis/Hydrolysis ratio	Methyl glucoside (mg/ml)
Wt	33	7.5, ±0.6
H222D	78	10.6, ±0.4
H222E	51	9.7, ±1.8
H222Q	68	12.5, ±0.8

Fig. 7. Alcoholysis reaction yields of the wild-type α-amylase from *Thermotoga maritima* and some of the mutants generated
Quantification of alcoholysis reactions generated by wild-type α-amylase from *Thermotoga maritima* and the H222 residue mutants. (A) Alcoholysis and hydrolysis events from 6% starch - 20% methanol obtained with 20 U/ml of the enzymes shown; (B) alcoholysis/hydrolysis ratios and methyl-glucoside yields.

The comparison of liquefying and saccharifying α-amylases, neopollulanases, CGTases and maltogenic amylases through a multiple sequence alignment (Fig. 3) has also made possible the identification of other residues potentially involved in the transglycosylation activity (Fig. 6). In the CGTase from *Bacillus circulans*, residue F260 has been identified as part of a switch for the transglycosylation and hydrolysis reactions (van der Veen et al., 2001). Mutants formed by changing the equivalent residue in wild-type AmyA (F260) to W and G and the H222Q mutant showed opposite behaviors. In the presence of soluble starch as substrate, mutants H222Q and F260G leave higher amounts of high-molecular weight oligosaccharides, while the wild-type enzyme and mutant F260W show a higher proportion of glucose. These differences were seen as changes in the transglycolyslation/hydrolysis ratios. In the double mutant H222Q-F260W, the more transglycosidic pattern of H222Q was recessive, thus eliminating or reducing the presence of longer oligossacharides (Damián-Almazo et al., 2008).

4. Conclusions

Site-directed mutagenesis is a powerful tool for both the study of protein function and the design of novel proteins. Using several approaches to identify phylogenetically conserved residues or residues involved in binding, it has been possible to modify the properties of enzymes that have industrial and biotechnological applications. In order to increase the transglycosylation reactions carried out by α-amylases, we have applied site-directed mutagenesis to residues close to the active site. Based on multiple sequence alignments of natural transferases, like CGTases, we identified conserved residues involved in the transferase reactions of fungal and bacterial α-amylases. Changes to these residues in α-amylases that originally were unable to perform the translycosylation reactions altered the product profiles and increased the translgycosylation/hydrolysis ratios. Furthermore, it was possible to increase the alcoholysis reactions in the α-amylase from *Thermotoga maritima*, which was already capable of carrying out this kind of reaction at a low level.

5. Acknowledgements

This work was funded by PAPIIT [Grant number IN206311 to GSR]; and Consejo Nacional de Ciencia y Tecnología [Grant number 154194]

6. References

Alcolombri, U., Elias, M. & Tawfik, D.S. (2011). Directed evolution of sulfotransferases and paraoxonases by ancestral libraries. *J Mol Biol*, Vol. 411, No. 4, pp. 837-853, 1089-8638 (Electronic) 0022-2836 (Linking).

Ashkenazi, A., Presta, L.G., Marsters, S.A., Camerato, T.R., Rosenthal, K.A., Fendly, B.M. & Capon, D.J. (1990). Mapping the CD4 binding site for human immunodeficiency virus by alanine-scanning mutagenesis. *Proc Natl Acad Sci U S A*, Vol. 87, No. 18, pp. 7150-7154, 0027-8424 (Print) 0027-8424 (Linking).

Ben Mabrouk, S., Aghajari, N., Ben Ali, M., Ben Messaoud, E., Juy, M., Haser, R. & Bejar, S. (2011). Enhancement of the thermostability of the maltogenic amylase MAUS149 by Gly312Ala and Lys436Arg substitutions. *Bioresour Technol*, Vol. 102, No. 2, pp. 1740-1746, 1873-2976 (Electronic) 0960-8524 (Linking).

Blaber, M., Baase, W.A., Gassner, N. & Matthews, B.W. (1995). Alanine scanning mutagenesis of the alpha-helix 115-123 of phage T4 lysozyme: effects on structure, stability and the binding of solvent. *J Mol Biol*, Vol. 246, No. 2, pp. 317-330, 0022-2836 (Print) 0022-2836 (Linking).

Boel, E., Brady, L., Brzozowski, A.M., Derewenda, Z., Dodson, G.G., Jensen, V.J., Petersen, S.B., Swift, H., Thim, L. & Woldike, H.F. (1990). Calcium binding in alpha-amylases: an X-ray diffraction study at 2.1-A resolution of two enzymes from Aspergillus. *Biochemistry*, Vol. 29, No. 26, pp. 6244-6249.

Bolon, D.N. & Mayo, S.L. (2001). Enzyme-like proteins by computational design. *Proc Natl Acad Sci U S A*, Vol. 98, No. 25, pp. 14274-14279.

Brzozowski, A.M., Lawson, D.M., Turkenburg, J.P., Bisgaard-Frantzen, H., Svendsen, A., Borchert, T.V., Dauter, Z., Wilson, K.S. & Davies, G.J. (2000). Structural analysis of a chimeric bacterial alpha-amylase. High-resolution analysis of native and ligand complexes. *Biochemistry*, Vol. 39, No. 31, pp. 9099-9107.

Buchholz, K. & Seibel, J. (2008). Industrial carbohydrate biotransformations. *Carbohydr Res*, Vol. 343, No. 12, pp. 1966-1979, 0008-6215 (Print) 0008-6215 (Linking).

Cantarel, B.L., Coutinho, P.M., Rancurel, C., Bernard, T., Lombard, V. & Henrissat, B. (2009). The Carbohydrate-Active EnZymes database (CAZy): an expert resource for Glycogenomics. *Nucleic Acids Res*, Vol. 37, No. Database issue, pp. D233-238, 1362-4962 (Electronic) 0305-1048 (Linking).

Castle, L.A., Siehl, D.L., Gorton, R., Patten, P.A., Chen, Y.H., Bertain, S., Cho, H.J., Duck, N., Wong, J., Liu, D. & Lassner, M.W. (2004). Discovery and directed evolution of a glyphosate tolerance gene. *Science*, Vol. 304, No. 5674, pp. 1151-1154, 1095-9203 (Electronic) 0036-8075 (Linking).

Clark, L.A., Boriack-Sjodin, P.A., Day, E., Eldredge, J., Fitch, C., Jarpe, M., Miller, S., Li, Y., Simon, K. & van Vlijmen, H.W. (2009). An antibody loop replacement design feasibility study and a loop-swapped dimer structure. *Protein Eng Des Sel*, Vol. 22, No. 2, pp. 93-101, 1741-0134 (Electronic) 1741-0126 (Linking).

Clouthier, C.M., Kayser, M.M. & Reetz, M.T. (2006). Designing new Baeyer-Villiger monooxygenases using restricted CASTing. *J Org Chem*, Vol. 71, No. 22, pp. 8431-8437, 0022-3263 (Print) 0022-3263 (Linking).

Cohen, H.M., Tawfik, D.S. & Griffiths, A.D. (2004). Altering the sequence specificity of HaeIII methyltransferase by directed evolution using in vitro compartmentalization. *Protein Eng Des Sel*, Vol. 17, No. 1, pp. 3-11, 1741-0126 (Print).

Crabb, W.D. & Shetty, J.K. (1999). Commodity scale production of sugars from starches. *Curr Opin Microbiol*, Vol. 2, No. 3, pp. 252-256, 1369-5274 (Print) 1369-5274 (Linking).

Crameri, A., Raillard, S.A., Bermudez, E. & Stemmer, W.P. (1998). DNA shuffling of a family of genes from diverse species accelerates directed evolution. *Nature*, Vol. 391, No. 6664, pp. 288-291.

Cunningham, B.C. & Wells, J.A. (1989). High-resolution epitope mapping of hGH-receptor interactions by alanine-scanning mutagenesis. *Science*, Vol. 244, No. 4908, pp. 1081-1085, 0036-8075 (Print) 0036-8075 (Linking).

Chahid, Z., Montet, D., Pina, M. & Graille, J. (1992). Effect of water activity on enzymatic synthesis of alkylglycosides. *Biotechnol Lett*, Vol. 14, No., pp. 281-284.

Chen, F., Gaucher, E.A., Leal, N.A., Hutter, D., Havemann, S.A., Govindarajan, S., Ortlund, E.A. & Benner, S.A. (2010). Reconstructed evolutionary adaptive paths give polymerases accepting reversible terminators for sequencing and SNP detection. *Proc Natl Acad Sci U S A*, Vol. 107, No. 5, pp. 1948-1953, 1091-6490 (Electronic) 0027-8424 (Linking).

Cherry, J.R. & Fidantsef, A.L. (2003). Directed evolution of industrial enzymes: an update. *Curr Opin Biotechnol*, Vol. 14, No. 4, pp. 438-443, 0958-1669 (Print) 0958-1669 (Linking).

Christians, F.C., Scapozza, L., Crameri, A., Folkers, G. & Stemmer, W.P. (1999). Directed evolution of thymidine kinase for AZT phosphorylation using DNA family shuffling. *Nat Biotechnol*, Vol. 17, No. 3, pp. 259-264.

Damián-Almazo, J., López-Munguía, A., Soberón-Mainero, X. & Saab-Rincón, G. (2008). Role of the phenylalanine 260 residue in defining product profile and alcoholytic activity of the α-amylase AmyA from Thermotoga maritima. *Biologia*, Vol. 63, No. 6, pp. 1035-1043.

Damian-Almazo, J.Y., Moreno, A., Lopez-Munguia, A., Soberon, X., Gonzalez-Munoz, F. & Saab-Rincon, G. (2008). Enhancement of the alcoholytic activity of alpha-amylase AmyA from Thermotoga maritima MSB8 (DSM 3109) by site-directed mutagenesis. *Appl Environ Microbiol*, Vol. 74, No. 16, pp. 5168-5177, 1098-5336 (Electronic) 0099-2240 (Linking).

Davies, G. & Henrissat, B. (1995). Structures and mechanisms of glycosyl hydrolases. *Structure*, Vol. 3, No. 9, pp. 853-859, 0969-2126 (Print) 0969-2126 (Linking).

Davies, G.J., Wilson, K.S. & Henrissat, B. (1997). Nomenclature for sugar-binding subsites in glycosyl hydrolases. *Biochem J*, Vol. 321 (Pt 2), No., pp. 557-559, 0264-6021 (Print) 0264-6021 (Linking).

Di Giulio, M. (2001). The universal ancestor was a thermophile or a hyperthermophile. *Gene*, Vol. 281, No. 1-2, pp. 11-17.

Di Giulio, M. (2003). The Universal Ancestor was a Thermophile or a Hyperthermophile: Tests and Further Evidence. *Journal of Theoretical Biology*, Vol. 221, No. 3, pp. 425-436.

Ditursi, M.K., Kwon, S.J., Reeder, P.J. & Dordick, J.S. (2006). Bioinformatics-driven, rational engineering of protein thermostability. *Protein Eng Des Sel*, Vol. 19, No. 11, pp. 517-524, 1741-0126 (Print) 1741-0126 (Linking).

Flores, H. & Ellington, A.D. (2005). A modified consensus approach to mutagenesis inverts the cofactor specificity of Bacillus stearothermophilus lactate dehydrogenase. *Protein Eng Des Sel*, Vol. 18, No. 8, pp. 369-377, 1741-0126 (Print) 1741-0126 (Linking).

Fukumoto, J. & Okada, S. (1963). Studies on bacterial amylase. Amylase types of Bacillus subtilis species. *J Ferm Technol*, Vol. 41, No., pp. 427.

Funke, S.A., Otte, N., Eggert, T., Bocola, M., Jaeger, K.E. & Thiel, W. (2005). Combination of computational prescreening and experimental library construction can accelerate enzyme optimization by directed evolution. *Protein Eng Des Sel*, Vol. 18, No. 11, pp. 509-514, 1741-0126 (Print) 1741-0126 (Linking).

Ghollasi, M., Khajeh, K., Naderi-Manesh, H. & Ghasemi, A. (2010). Engineering of a Bacillus alpha-amylase with improved thermostability and calcium independency. *Appl Biochem Biotechnol*, Vol. 162, No. 2, pp. 444-459, 1559-0291 (Electronic) 0273-2289 (Linking).

Gibbs, C.S. & Zoller, M.J. (1991). Identification of electrostatic interactions that determine the phosphorylation site specificity of the cAMP-dependent protein kinase. *Biochemistry*, Vol. 30, No. 22, pp. 5329-5334, 0006-2960 (Print) 0006-2960 (Linking).

Giver, L., Gershenson, A., Freskgard, P.O. & Arnold, F.H. (1998). Directed evolution of a thermostable esterase. *Proc Natl Acad Sci U S A*, Vol. 95, No. 22, pp. 12809-12813.

Grishin, N.V. (2001). Fold change in evolution of protein structures. *J Struct Biol*, Vol. 134, No. 2-3, pp. 167-185, 1047-8477 (Print) 1047-8477 (Linking).

Gromiha, M.M., Oobatake, M. & Sarai, A. (1999). Important amino acid properties for enhanced thermostability from mesophilic to thermophilic proteins. *Biophysical Chemistry*, Vol. 82, No. 1, pp. 51-67.

Guzman-Maldonado, H. & Paredes-Lopez, O. (1995). Amylolytic enzymes and products derived from starch: a review. *Crit Rev Food Sci Nutr*, Vol. 35, No. 5, pp. 373-403.

Gyemant, G., Hovanszki, G. & Kandra, L. (2002). Subsite mapping of the binding region of alpha-amylases with a computer program. *Eur J Biochem*, Vol. 269, No. 21, pp. 5157-5162.

Halabi, N., Rivoire, O., Leibler, S. & Ranganathan, R. (2009). Protein sectors: evolutionary units of three-dimensional structure. *Cell*, Vol. 138, No. 4, pp. 774-786, 1097-4172 (Electronic) 0092-8674 (Linking).

Hancock, S.M., Corbett, K., Fordham-Skelton, A.P., Gatehouse, J.A. & Davis, B.G. (2005). Developing promiscuous glycosidases for glycoside synthesis: residues W433 and E432 in Sulfolobus solfataricus beta-glycosidase are important glucoside- and galactoside-specificity determinants. *Chembiochem*, Vol. 6, No. 5, pp. 866-875, 1439-4227 (Print) 1439-4227 (Linking).

Henrissat, B. (1991). A classification of glycosyl hydrolases based on amino acid sequence similarities. *Biochem J*, Vol. 280 (Pt 2), No., pp. 309-316, 0264-6021 (Print) 0264-6021 (Linking).

Henrissat, B. & Bairoch, A. (1993). New families in the classification of glycosyl hydrolases based on amino acid sequence similarities. *Biochem J*, Vol. 293 (Pt 3), No., pp. 781-788, 0264-6021 (Print) 0264-6021 (Linking).

Henrissat, B. & Bairoch, A. (1996). Updating the sequence-based classification of glycosyl hydrolases. *Biochem J*, Vol. 316 (Pt 2), No., pp. 695-696, 0264-6021 (Print).

Henrissat, B. & Davies, G. (1997). Structural and sequence-based classification of glycoside hydrolases. *Curr Opin Struct Biol*, Vol. 7, No. 5, pp. 637-644.

Hibbert, E.G., Baganz, F., Hailes, H.C., Ward, J.M., Lye, G.J., Woodley, J.M. & Dalby, P.A. (2005). Directed evolution of biocatalytic processes. *Biomol Eng*, Vol. 22, No. 1-3, pp. 11-19, 1389-0344 (Print) 1389-0344 (Linking).

Hirata, A., Adachi, M., Sekine, A., Kang, Y.N., Utsumi, S. & Mikami, B. (2004a). Structural and enzymatic analysis of soybean beta-amylase mutants with increased pH optimum. *J Biol Chem*, Vol. 279, No. 8, pp. 7287-7295, 0021-9258 (Print) 0021-9258 (Linking).

Hirata, A., Adachi, M., Utsumi, S. & Mikami, B. (2004b). Engineering of the pH optimum of Bacillus cereus beta-amylase: conversion of the pH optimum from a bacterial type to a higher-plant type. *Biochemistry*, Vol. 43, No. 39, pp. 12523-12531, 0006-2960 (Print) 0006-2960 (Linking).

Jochens, H., Aerts, D. & Bornscheuer, U.T. (2010). Thermostabilization of an esterase by alignment-guided focussed directed evolution. *Protein Eng Des Sel*, Vol. 23, No. 12, pp. 903-909, 1741-0134 (Electronic) 1741-0126 (Linking).

Jochens, H. & Bornscheuer, U.T. (2010). Natural diversity to guide focused directed evolution. *Chembiochem*, Vol. 11, No. 13, pp. 1861-1866, 1439-7633 (Electronic) 1439-4227 (Linking).

Joerger, A.C., Mayer, S. & Fersht, A.R. (2003). Mimicking natural evolution in vitro: an N-acetylneuraminate lyase mutant with an increased dihydrodipicolinate synthase activity. *Proc Natl Acad Sci U S A*, Vol. 100, No. 10, pp. 5694-5699.

Jurgens, C., Strom, A., Wegener, D., Hettwer, S., Wilmanns, M. & Sterner, R. (2000). Directed evolution of a (beta alpha)8-barrel enzyme to catalyze related reactions in two different metabolic pathways. *Proc Natl Acad Sci U S A*, Vol. 97, No. 18, pp. 9925-9930.

Kadziola, A., Sogaard, M., Svensson, B. & Haser, R. (1998). Molecular structure of a barley alpha-amylase-inhibitor complex: implications for starch binding and catalysis. *J Mol Biol*, Vol. 278, No. 1, pp. 205-217.

Kandra, L., Gyemant, G., Remenyik, J., Hovanszki, G. & Liptak, A. (2002). Action pattern and subsite mapping of Bacillus licheniformis alpha- amylase (BLA) with modified maltooligosaccharide substrates. *FEBS Lett*, Vol. 518, No. 1-3, pp. 79-82.

Kaplan, J. & DeGrado, W.F. (2004). De novo design of catalytic proteins. *Proc Natl Acad Sci U S A*, Vol. 101, No. 32, pp. 11566-11570.

Kim, T.J., Park, C.S., Cho, H.Y., Cha, S.S., Kim, J.S., Lee, S.B., Moon, T.W., Kim, J.W., Oh, B.H. & Park, K.H. (2000). Role of the glutamate 332 residue in the transglycosylation activity of ThermusMaltogenic amylase. *Biochemistry*, Vol. 39, No. 23, pp. 6773-6780, 0006-2960 (Print).

Kortemme, T., Kim, D.E. & Baker, D. (2004). Computational alanine scanning of protein-protein interfaces. *Sci STKE*, Vol. 2004, No. 219, pp. pl2, 1525-8882 (Electronic) 1525-8882 (Linking).

Kristan, K., Stojan, J., Adamski, J. & LaniÅ¡nik RiÅ¾ner, T. (2007). Rational design of novel mutants of fungal 17Î²-hydroxysteroid dehydrogenase. *J Biotechnol*, Vol. 129, No. 1, pp. 123-130.

Kuhlman, B., Dantas, G., Ireton, G.C., Varani, G., Stoddard, B.L. & Baker, D. (2003). Design of a novel globular protein fold with atomic-level accuracy. *Science*, Vol. 302, No. 5649, pp. 1364-1368, 1095-9203 (Electronic) 0036-8075 (Linking).

Kuipers, R.K., Joosten, H.J., van Berkel, W.J., Leferink, N.G., Rooijen, E., Ittmann, E., van Zimmeren, F., Jochens, H., Bornscheuer, U., Vriend, G., dos Santos, V.A. & Schaap, P.J. (2010). 3DM: systematic analysis of heterogeneous superfamily data to discover protein functionalities. *Proteins*, Vol. 78, No. 9, pp. 2101-2113, 1097-0134 (Electronic) 0887-3585 (Linking).

Kumar, S. & Nussinov, R. (2001). How do thermophilic proteins deal with heat? *Cell Mol Life Sci*, Vol. 58, No. 9, pp. 1216-1233, 1420-682X (Print) 1420-682X (Linking).

Kumar, S., Tsai, C.J. & Nussinov, R. (2000). Factors enhancing protein thermostability. *Protein Eng*, Vol. 13, No. 3, pp. 179-191, 0269-2139 (Print) 0269-2139 (Linking).

Kuriki, T., Kaneko, H., Yanase, M., Takata, H., Shimada, J., Handa, S., Takada, T., Umeyama, H. & Okada, S. (1996). Controlling substrate preference and transglycosylation activity of neopullulanase by manipulating steric constraint and hydrophobicity in active center. *J Biol Chem*, Vol. 271, No. 29, pp. 17321-17329.

Ladenstein, R. & Antranikian, G. (1998). Proteins from hyperthermophiles: stability and enzymatic catalysis close to the boiling point of water. *Adv Biochem Eng Biotechnol*, Vol. 61, No., pp. 37-85.

Landwehr, M., Carbone, M., Otey, C.R., Li, Y. & Arnold, F.H. (2007). Diversification of catalytic function in a synthetic family of chimeric cytochrome p450s. *Chem Biol*, Vol. 14, No. 3, pp. 269-278, 1074-5521 (Print).

Leemhuis, H., Rozeboom, H.J., Dijkstra, B.W. & Dijkhuizen, L. (2004). Improved thermostability of bacillus circulans cyclodextrin glycosyltransferase by the introduction of a salt bridge. *Proteins*, Vol. 54, No. 1, pp. 128-134, 1097-0134 (Electronic) 0887-3585 (Linking).

Lehmann, M., Kostrewa, D., Wyss, M., Brugger, R., D'Arcy, A., Pasamontes, L. & van, L.A. (2000). From DNA sequence to improved functionality: using protein sequence comparisons to rapidly design a thermostable consensus phytase. *Protein Eng*, Vol. 13, No. 1, pp. 49-57.

Lehmann, M. & Wyss, M. (2001). Engineering proteins for thermostability: the use of sequence alignments versus rational design and directed evolution. *Curr Opin Biotechnol*, Vol. 12, No. 4, pp. 371-375.

Levy, M. & Ellington, A.D. (2001). Selection of deoxyribozyme ligases that catalyze the formation of an unnatural internucleotide linkage. *Bioorg Med Chem*, Vol. 9, No. 10, pp. 2581-2587.

Li, Y., Drummond, D.A., Sawayama, A.M., Snow, C.D., Bloom, J.D. & Arnold, F.H. (2007). A diverse family of thermostable cytochrome P450s created by recombination of stabilizing fragments. *Nat Biotechnol*, Vol. 25, No. 9, pp. 1051-1056, 1087-0156 (Print).

Lichtarge, O., Bourne, H.R. & Cohen, F.E. (1996). An evolutionary trace method defines binding surfaces common to protein families. *J Mol Biol*, Vol. 257, No. 2, pp. 342-358, 0022-2836 (Print) 0022-2836 (Linking).

Lichtarge, O., Yao, H., Kristensen, D.M., Madabushi, S. & Mihalek, I. (2003). Accurate and scalable identification of functional sites by evolutionary tracing. *J Struct Funct Genomics*, Vol. 4, No. 2-3, pp. 159-166, 1345-711X (Print) 1345-711X (Linking).

Liebl, W., Stemplinger, I. & Ruile, P. (1997). Properties and gene structure of the Thermotoga maritima alpha-amylase AmyA, a putative lipoprotein of a hyperthermophilic bacterium. *J Bacteriol*, Vol. 179, No. 3, pp. 941-948.

Liu, Y.H., Lu, F.P., Li, Y., Yin, X.B., Wang, Y. & Gao, C. (2008). Characterisation of mutagenised acid-resistant alpha-amylase expressed in Bacillus subtilis WB600. *Appl Microbiol Biotechnol*, Vol. 78, No. 1, pp. 85-94, 0175-7598 (Print) 0175-7598 (Linking).

MacGregor, E.A. (1988). Alpha-amylase structure and activity. *J Protein Chem*, Vol. 7, No. 4, pp. 399-415, 0277-8033 (Print) 0277-8033 (Linking).

Mackay, R.M., Baird, S., Dove, M.J., Erratt, J.A., Gines, M., Moranelli, F., Nasim, A., Willick, G.E., Yaguchi, M. & Seligy, V.L. (1985). Glucanase gene diversity in prokaryotic and eukaryotic organisms. *Biosystems*, Vol. 18, No. 3-4, pp. 279-292, 0303-2647 (Print) 0303-2647 (Linking).

Machius, M., Declerck, N., Huber, R. & Wiegand, G. (1998). Activation of Bacillus licheniformis alpha-amylase through a disorder-- >order transition of the substrate-binding site mediated by a calcium- sodium-calcium metal triad. *Structure*, Vol. 6, No. 3, pp. 281-292.

Machius, M., Wiegand, G. & Huber, R. (1995). Crystal structure of calcium-depleted *Bacillus licheniformis* alpha- amylase at 2.2 A resolution. *J Mol Biol*, Vol. 246, No. 4, pp. 545-559.

Mansfeld, J., Vriend, G., Dijkstra, B.W., Veltman, O.R., Van den Burg, B., Venema, G., Ulbrich-Hofmann, R. & Eijsink, V.G. (1997). Extreme stabilization of a thermolysin-like protease by an engineered disulfide bond. *J Biol Chem*, Vol. 272, No. 17, pp. 11152-11156.

Matsubara, S. (1961). Studies on Taka-amylase A. VII. Transmaltosidation by Taka-amylase A. *J Biochem*, Vol. 49, No. 3, pp. 226-231.

Matsui, I., Ishikawa, K., Miyairi, S., Fukui, S. & Honda, K. (1992a). Alteration of bond-cleavage pattern in the hydrolysis catalyzed by Saccharomycopsis alpha-amylase altered by site-directed mutagenesis. *Biochemistry*, Vol. 31, No. 22, pp. 5232-5236.

Matsui, I., Ishikawa, K., Miyairi, S., Fukui, S. & Honda, K. (1992b). A mutant alpha-amylase with enhanced activity specific for short substrates. *FEBS Lett*, Vol. 310, No. 3, pp. 216-218.

Matsumura, I. & Ellington, A.D. (2001). In vitro evolution of beta-glucuronidase into a beta-galactosidase proceeds through non-specific intermediates. *J Mol Biol*, Vol. 305, No. 2, pp. 331-339.

Maxwell, K.L. & Davidson, A.R. (1998). Mutagenesis of a Buried Polar Interaction in an SH3 Domain:â€‰ Sequence Conservation Provides the Best Prediction of Stability Effectsâ€ *Biochemistry*, Vol. 37, No. 46, pp. 16172-16182.

McCarter, J.D. & Withers, S.G. (1994). Mechanisms of enzymatic glycoside hydrolysis. *Curr Opin Struct Biol*, Vol. 4, No., pp. 885-892.

Meiler, J. & Baker, D. (2006). ROSETTALIGAND: protein-small molecule docking with full side-chain flexibility. *Proteins*, Vol. 65, No. 3, pp. 538-548, 1097-0134 (Electronic) 0887-3585 (Linking).

Minshull, J. & Stemmer, W.P. (1999). Protein evolution by molecular breeding. *Curr Opin Chem Biol*, Vol. 3, No. 3, pp. 284-290.

Moreno-Beltran, A., Salgado, L., Vazquez-Duhalt, R. & López-Munguía, A. (1999). Modelling the alcoholysis reaction of beta-galactosidase in reverse micelles. *J Mol Catalysis B: Enzymatic*, Vol. 6, No. 1-2, pp. 1-10.

Moreno, A., Damian-Almazo, J.Y., Miranda, A., Saab-Rincon, G., Gonzalez, F. & Lopez-Munguia, A. (2010). Transglycosylation reactions of Thermotoga maritima [alpha]-amylase. *Enzyme and Microbial Technology*, Vol. 46, No. 5, pp. 331-337.

Morley, K.L. & Kazlauskas, R.J. (2005). Improving enzyme properties: when are closer mutations better? *Trends Biotechnol*, Vol. 23, No. 5, pp. 231-237, 0167-7799 (Print) 0167-7799 (Linking).

Nagi, A.D. & Regan, L. (1997). An inverse correlation between loop length and stability in a four-helix-bundle protein. *Fold Des*, Vol. 2, No. 1, pp. 67-75, 1359-0278 (Print) 1359-0278 (Linking).

Nakajima, R., Imanaka, T. & Aiba, S. (1986). Comparison of amino acid sequences of eleven different α-amylases. *Appl Microbiol Biotechnol*, Vol. 23, No. 5, pp. 355-360.

Nielsen, J.E. & Borchert, T.V. (2000). Protein engineering of bacterial alpha-amylases. *Biochim Biophys Acta*, Vol. 1543, No. 2, pp. 253-274.

Nikolova, P.V., Henckel, J., Lane, D.P. & Fersht, A.R. (1998). Semirational design of active tumor suppressor p53 DNA binding domain with enhanced stability. *Proc Natl Acad Sci U S A*, Vol. 95, No. 25, pp. 14675-14680, 0027-8424 (Print) 0027-8424 (Linking).

Ochoa-Leyva, A., Soberon, X., Sanchez, F., Arguello, M., Montero-Moran, G. & Saab-Rincon, G. (2009). Protein design through systematic catalytic loop exchange in the (beta/alpha)8 fold. *J Mol Biol*, Vol. 387, No. 4, pp. 949-964, 1089-8638 (Electronic) 0022-2836 (Linking).

Palackal, N., Brennan, Y., Callen, W.N., Dupree, P., Frey, G., Goubet, F., Hazlewood, G.P., Healey, S., Kang, Y.E., Kretz, K.A., Lee, E., Tan, X., Tomlinson, G.L., Verruto, J., Wong, V.W., Mathur, E.J., Short, J.M., Robertson, D.E. & Steer, B.A. (2004). An evolutionary route to xylanase process fitness. *Protein Sci*, Vol. 13, No. 2, pp. 494-503, 0961-8368 (Print) 0961-8368 (Linking).

Pantazes, R.J., Saraf, M.C. & Maranas, C.D. (2007). Optimal protein library design using recombination or point mutations based on sequence-based scoring functions. *Protein Eng Des Sel*, Vol. 20, No. 8, pp. 361-373, 1741-0126 (Print).

Paramesvaran, J., Hibbert, E.G., Russell, A.J. & Dalby, P.A. (2009). Distributions of enzyme residues yielding mutants with improved substrate specificities from two different directed evolution strategies. *Protein Eng Des Sel*, Vol. 22, No. 7, pp. 401-411, 1741-0134 (Electronic) 1741-0126 (Linking).

Park, H.S., Nam, S.H., Lee, J.K., Yoon, C.N., Mannervik, B., Benkovic, S.J. & Kim, H.S. (2006). Design and evolution of new catalytic activity with an existing protein scaffold. *Science*, Vol. 311, No. 5760, pp. 535-538.

Park, S., Morley, K.L., Horsman, G.P., Holmquist, M., Hult, K. & Kazlauskas, R.J. (2005). Focusing mutations into the P. fluorescens esterase binding site increases enantioselectivity more effectively than distant mutations. *Chem Biol*, Vol. 12, No. 1, pp. 45-54, 1074-5521 (Print) 1074-5521 (Linking).

Pascarella, S. & Argos, P. (1992). Analysis of insertions/deletions in protein structures. *J Mol Biol*, Vol. 224, No. 2, pp. 461-471, 0022-2836 (Print) 0022-2836 (Linking).

Pavelka, A., Chovancova, E. & Damborsky, J. (2009). HotSpot Wizard: a web server for identification of hot spots in protein engineering. *Nucleic Acids Res*, Vol. 37, No. Web Server issue, pp. W376-383, 1362-4962 (Electronic) 0305-1048 (Linking).

Perl, D., Mueller, U., Heinemann, U. & Schmid, F.X. (2000). Two exposed amino acid residues confer thermostability on a cold shock protein. *Nat Struct Biol*, Vol. 7, No. 5, pp. 380-383, 1072-8368 (Print) 1072-8368 (Linking).

Reetz, M.T., Bocola, M., Carballeira, J.D., Zha, D. & Vogel, A. (2005). Expanding the range of substrate acceptance of enzymes: combinatorial active-site saturation test. *Angew Chem Int Ed Engl*, Vol. 44, No. 27, pp. 4192-4196, 1433-7851 (Print) 1433-7851 (Linking).

Reetz, M.T., Carballeira, J.D., Peyralans, J., Hobenreich, H., Maichele, A. & Vogel, A. (2006a). Expanding the substrate scope of enzymes: combining mutations obtained by CASTing. *Chemistry*, Vol. 12, No. 23, pp. 6031-6038, 0947-6539 (Print).

Reetz, M.T., Carballeira, J.D. & Vogel, A. (2006b). Iterative saturation mutagenesis on the basis of B factors as a strategy for increasing protein thermostability. *Angew Chem Int Ed Engl*, Vol. 45, No. 46, pp. 7745-7751, 1433-7851 (Print).

Reetz, M.T., Soni, P. & Fernandez, L. (2009). Knowledge-guided laboratory evolution of protein thermolability. *Biotechnol Bioeng*, Vol. 102, No. 6, pp. 1712-1717, 1097-0290 (Electronic) 0006-3592 (Linking).

Reetz, M.T., Soni, P., Fernandez, L., Gumulya, Y. & Carballeira, J.D. (2010). Increasing the stability of an enzyme toward hostile organic solvents by directed evolution based on iterative saturation mutagenesis using the B-FIT method. *Chem Commun (Camb)*, Vol. 46, No. 45, pp. 8657-8658, 1364-548X (Electronic) 1359-7345 (Linking).

Rivera, M.H., Lopez-Munguia, A., Soberon, X. & Saab-Rincon, G. (2003). Alpha-amylase from Bacillus licheniformis mutants near to the catalytic site: effects on hydrolytic and transglycosylation activity. *Protein Eng*, Vol. 16, No. 7, pp. 505-514.

Rodríguez-Zavala, J.S. (2008). Enhancement of coenzyme binding by a single point mutation at the coenzyme binding domain of E. coli lactaldehyde dehydrogenase. *Protein Science*, Vol. 17, No. 3, pp. 563-570.

Rodríguez, M., Gómez, A., González, F., Bárzana, E. & López-Munguía, A. (1996). Selectivity of methyl-fructoside synthesis with β-fructufuranosidase. *Appl Biochem Biotechnol*, Vol. 59, No., pp. 167-175.

Rogers, J.C. (1985). Conserved amino acid sequence domains in alpha-amylases from plants, mammals, and bacteria. *Biochemical and Biophysical Research Communications*, Vol. 128, No. 1, pp. 470-476.

Rosell, A., Valencia, E., Ochoa, W.F., Fita, I., Pares, X. & Farres, J. (2003). Complete Reversal of Coenzyme Specificity by Concerted Mutation of Three Consecutive Residues in Alcohol Dehydrogenase. *J. Biol. Chem.*, Vol. 278, No. 42, pp. 40573-40580.

Saab-Rincon, G., del-Rio, G., Santamaria, R.I., Lopez-Munguia, A. & Soberon, X. (1999). Introducing transglycosylation activity in a liquefying alpha-amylase. *FEBS Lett*, Vol. 453, No. 1-2, pp. 100-106.

Sakamoto, T., Joern, J.M., Arisawa, A. & Arnold, F.H. (2001). Laboratory evolution of toluene dioxygenase to accept 4-picoline as a substrate. *Appl Environ Microbiol*, Vol. 67, No. 9, pp. 3882-3887.

Sali, A., Potterton, L., Yuan, F., van, V.H. & Karplus, M. (1995). Evaluation of comparative protein modeling by MODELLER. *Proteins Struct Funct Genet*, Vol. 23, No. 3, pp. 318-326.

Sanchez, R. & Sali, A. (1997). Evaluation of comparative protein structure modeling by MODELLER-3. *Proteins Struct Funct Genet*, Vol. 1997, No. 50, pp. 50-58.

Santamaria, R.I., Del Rio, G., Saab, G., Rodriguez, M.E., Soberon, X. & Lopez-Manguia, A. (1999). Alcoholysis reactions from starch with alpha-amylases. *FEBS Lett*, Vol. 452, No. 3, pp. 346-350.

Saravanan, M., Vasu, K. & Nagaraja, V. (2008). Evolution of sequence specificity in a restriction endonuclease by a point mutation. *Proc Natl Acad Sci U S A*, Vol. 105, No. 30, pp. 10344-10347, 1091-6490 (Electronic) 0027-8424 (Linking).

Savile, C.K., Janey, J.M., Mundorff, E.C., Moore, J.C., Tam, S., Jarvis, W.R., Colbeck, J.C., Krebber, A., Fleitz, F.J., Brands, J., Devine, P.N., Huisman, G.W. & Hughes, G.J. (2010). Biocatalytic asymmetric synthesis of chiral amines from ketones applied to sitagliptin manufacture. *Science*, Vol. 329, No. 5989, pp. 305-309, 1095-9203 (Electronic) 0036-8075 (Linking).

Schmitzer, A.R., Lepine, F. & Pelletier, J.N. (2004). Combinatorial exploration of the catalytic site of a drug-resistant dihydrofolate reductase: creating alternative functional configurations. *Protein Eng Des Sel*, Vol. 17, No. 11, pp. 809-819, 1741-0126 (Print) 1741-0126 (Linking).

Shinoyama, H., Kamiyama, Y. & Yasui, T. (1988). Enzymatic synthesis of alkyl β–xylosides from xylobiose by application of the transxylosyl reaction of *Aspergillus niger* β-xylosidase. *Agric Biol Chem*, Vol. 52, No., pp. 2197-2202.

Siegel, J.B., Zanghellini, A., Lovick, H.M., Kiss, G., Lambert, A.R., St Clair, J.L., Gallaher, J.L., Hilvert, D., Gelb, M.H., Stoddard, B.L., Houk, K.N., Michael, F.E. & Baker, D. (2010). Computational design of an enzyme catalyst for a stereoselective bimolecular Diels-Alder reaction. *Science*, Vol. 329, No. 5989, pp. 309-313, 1095-9203 (Electronic) 0036-8075 (Linking).

Sinnott, M.L. (1990). Catalytic Mechanisms of Enzymic Glycosyl Transfer. *Chem Rev*, Vol. 90, No., pp. 1171-1202.

Song, J.K., Chung, B., Oh, Y.H. & Rhee, J.S. (2002). Construction of DNA-shuffled and incrementally truncated libraries by a mutagenic and unidirectional reassembly method: changing from a substrate specificity of phospholipase to that of lipase. *Appl Environ Microbiol*, Vol. 68, No. 12, pp. 6146-6151.

Song, J.K. & Rhee, J.S. (2000). Simultaneous enhancement of thermostability and catalytic activity of phospholipase A(1) by evolutionary molecular engineering. *Appl Environ Microbiol*, Vol. 66, No. 3, pp. 890-894.

Spector, S., Wang, M., Carp, S.A., Robblee, J., Hendsch, Z.S., Fairman, R., Tidor, B. & Raleigh, D.P. (2000). Rational modification of protein stability by the mutation of charged surface residues. *Biochemistry*, Vol. 39, No. 5, pp. 872-879, 0006-2960 (Print) 0006-2960 (Linking).

Stemmer, W.P. (1994a). DNA shuffling by random fragmentation and reassembly: in vitro recombination for molecular evolution. *Proc Natl Acad Sci U S A*, Vol. 91, No. 22, pp. 10747-10751.

Stemmer, W.P. (1994b). Rapid evolution of a protein in vitro by DNA shuffling. *Nature*, Vol. 370, No. 6488, pp. 389-391.

Straathof, A.J.J., Vrijenhoef, J.P., Sprangers, E.P.A.T., Bekkum, H.v. & Kieboom, A.P.G. (1988). Enzymic formation of β-D-fructofuranosides from sucrose: activity and selectivity of invertase in mixtures of water and alcohol. *J Carbohydr Chem*, Vol. 7, No., pp. 223-238.

Suel, G.M., Lockless, S.W., Wall, M.A. & Ranganathan, R. (2003). Evolutionarily conserved networks of residues mediate allosteric communication in proteins. *Nat Struct Biol*, Vol. 10, No. 1, pp. 59-69, 1072-8368 (Print) 1072-8368 (Linking).

Sullivan, B.J., Durani, V. & Magliery, T.J. (2011). Triosephosphate isomerase by consensus design: dramatic differences in physical properties and activity of related variants. *J Mol Biol*, Vol. 413, No. 1, pp. 195-208, 1089-8638 (Electronic) 0022-2836 (Linking).

Svensson, B. (1994). Protein engineering in the alpha-amylase family: catalytic mechanism, substrate specificity, and stability. *Plant Mol Biol*, Vol. 25, No. 2, pp. 141-157.

Turner, N.J. (2009). Directed evolution drives the next generation of biocatalysts. *Nat Chem Biol*, Vol. 5, No. 8, pp. 567-573, 1552-4469 (Electronic) 1552-4450 (Linking).

Uchiyama, H., Inaoka, T., Ohkuma-Soyejima, T., Togame, H., Shibanaka, Y., Yoshimoto, T. & Kokubo, T. (2000). Directed evolution to improve the thermostability of prolyl endopeptidase. *J Biochem (Tokyo)*, Vol. 128, No. 3, pp. 441-447.

Ulmer, K.M. (1983). Protein engineering. *Science*, Vol. 219, No. 4585, pp. 666-671, 0036-8075 (Print) 0036-8075 (Linking).

Vallee, B.L., Stein, E.A., Sumerwell, W.N. & Fischer, E.H. (1959). Metal content of alpha-amylases of various origins. *J Biol Chem*, Vol. 234, No., pp. 2901-2905, 0021-9258 (Print) 0021-9258 (Linking).

van der Veen, B.A., Leemhuis, H., Kralj, S., Uitdehaag, J.C.M., Dijkstra, B.W. & Dijkhuizen, L. (2001). Hydrophobic Amino Acid Residues in the Acceptor Binding Site Are Main Determinants for Reaction Mechanism and Specificity of Cyclodextrin-glycosyltransferase. *J. Biol. Chem.*, Vol. 276, No. 48, pp. 44557-44562.

Vihinen, M. & Mantsala, P. (1989). Microbial amylolytic enzymes. *Crit Rev Biochem Mol Biol*, Vol. 24, No. 4, pp. 329-418.

Vihinen, M. & Mantsala, P. (1990). Conserved residues of liquefying alpha-amylases are concentrated in the vicinity of active site. *Biochem Biophys Res Commun*, Vol. 166, No. 1, pp. 61-65.

Vulfson, E.N., Patel, R., Beecher, J.E., Andrews, A.T. & Law, B.A. (1990). Glycosidases in organic solvents: I. Alkyl −β−glucoside synthesis in a water-organic two-phase system. *Enzyme Microb Technol*, Vol. 12, No., pp. 950-954.

Wilming, M., Iffland, A., Tafelmeyer, P., Arrivoli, C., Saudan, C. & Johnsson, K. (2002). Examining reactivity and specificity of cytochrome c peroxidase by using combinatorial mutagenesis. *Chembiochem*, Vol. 3, No. 11, pp. 1097-1104, 1439-4227 (Print) 1439-4227 (Linking).

Woodyer, R., van der Donk, W.A. & Zhao, H. (2003). Relaxing the nicotinamide cofactor specificity of phosphite dehydrogenase by rational design. *Biochemistry*, Vol. 42, No. 40, pp. 11604-11614, 0006-2960 (Print) 0006-2960 (Linking).

Yamashiro, K., Yokobori, S., Koikeda, S. & Yamagishi, A. (2010). Improvement of Bacillus circulans beta-amylase activity attained using the ancestral mutation method. *Protein Eng Des Sel*, Vol. 23, No. 7, pp. 519-528, 1741-0134 (Electronic) 1741-0126 (Linking).

Yin, Q., Teng, Y., Ding, M. & Zhao, F. (2011). Site-directed mutagenesis of aromatic residues in the carbohydrate-binding module of Bacillus endoglucanase EGA decreases enzyme thermostability. *Biotechnol Lett*, Vol. 33, No. 11, pp. 2209-2216, 1573-6776 (Electronic) 0141-5492 (Linking).

Zanghellini, A., Jiang, L., Wollacott, A.M., Cheng, G., Meiler, J., Althoff, E.A., Rothlisberger, D. & Baker, D. (2006). New algorithms and an in silico benchmark for computational enzyme design. *Protein Sci*, Vol. 15, No. 12, pp. 2785-2794, 0961-8368 (Print) 0961-8368 (Linking).

Zhang, J.H., Dawes, G. & Stemmer, W.P. (1997). Directed evolution of a fucosidase from a galactosidase by DNA shuffling and screening. *Proc Natl Acad Sci U S A*, Vol. 94, No. 9, pp. 4504-4509.

Zhao, H. & Arnold, F.H. (1999). Directed evolution converts subtilisin E into a functional equivalent of thermitase. *Protein Eng*, Vol. 12, No. 1, pp. 47-53.

Zhao, H., Giver, L., Shao, Z., Affholter, J.A. & Arnold, F.H. (1998). Molecular evolution by staggered extension process (StEP) in vitro recombination. *Nat Biotechnol*, Vol. 16, No. 3, pp. 258-261.

6

Biological Activity of Insecticidal Toxins: Structural Basis, Site-Directed Mutagenesis and Perspectives

Silvio Alejandro López-Pazos[1,3] and Jairo Cerón[2]
[1]Facultad de Ciencias de la Salud, Universidad Colegio Mayor de Cundinamarca,
[2]Instituto de Biotecnología, Universidad Nacional de Colombia, Santafé de Bogotá DC,
[3]Biology & Rural Ecology research group, Corporación Ramsar Guamuez,
El Encano (Pasto-Nariño)
[1,2,3]Colombia

1. Introduction

Insect pests destroy about 18% of crop production each year and transmit disease agents (Oerke & Dehn, 2004). Beetles (order Coleoptera) are the largest and most diverse group of eukaryotes. They contain species of harvest pests that produce major losses around the world (Wang et al., 2007). Some examples of coleopteran pests follow: *Dectes texanus* [Coleoptera (order): Cerambycidae (family)], attacks soybeans; *Tribolium castaneum* (Coleoptera: Tenebrionidae), a biological problem of stored products; *Hypothenemus hampei* (Coleoptera: Scolytidae), an entomological problem of coffee crops; and *Premnotrypes vorax* (Coleoptera: Curculionidae), a potato pest in South America (Abdelghany et al., 2010; Tindall et al., 2010; López-Pazos et al., 2009b; Pai & Bernasconi, 2008; Damon, 2000). Lepidopteran species constitute an important group of harmful harvest pests that affect commercial agriculture. Among them are the following: the cotton bollworms, *Helicoverpa armigera* and *H. zea* (both Lepidoptera: Noctuidae); *Tecia solanivora* (Lepidoptera: Gelechiidae), a pest in potato crops of the Americas; *Plutella xylostella* (Lepidoptera: Plutellidae), of great importance in cruciferous crops; and the fall armyworm, *Spodoptera frugiperda* (Lepidoptera: Noctuidae), which causes losses in corn, cotton and rice (Keszthelyi et al., 2011; Du et al., 2011; Chagas et al., 2010; Suckling & Brockerhoff, 2010; Bosa et al., 2006; Monnerat et al., 2006).

The biological control of insect pests is an important alternative to the management of insects (or Integrated Pest Management-IPM). Unfortunately insect pests have been attacked primarily with chemical products, which cause huge environmental losses and adverse effects on human health. However, biological control and IPM-compatible chemicals can be used together [as outlined in a recent review by Gentz et al., (2010). Extensive research has centred on the search for an appropriate insecticidal peptide or polypeptide with toxicity to pest organisms, but not to flora and fauna. Researchers also hope to establish the most appropriate means of delivering the biological molecule to its site of action (De Lima et al., 2007). Recombinant DNA technology allows the exploitation of the insecticidal properties of entomopathogenic organisms. It offers environmentally friendly options for the cost-effective control of insect pests (St Leger & Wang, 2010). Bioinsecticides include microbial agents, natural enemies, plant defences, metabolites, pheromones and genes that transcribe toxic peptides or proteins. The number and variety of toxins is extensive. For example, there are at least 0.5 million insecticidal toxins from arachnids, and evidence suggests that the use of novel toxic factors is likely to be extensive (Whetstone & Hammock 2007).

2. Typical anti-insect toxins

There are two classes of insecticidal toxins: (1) peptide-like toxins (3-10 kDa) from some scorpion and spider venoms and (2) the high molecular mass toxins (*i.e.*, about 1000 residues), such as the latrotoxins from the venom of the spider *Latrodectus* or the crystal proteins of *Bacillus thuringiensis* (De Lima et al., 2007; Schnepf et al., 1998). The toxins of the first group consist of one chain that contains many cysteine residues and intramolecular disulfide bridges. These peptides interact with ion channels (*i.e.*, those for Na^+, K^+, $Ca2^+$ and Cl^-) on cellular membranes (De Lima et al., 2007). Recently a peptide-like toxin nomenclature has been proposed that takes into account the basis of activity, the biological source and the relationship with other toxins (King et al., 2008). The primary sources of entomopathogenic proteins in the second group of toxins are several organisms, including spiders, snakes, scorpions, anemones, snails, lacewings, insects, fungi and bacteria (De Lima et al., 2007; Schnepf et al., 1998).

Toxins from arthropod venoms consist of combinations of biologically active compounds (peptides, proteins, nucleotides, lipids and other molecules). They are used for paralysing insects and for defence against natural enemies. They interact with ion channels and/or receptors from neurological systems in the target organism (De Lima et al., 2007).Venom-derived peptide toxins target voltage-gated Na^+, K^+, Ca^{2+}, or Cl^- channels. Proteins, such as neuropeptides and hormones, are analogous. Their effects depend upon their specific activities (Whetstone & Hammock, 2007). Antagonists disrupt and interfere with development and behaviour. Spiders and scorpions maybe the most important arthropods having insecticidal toxins. Many spider venoms contain a complex mixture of both neurotoxic and cytolytic toxins (see: www.arachnoserver.org). Virtually all insecticidal spider toxins contain a cystine-knot motif that provides them with chemical and biological stability (King et al., 2002; Tedford et al., 2004). These types of venoms contain acylpolyamines (from the Araneidae family), cytolytic toxins (from the Zodariidae family) and neurotoxic peptides (J-atratoxins), and neurotoxins (>10 kDa) and enzymes (~35 kDa) in the Sicariidae and Theridiidae families respectively (Vassilevski et al., 2009; Gunning et al., 2008). *Agelenopsis aperta* employs venom that is very active against insects. It is composed of toxins (agatoxins) that attack transmitter-activated cation channels, voltage-activated sodium channels and voltage-activated calcium channels. The α-agatoxins, μ-agatoxins and ω-agatoxins alter insect ion channels (Adams, 2004). Australian funnel-web spiders [Mygalomorphae (order): Hexathelidae (family): Atracinae (subfamily)] have ω-atracotoxins (36–37 residues with six cysteines in a disulfide pattern), which slow insect cation voltage-dependent channels (Chong et al., 2007).

Scorpions are a special group of organisms that have interesting toxins. These toxins have 23-78 residues. Generally the conformation has an α-helix packed against a three-stranded β-sheet stabilized by four disulfide bonds. Scorpion toxins recognize the face of voltage-dependent sodium channels and alter their gating. They are defined as α-or β-toxins, based on their mechanism of action (Rodríguez de la Vega et al., 2010; Gurevitz et al., 2007; Karbat et al., 2004). Anti-insect α-toxins bind to voltage-dependent sodium channels with high affinity (Gordon et al., 2007). Scorpion β-toxins change the voltage dependence of channel activation. The first class of entomopathogenic scorpion β-toxins is comprised of excitatory toxins. They are composed of 70-76 amino acids. These toxins may induce spastic paralysis by the activation of sodium flux at negative membrane potential. A second group consists of depress ant toxins, which induce flaccid paralysis by depolarization of the axonal membrane. A third set is composed of active toxins, which act on both insect and mammalian sodium channels, with typical depressant effects on insects (Gurevitz et al., 2007).

Surprisingly, some insects (such as the tobacco hornworm *Manduca sexta*) produce insecticidal peptides (each peptide has 23 amino acids) from haemolymph. These molecules can cause paralysis in the larvae of many insects (Skinner et al., 1991). For example, a dose of 105 plaque-forming units of baculovirus containing a poneratoxin DNA sequence from the ant, *Paraponera clavata*, was adequate for controlling lepidopteran individuals (*S. frugiperda*) (Szolajska et al., 2004).

Microorganisms possess toxins for the biological control of insects. Fungus is an entomopathogenic option. *Beauveria bassiana* has a long history in relation to the control of lepidopteran, coleopteran and dipteran species (Howard et al., 2010; Qin et al., 2010; Cruz et al., 2006; Shah & Pell, 2003). *Metarhizium anisopliae* has been used against ticks and insects, this fungus has a wide set of virulent factors, such as lipolytic enzymes, proteases, chitinases and toxins (destruxins) (Schrank & Vainstein, 2010; Pava-Ripoll et al., 2008). Ascomycota (genera *Cordyceps, Hypocrella* and *Torrubiella*), Zygomycota (genera *Conidiobolus* and *Entomophaga*), Deuteromycota (genus *Aschersonia*), Zygomycetes (genus *Entomophthora*) and Hyphomycetes (genus *Hirsutella*), which have activity against lepidopterans and coleopterans (Shah & Pell, 2003). Many bacteria, such as *Serratia marcescens, Photorhabdus luminescens, B. thuringiensis* and *Xenorhabdus nematophilus*, can produce entomopathogenic toxins (Roh et al., 2010; Whetstone & Hammock, 2007). Baculoviruses have been used as safe and effective biopesticides for the protection of crops and forests in the Americas, Europe and Asia. The *oryctes* virus has also demonstrated insecticidal activity against the rhinoceros beetle. The entomopathogenic parvoviruses are an insecticidal option. The *H. armigera* stunt virus (a tetravirus) has been isolated from pests and may be useful for the development of genetically modified plants (Whetstone & Hammock, 2007).

Plants produce a great variety of toxic compounds that are responsible for insect self-defense mechanisms. Plant cyclotides contain 30 amino acids with acyclic peptide backbone and a knotted alignment of three conserved disulphide bonds connected in a "cystine knot" motif. Members of Lepidoptera and Coleoptera are susceptible to plant cyclotides from the Violaceae, Rubiaceae and Cucurbitaceae families (Gruber et al., 2007). Plant cysteine proteases are accumulated after lepidopteran infestation affecting insect growth (Pechan et al., 2002). Plant defensins are antimicrobial proteins with eight conserved cysteines and four disulfide bridges. Defensins attack lepidopteran α-amylases, causing feeding inhibition (Kanchiswamy et al., 2010; Rayapuram & Baldwin, 2008). Plant glucanases, chitinases, lectins and dehydrins are induced after attack by lepidopteran and coleopteran pests (Ralph et al., 2006).

3. The phylogenetic relationship of insecticidal toxins and their comparison with lepidopteran- and coleopteran-specific molecules

Twenty-seven amino acid sequences from the RCSB Protein Data Bank (PDB) (http://www.pdb.org/pdb/home/home.do) were selected by a bibliographical revision, using the criteria of established insect-specific toxicity. Next a phylogenetic analysis of insect-specific toxins was performed (Figure 1) by means of Phylogeny.fr platform (http://www.phylogeny.fr/) (Dereeeper et al., 2008). The available data from a bibliographical search,show insecticidal protein sequences from a large variety of organisms with toxicity against several orders of targets, including 11 anti-lepidopteran toxins and five coleopteran-specific toxins (Table 1).

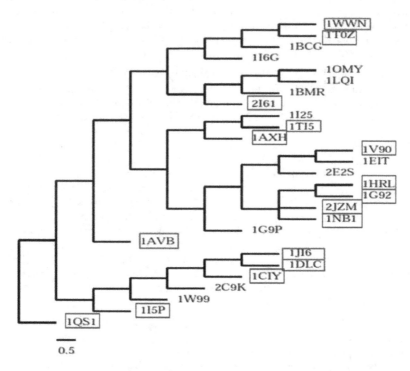

Fig. 1. Phylogenetic tree for insecticidal toxins. The blue squares indicate the coleopteran-specific amino acid sequences and the red squares show antilepidopteran toxins. The analysis of the toxins was done by the parsimony method with the TNT 1.1 program, using the alignment previously obtained with MUSCLE 3.7. The analysis was carried out 1000 times in order to obtain a strict consensus tree by using the bootstrapping tool. The consensus phylogenetic tree was computed by the TreeDyn 198.3. See the text for an analysis.

ID PDB	TOXIN	SOURCE	ORDER TARGET	REFERENCES
1AVB	Arcelin 1	*Phaseolus vulgaris*	Coleoptera	Fabre et al., 1998; Mourey et al., 1998
1AXH	ω-ACTX-HV1	*Hadronyche versuta*	Lepidoptera, Diptera, Ixodida	Chong et al., 2007; Fletcher et al., 1997
1BCG	Bjxtr-IT	*Buthotus judaicus*	Blattaria	Possani et al., 1999; Oren et al., 1998
1BMR	Lqh III	*Leiurus quinquestriatushebraeus*	Blattaria	Krimm et al., 1999
1CIY	Cry1Aa	*Bacillus thuringiensis*	Lepidoptera	Grochulski et al., 1995; López-Pazos & Cerón, 2007
1DLC	Cry3A	*Bacillus thuringiensis*	Coleoptera	Li et al., 1991; López-Pazos & Cerón, 2007
1EIT	μ-agatoxin	*Agelenopsis aperta*	Diptera	Adams, 2004; Omecinsky et al., 1996
1G92	Poneratoxin	*Paraponera clavata*	Lepidoptera	Szolajska et al., 2004
1G9P	ω-Atracotoxin-HV2A	*Hadronyche versuta*	Orthoptera	Chong et al., 2007; Wang et al., 2001
1HRL	PP1	*Manduca sexta*	Lepidoptera	Yu et al., 1999; Skinner et al., 1991
1I5P	Cry2Aa	*Bacillus thuringiensis*	Lepidoptera, Diptera	Morse et al., 2001; López-Pazos & Cerón, 2007
1I6G	CsE-v5	*Centruroides sculpturatus Ewing*	Blattaria	Jablonsky et al., 2001; Possani et al., 1999; Lee et al., 1994
1J16	Cry3Bb1	*Bacillus thuringiensis*	Coleoptera	Galitsky et al., 2001; López-Pazos & Cerón, 2007
1LQI	Lqh(α)IT	*Leiurus quinquestriatus hebraeus*	Diptera	Tugarinov et al., 1997; Zilberberg et al., 1997
1I25	Huwentoxin-II	*Selenocosmia huwena*	Blattaria	Liang., 2004; Shu et al., 2002
1NB1	Kalata B1	*Oldenlandia affinis*	Lepidoptera	Rosengren et al., 2003; Gruber et al., 2007
1OMY	BmKaIT1	*Buthus martensii* Karsch	Diptera, Orthoptera	Ji et al., 1996

ID PDB	TOXIN	SOURCE	ORDER TARGET	REFERENCES
1QS1	VIP2	*Bacillus thuringiensis*	Lepidoptera	Han et al., 1999
1TI5	VrD1	*Vigna radiata*	Coleoptera	Liu et al., 2006
1T0Z	BmK IT-AP	*Buthus martensii* Karsch	Lepidoptera	Li et al., 2005; Hao et al., 2005
1V90	δ-palutoxin IT1	*Paracoelotes luctuosus*	Lepidoptera	De Lima et al., 2007; Ferrat et al., 2005
1WWN	BmK-βIT	*Buthus martensii* Karsch	It displays toxicity against Diptera and is related with AaIT from *Androctonus australis* Hector with activity against Blattaria, Orthoptera, Diptera and Coleoptera	Pava-Ripoll et al., 2008; Zlotkin et al., 2000
1W99	Cry4Ba	*Bacillus thuringiensis*	Diptera	Boonserm et al., 2005; López-Pazos & Cerón, 2007
2C9K	Cry4Aa	*Bacillus thuringiensis*	Diptera	van Frankenhuyzen, 2009; Boonserm et al., 2006
2E2S	Agelenin	*Agelena opulenta*	Orthoptera	Yamaji et al., 2007
2I61	LqhIT2	*Leiurus quinquestriatushebraeus*	Lepidoptera, Diptera	Karbat et al., 2007; De Lima et al., 2007
2JZM	Chymotrypsin inhibitor C1	*Nicotiana alata*	Lepidoptera	Schirra et al., 2008; Schirra et al., 2001; Miller et al., 2000

Table 1. Some toxins from several sources for which **experimentally determined structures** are available in the Protein Data Bank (PDB).

The observed toxin phylogenies - specifically active against lepidopteran species - have several relationships among them and are distributed along all of the branches (Figure 1). *B. thuringiensis* proteins (Cry and vegetative insecticidal protein (VIP)) are closely related in a separated branch, containing three lepidopteran-specific proteins (Cry1Aa, Cry2Aa and VIP2). BmK IT-AP is related with BmK-βIT, Bjxtr-IT and CsE-v5. The antilepidopteran structure 2I61 is in the same group as 1BMR, 1LQI and 1OMY. The Hadronyche versuta toxin (ω-ACTX-Hv1a) has proximity with Huwentoxin-II (*Ornithoctonus huwena*) and the coleopteran-specific VrD1 from the wild mung bean. 1V90 (a lepidopteran-specific toxin), 1EIT and 2E2S are close. The antilepidopteran toxic factors PP1, Poneratoxin, Kalata B1 and chymotrypsin inhibitor C1, have proximity with ω-Atracotoxin-Hv2A from *H. versuta*. Only arcelin1 is in a different site. One might ask whether the amino acid sequences associated with antilepidopteran toxins could have the same biological role, such as 1G9P, 2E2S, 1EIT, 1I25 or 1WWN. Moreover, the phylogenetic tree showed no relationship among Coleopteran-specific sequences, except for 1DLC and 1JI6, which belong to the family of *B. thuringiensis* Cry toxins (Figure 1, Table 1). However, this analysis indicates that 1T0Z (from the Asian scorpion *Buthus martensi* Karsch) and 1I25 (from the Chinese bird spider *O. huwena*) may have anti-coleopteran properties due to the fact that they are in the same branch as 1WWN and 1TI5, respectively (Figure 1). Studies have shown that insecticidal toxins purified from arthropod venoms exert their effects via specific interactions with ion channels and receptors in the central or peripheral nervous system (De Lima et al., 2007; Bloomquist, 2003; Johnson et al., 1998; Fletcher et al., 1997). *B. martensi* Karsch venom has four peptides related to the excitatory insect toxin family and 10 related to the depressant insect toxin (Goudet et al., 2002). Huwentoxin-II (from the spider *O. huwena*) can paralyse cockroaches for hours (ED50 of 29 ± 12 nmol/g) and increase the activity of Huwentoxin-I (a toxin targeting ion channels) (Liang, 2004).

4. Insecticidal toxins and site-directed mutagenesis: case reports

Site-directed mutagenesis is a powerful methodology for studying function and protein structure through manipulation at the level of the DNA molecule. Advances in site-directed mutagenesis have allowed the transfer of new or improved gene roles between organisms, such as bacteria, plants and animals (Adair & Wallace, 1998; James & Dickinson, 1998). In this section, we describe several experiences of the application of site-directed mutagenesis on insecticidal toxin sequences.

4.1 Mutagenesis exposes essential residues in the anti-insect toxin Av2 from *Anemonia viridis*

Sea anemones (Metazoa, Cnidaria, Anthozoa, and Hexacorallia) are sessile predators that are highly dependent on their venom for prospering in a wide range of ecological environments. Venom analysis shows a significant collection of low molecular weight toxins: ~20 kDa pore-forming toxins, 3.5–6.5 kDa voltage-gated potassium channel-active toxins and 3–5 kDa polypeptide toxins active on voltage-gated sodium channels (Navs) (Moran et al., 2009). [A Nav has a central role in the excitability of animals. It functions in the initiation and propagation of action potentials (Goldin, 2002).]

The *Anemonia viridis* toxin 2 (Av2) is a lethal neurotoxin. Av2 has shown a clear preference for insect Nav from the assessment of toxin effects on the *Drosophila melanogaster* sodium channel (DmNav1) expressed in *Xenopus laevis* oocytes (Moran et al., 2009; Warmke et al., 1997). Hence, mutagenesis offers a means of examining residues thought to be important for Av2 activity on insect Navs. A synthetic gene coding for Av2 was designed. It was cloned into the expression vector pET-14b and used to transform appropriate *Escherichia coli* cells (strain BL21). Av2 point mutations (Note: amino acid abbreviations and single-letter designations are given in Table 1 of the chapter by Figurski et al.) [V2A (*i.e.*, residue 2 changed from V to A), P3A, L5A, D7A, S8A, D9A, G10A, G10P, S12A, V13A, R14A, G15A, G15P, N16A, T17A, L18A, G20P, I21A, P28A, S29A, W31A, I32A, N33A, K35A, K36A,

H37A, P39A, T40A, I41A, W43A and Q47A] were established by means of PCR (Polymerase Chain Reaction) using the appropriate primers and the synthetic Av2 gene as the DNA template. The mutant proteins were purified by reverse-phase high performance liquid chromatography. Toxicity assays were done on *Sarcophaga falculata* blowfly larvae. (They were scrutinized for immobilization and contraction). Competition binding assays were done with the neuronal membranes of adult cockroaches (*Periplaneta americana*). The toxicity correlated well with the results of the binding assays. This study indicated that N-terminal aliphatic residues (V2 and L5) play a role in such activity. The central region of the toxin is not involved in the toxic activity. W23 and L24 are important residues in toxin structure. At the C-terminus, it is noteworthy that residue I41 is involved in the bioactive surface of Av2. Residues V2, L5, D9, N16, L18 and I41 are pivotal amino acids for toxicity to blowfly larvae and for binding to cockroach neuronal membranes. The information from these mutants may be applicable to other insect orders (Moran et al., 2006).

4.2 Mutagenesis demonstrates that N183 is a key residue for the mode of action of the Cry4Ba protein

B. thuringiensis is a biopesticide bacterium. Its insecticidal properties are attributed (predominantly) to Cry toxins (a protein family), which are synthesized during the sporulation phase of the organism (Roh et al., 2007). The Cry protein is ingested by the susceptible insect, solubilized in the gut lumen, and cleaved by proteases to yield the activated 60 kDa toxin. Next Cry toxins are recognized by cadherin-like receptors (CADR) to assemble oligomeric forms of the toxin. The toxin oligomers have binding affinities to the secondary receptors: aminopeptidase N (APN), alkaline phosphatase (ALP), ADAM metalloprotease or glycosylphosphatidyl-inositol (GPI)-anchored proteins. The oligomers insert into the apical membrane of midgut-generating pores to cause osmotic lysis and insect death (Ochoa-Campuzano et al. 2007; Pigott & Ellar, 2007). Cry toxin is composed of three functional domains. Domain I comprises seven hydrophobic and amphipathic α-helices and is capable of forming pores in the apical membrane of the insect midgut. Domain II is made of three variable anti-parallel β-sheets, which are responsible for receptor recognition. Domain III has two anti-parallel β-strands involved in structural stability and receptor binding (Schnepf et al., 1998). Site-directed mutagenesis on Cry proteins revealed the function of each domain in the toxicity to the target insect. This fact provides a perspective on the generation of toxins with enhanced toxicity or new specificities.

A collection of Cry4Ba mutants (Figure 2), which are modified in polar uncharged residues (Y178, Q180, N183, N185, and N195) within α-helix 5, were developed to observe their effects on biological activity. All mutant toxins were generated using PCR-based site-directed mutagenesis, and each mutant was expressed from the *lac* promoter in *E. coli* upon IPTG (isopropyl β-D-thiogalactopyranoside) induction. The Cry4Ba-N183A mutant does not display lethality, while alanine substitutions for other residues (Y178, Q180, N185, and N195) still maintained more than 70% of the insect toxicity of the Cry4Ba standard (Figure 2). This result indicated that N183 plays an important role in the functionality of the Cry4Ba toxin (Likitvivatanavong et al., 2006).

Other studies indicated that N183 plays a crucial role in both toxic and structural properties. Mutants N183Q and N183K were made so as to be insoluble at alkaline pH. Mutations at N183 using several residues (with different structural characteristics) revealed that substitutions with a polar amino acid still retained lethal activity similar to the Cry4Ba standard. Nevertheless, changes to charged or nonpolar residues suppressed biological activity (Figure 2). In conclusion, N183 polarity and α-helix 5 localization (in the middle of domain I) are very important to the toxicity of the Cry4Ba protein (Likitvivatanavong et al., 2006).

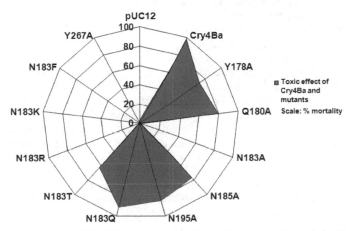

Fig. 2. Biological activity of Cry4Ba and mutants. The red colour indicates lethality and level. Bioassays for mosquito-larvicidal activity were performed using 2-day-old *Stegomyia* (*Aedes*) *aegypti* (mosquito) larvae. The altered residues in the mutant proteins are given on the outside of the graph. The gene for the mutant protein was inserted into the plasmid expression vector pUC12 and induced from the *lac* promoter. pUC12 on the graph depicts the toxicity of the vector alone.

4.3 A Juvenile hormone esterase with a mutated α helix shows improved insecticidal effects

Juvenile hormone (JH) regulates several physiological events in insects (development, metamorphosis, reproduction, diapause, migration, polyphenism and metabolism). JH esterase (JHE) is a hydrolytic enzyme from the α/β-hydrolase fold family, which metabolizes JH (Kamita et al., 2003). When JHE is injected into lepidopteran larval states, it causes a darkening and a decrease in feeding (Hammock et al., 1990; Philpott & Hammock, 1990). JHE is rapidly cleared from the haemolymph following inoculation, suggesting a discriminatory system for its elimination (El-Sayed et al., 2011). In testing, it was revealed that the double histidine mutated JHE [JHE K204H and R208H (in an amphipathic α helix)] is capable of blocking clearance from the haemolymph by reducing its binding to the JHE receptor. These experiments used *Autographa californica* NPV (AcMNPV, a baculovirus with pathogenic activity towards insect pests) as an expression vehicle. JHE shows enhanced insecticidal activity against the lepidopteran larvae of *M. sexta* (tobacco hornworm), *Heliothis virescens* (tobacco budworm) and *Agrotis ipsilon* (black cutworm) (El-Sayed et al., 2011).

Mutant and wild-type JHEs were produced and purified from insect cells, and their activities were found in the culture supernatants of insect cells. The specific activity of mutant JHE was 6.5 nmol of JH III acid (a metabolism product of JH by JHE) formed min^{-1} mg^{-1}. The specific activity of wild-type JHE was 61.3 nmol of JH III acid formed min^{-1} mg^{-1}. The K204H and/or R208H alterations, although far-removed from the catalytic site of the protein, induced allosteric properties that led to a decrease in activity. No statistically significant differences were seen in the clearance of JH hydrolysis activity in the fourth instars of *H. virescens, A. ipsilon* and *M. sexta*. Bioassays (using the first instars of *H. virescens* and *A. ipsilon*) were done to establish the lethal concentration and the lethal time and to determine the result of the expression of mutant JHE on the insecticidal lethality of the baculovirus. The results showed that the median lethal concentration of mutant JHE was 3.2-fold lower in *H. virescens*, in contrast to the effect of AcMNPV. There is no effect on *A. ipsilon*, as observed by the bioassay (Table 2). The most notable difference between the esterases was the higher median lethal concentration (1.9-fold) of mutant JHE compared to a non-mutant JHE against *A. ipsilon* (Table 2). The median lethal concentration of mutant JHE in *H. virescens* was 3.5-fold lower than mutant JHE in *A. ipsilon*. The median lethal time of *H. virescens* and *A. ipsilon* treated with mutant JHE was about 4.8 and 5.3 days, respectively. It was about the same for non-mutant JHE. In addition, feeding assays were carried out using the first instars of *M. sexta* (for 4 days on an artificial diet or on a tomato leaf). The results showed 41–90% lower mass for the mutant than for the JHE wild type (non-mutant) at the end of the experiment. The study showed that point mutations of the amphipathic α-helix were sufficient for improving insecticidal activity (El-Sayed et al., 2011).

Insect	Esterase	Median lethal concentration (x10⁵) (95% Confidence Limits)
H. virescens	Mutant JHE	1.8 (1.0-2.6)
	Wild type JHE	**2.7** (1.8-3.8)
A. ipsilon	Mutant JHE	6.3 (3.6-13)
	Wild type JHE	3.3 (2.3-4.6)

Table 2. Lethal concentrations of mutant and wild-type versions of JHE in the first instar larvae of *H. virescens* and *A. ipsilon*. Insects were inoculated with recombinant JHEs in a polyhedral virus vehicle. The median lethal concentration is expressed as polyhedra per ml (modified of El-Sayed et al., 2011).

4.4 Predicting important residues responsible for the capacity of scorpion α-toxins to discriminate between insect and mammalian voltage-gated sodium channels

Scorpion toxins are poison molecules (61–67 amino acids). Scorpion α-toxins recognize voltage-gated sodium channels (NaCh). NaChs mediate the temporary increase in sodium ion permeability thereby generating action potentials. The toxin expands the action potential by delaying the inactivation stage (Gordon et al., 2007). LqhαIT, from the scorpion *Leiurus quinquestriatus hebraeus*, is an α-toxin that is highly active on insect NaChs. A mutagenic analysis of LqhαIT was performed, revealing that the residues important for function are grouped into two different domains. A new toxin made by putting the efficient region of LqhαIT onto Aah2 (an anti-mammalian α-toxin from the scorpion *Androctonus australis* Hector) proved to be anti-insect (Karbat et al., 2004).

Mutations in the cDNAs of *L. quinquestriatus hebraeus* encoding LqhαIT were generated by PCR (Gurevitz et al., 1991). A CD (Circular Dichroism) Spectroscopy analysis was recorded at 25°C (Karbat et al., 2004). Some residues (Y14, E15, D19, Y21, E24, L25, K28, A39, N54 and P56) had no effect on the biological action or alteration of the CD spectrum. N44 and mutants F17G/A, R18A, W38A had decreased lethality and an unchanged CD spectrum. The F17W and W38Y mutants had activities similar to wild-type LqhαIT, so aromatic side chains affect toxin function. The substitutions I57A/T, R58K, V59A/G, R58K/V59A, K62A/L/R and R64N in the C-terminal region reduced biological activity. The substitution R58N had a marked negative effect on biological activity. This result implies that both charged amine groups and the aliphatic moiety in R58 are principal determinants in functionality. Biologically important residues appear in two domains. The first domain (core-domain) consists of F17, R18, W38 and N44. The second domain (NC-domain) is formed by residues K8, Y10, P56, I57, R58, V59, K62 and R64 (Karbat et al., 2004). LqhαIT and Aah2 have an overall similarity of 70%, although the similarity varies in the NC-domain. The core-domain and the NC-domain of Aah2 were replaced by the LqhαIT counterparts to generate four hybrids (Table 3). The constructs were evaluated with biological assays using *S. falculata* blowfly larvae. Immobilization and contraction were measured, and an effective dose of 50% (ED50) was calculated (Table 3) (Karbat et al., 2004).

Toxin	ED50/100 mg of *S. falculata* body weight
Parental	
LqhαIT	13 ng
Aah2	> 10 µg
Mutant toxin	
Aah2$^{LqhαIT(8-10)}$	> 10 µg
Aah2$^{LqhαIT(56-64)}$	> 10 µg
Aah2$^{LqhαIT(8-10, 56-64)}$	64 ng
Aah2$^{LqhαIT(8-10, G17F, 56-64)}$	37 ng

Table 3. Toxicity assays of Aah2 and its counterpart mutants (Karbat et al., 2004).

The similar activities of Aah2$^{LqhαIT(8-10, G17F, 56-64)}$ and LqhαIT indicate that their functional NC-domains are equally oriented. This indicates that the increase of insecticidal activity is related to the arrangement of the NC-domain in a structure that projects into the solvent. Remarkably this conformation is universal to all scorpion α-toxins with lethality on insects, in contrast with the flat face in α-toxins that are toxic to mammals (Karbat et al., 2004).

5. Final remarks

5.1 Novel sources?

Whole-genome sequencing projects are a resource of biological functions and their annotation allows for the detection of proteins through orthologous sequences (common ancestry), searches and primary and tertiary structure correlation - a process named "comparative genomics" (Lee et al. 2007; Ellegren, 2008). This theoretical approach makes it possible to find candidate toxins in sequenced genomes. An appropriate criterion for the identification of novel lepidopteran and coleopteran candidate toxins can be understood in terms of the "guilt by association" principle (Gabaldon &Huynen, 2004; Aravind, 2000). For this reason, we applied a very basic protocol (Figure 3). BLAST (tblastn) searches from the National Centre for Biotechnology Information (NCBI) (http://blast.ncbi.nlm.nih.gov /Blast.cgi). Searches were done using each toxin (from Table 1) as a query. The iterative searches were done for proteins larger than 100 aminoacids with an inclusion threshold of 0.01 (the statistical significance limit for inclusion of a sequence in the process) and for proteins smaller than 100 aminoacids with an inclusion threshold of 0.1. The searches used the 881 completely sequenced bacterial and archaeal genomes available on the NCBI Microbial Genomes website at the time of this analysis (January 2011) and the entire NCBI environmental samples database (1.66 million Whole Genome Shotgun reads) (see http://www.ncbi.nlm.nih.gov/). The searches were done until either convergence was achieved or until the last iteration before the first known false positives appeared. Significant hits to proteins encoded in these genomes were further classified as possible insect-specific toxins. The BLAST analysis showed fourteen microbial sequences with a high similarity to insecticidal queries (Table 4). There is a version of Arcelin 1 encoded in the genome of the cyanobacterium *Acaryochloris marina* (Tables 1 and 4). Cry proteins from *B. thuringiensis* have a degree of correspondence to sequences in the genomes of four bacteria and one archaeon (Table 4). The VIP2 toxin from *B. thuringiensis* appears to be very diverse in nature. We found VIP-like toxins encoded by eleven bacterial genomes (Table 4). The identified lepidopteran-active toxins are associated with Cry1Aa, Cry2Aa and VIP2. Anti-coleopteran-like toxins were identified, and they are related to Arcelin 1 and Cry3A (Table 4). The search in the Environmental Sample Database showed seven most probable insecticidal sequences related with a Blattaria-active toxin, a coleopteran-specific toxin, four lepidopteran-active toxins and an anti-dipteran toxin (Table 4).

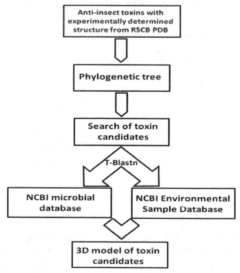

Fig. 3. Diagram of the work.The search for lepidopteran- and coleopteran-specific toxins was done through a basic strategy with the BLAST program on microbial and environmental genomes.

For our trial, the most important organisms harbouring lepidopteran- and coleopteran-active toxins are *A. marina*, *B. weihenstephanensis* and *Clostridium difficile*. First, *A. marina* is a unicellular cyanobacterium containing chlorophyll d as a major pigment (Ohashi et al., 2008). Second, *B. weihenstephanensis* is a Gram-positive, facultatively anaerobic, spore-forming bacterium. This organism has food poisoning potential and is able to grow aerobically at 7°C. *B. weihenstephanensis* has a 16s rDNA signature sequence [1003]TCTAGAGATAGA and the signature sequence [4]ACAGTT of the gene for CspA (a major cold shock protein) (Lechner et al., 1998). Third, *C. difficile* is a Gram-positive spore-forming anaerobic bacterium thought to be involved in diarrhoea and colitis. *C. difficile* codes for two potent toxins (A and B), which attach to specific receptors in the lumen of human colonic epithelium (Vaisnavi, 2010). It is interesting to note that "particular" organisms have versions of these kinds of toxins, such as *Methanosarcina acetivorans* (an acetate-using methanogen archaeon), *Dyadobacter fermentans* (a Gram-negative bacterium isolated from maize and related to *Runella slithyformis*), the marine bacterium *Microscilla furvescens*, and *Cupriavidus necator* - previously known as *Ralstonia eutropha*, a microorganism that can be isolated from several environmental sources, such as soil and water, and which is important in polyhydroxyalkanoate production and bioremediation by the degradation of chlorinated aromatic pollutants (Galagan et al., 2002;Chelius & Triplett, 2002; Lykidis et al., 2010). In addition, we detected other *Clostridium* and *Bacillus* species. The NCBI environmental samples database, a metagenome of the Sargasso Sea genetic diversity from the Venter et al. (2004) project, shows environmental sequences with anti-lepidopteran and anti-coleopteran potential (Table 4).

We built tertiary (3D) structures of some of the predicted toxins: a lepidopteran-active toxin, a coleopteran-specific toxin and a toxin from a metagenome sequence. Approximately 30% sequence identity in the primary sequence is required for the generation of useful structures (Forster, 2002; Paramasivan et al., 2006). Tertiary models of candidate insecticidal sequences were constructed by homology modelling using the crystal structure of homologous protein from the RCSB PDB database (http://www.pdb.org/pdb/home/home.do). We used SWISS-MODEL (http://swissmodel.expasy.org/) (Arnold et al., 2006) for the identification of templates (Table 4 footnotes). The structural alignments were generated with DeepView Swiss-PdbViewer 4.0 software (http://spdbv.vital-it.ch/) (Guex & Peitsch, 1997).

The final models (Figure 4) have a range of 33% to 37% identity with the templates. The toxins in Figure 4 correspond to the following (A) NCBI ID NC_009925.1 from the *A. marina* MBIC11017 genome (33% identity), (B) NCBI ID NC_010180 from the *B. weihenstephanensis* KBAB4 plasmid pBWB401 (37% identity) and (C) the hypothetical protein GOS_5670768 from the marine metagenome (33% identity) (Table 4). The most striking feature of the predicted structure of the candidate insect toxin from the *A. marina* genome consists of two large β-pleated sheets that form a scaffold on which is a possible a carbohydrate-binding region (Figure 4). These architectures and topologies are found in a wide variety of carbohydrate recognizing proteins, such as plant lectins, galactins and serum amyloid proteins (Loris et al., 1998). The model is structurally related to the jelly-roll topology, which facilitates viral entry into bacterial cells. Entry is mediated by interactions with sugar-modified proteins on the cell surface (Petrey & Honig, 2009). It has been postulated that the binding of the lectin to the sugar moiety of any of the glycosylated digestive enzymes is a potential factor of insecticidal activity (Peumans & Van Damme, 1995a, b). Based on the structural alignment of the aminoacid sequences of the toxin from *B. weihenstephanensis* with

INSECTICI DAL TOXIN (ID PDB)	ORGANISM TARGET GENOME/ENVIRONMEN TAL SOURCE	ID NCBI	E-VALUE	REGION
Microbial database				
1AVB[A]	*Acaryochloris marina* MBIC11017	NC_009925.1	3e-10	1669294-1669911
1CIY, 1DLC[B], 1I5P, 1JI6*, 1W99 and 2C9K**	*Bacillus weihenstephanensis* KBAB4 plasmid pBWB401	NC_010180	8e-97-2e-10	139296-138751
	Methanosarcina acetivorans C2A	NC_003552.1	4e-19-1e-04	3249335-3249832
	Dyadobacter fermentans DSM 18053	NC_013037.1	1e-15-4e-06	2869719-2870441
	Bacillus brevis NBRC 100599	NC_012491.1	5e-16-0.026[1]	4962833-4963585
	Ralstonia eutropha JMP134 Chromosome 1	NC_007347.1	1e-08-3.3[2]	411729-411409
1QS1	*Clostridium difficile*	ABHF02000033.1	2e-41	223624-224649
	Clostridium perfringens, E str. JGS1987	NZ_ABDW01000001 2.1	3e-39	66996-65971
	Clostridium botulinum, D str. 1873 plasmid pCLG1	NC_012946.1	1e-33	103322-104389
	Clostridium acetobutylicum ATCC 824	NC_003030.1	5e-17	398379-398876
	Bacillus cereus Rock4-18	NZ_ACMN0100016 2.1	1e-21	17703-17065
	Bacillus halodurans C-125	NC_002570.2	4e-15	3637460-3636978
	Streptomyces avermitilis MA-4680	NC_003155.4	5e-12	6590878-6591372
	Listeria monocytogenes FSL R2-561	AARS01000007.1	8e-12	72280-71786
	Lactobacillus brevis subsp. *gravesensis*	NZ_ACGG0100011 8.1	3e-11	220449-220006
	Aeromonas hydrophila subsp. *hydrophila*	NC_008570.1	2e-05	1214897-1215424
	Enterococcus faecalis V583	NC_004668.1	2e-05	311391-311870
Environmental database				
1BMR	hypothetical protein GOS_4202115 marine	gbIECA60195.1	0.057	88-243

INSECTICIDAL TOXIN (ID PDB)	ORGANISM TARGET GENOME/ENVIRONMENTAL SOURCE	ID NCBI	E-VALUE	REGION
1DLC[c]	metagenome hypothetical protein GOS_5670768 marine metagenome	gb I ECH33518.1	0.014	12-142
1QS1	hypothetical protein GOS_355881 marine metagenome	gb I EBA70908.1	6e-04	102-270
	hypothetical protein GOS_1734861 marine metagenome	gb I EDJ21677.1	8e-04	416-584
	hypothetical protein GOS_9568803 marine metagenome	gb I EBF61568.1	0.003	5-173
	hypothetical protein GOS_7854205 marine metagenome	gb I EBP79016.1	0.004	78-232
1W99	hypothetical protein GOS_6575573 marine metagenome	gb I EBX51304.1	0.010	29-95

Table 4. Results of the BLAST search in a microbial database (Blosum 62, E threshold 0.01) and Environmental Sample Database (Blosum 62, E threshold 0.01) (underlined by modelled sequences). * It is not compatible with B .weihenstephanensis. ** Only compatible with B. weihenstephanensis and M. acetivorans. [A] PDB template: 1G7Y chain C (lectin from the legume Dolichos biflorus). Model residues: 72-289.[B] PDB template: 3EB7 (Cry8Ea1). Model residues: 64-648.[C]PDB template: 2E58 (MnmC2 from Aquifex aeolicus). Model residues: 38-136. The ID PDB refers to code in Protein Data Bank; the ID NCBI refers to accession number in National Center for Biotechnology Information. The region column refers to the specific segment inside the DNA sequence from the ID NCBI column.

the Cry8Ea1 protein, a model of the toxin was obtained; and it corresponds to the general model for a Cry protein (Figure 4). The last structure corresponds to a sequence from the marine metagenome. It was built by homology to a possible transferase of Aquifex aeolicus, a hyperthermophilic microorganism that grows at 85-100°C. It has been suggested that this organism may be the earliest diverging eubacterium (Deckert et al., 1998). The model is composed of three α-helices and a large β-sheet, in which the first and second β-strands are arranged in parallel; and the third and fourth are anti-parallel. Interestingly, the model is somewhat similar to that of the aminoacyl-tRNA synthetase editing domain (Ribas de Pouplana & Schimmel, 2000; Naganuma et al., 2009). The phylogenetic relationships amongst these enzymes are clustered around substrate specificity (Guo et al., 2009). That the amino acid sequence from an ancient bacterium has identity with the Cry protein of B. thuringiensis, and that the toxin structure is similar to that of an aminoacyl-tRNA synthetase editing domain and that it has a helix-sheet formation, hints at the origin of these toxins and their specificities.

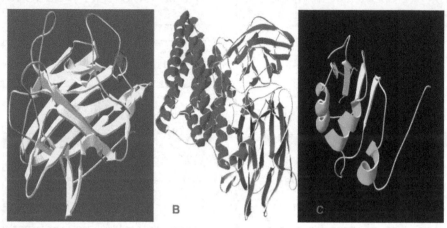

Fig. 4. Models of candidate toxins. (A) Insect toxin the from the *A. marina* genome (β-pleated sheets are in yellow); (B) Structure of the toxin from the *B. weihenstephanensis* genome (domain I is red; blue represents domain II; and domain III is green); and (C) model of the toxin from the marine metagenome (the helices are green, and the β-sheet is yellow). Also see the text.

5.2 *B. thuringiensis* vs. lepidopteran and coleopteran pests

The entomopathogenic bacterium *B. thuringiensis* has been used to help thwart the development of insect and plant resistance by using *cry* genes to construct lethal toxins against pest larvae. Some Cry proteins display biological activity against lepidopteran (Cry1, Cry2, Cry7, Cry8, Cry9, Cry15, Cry22, Cry32 and Cry51) and coleopteran (Cry1B, Cry1I, Cry3, Cry7, Cry8, Cry9, Cry14, Cry22, Cry23, Cry34, Cry35, Cry36, Cry37, Cry43 and Cry55) organisms (van Frankenhuyzen, 2009). Over the past fifteen years, research in our laboratory has focused on the study of the Cry proteins of the entomocidal bacterium *B. thuringiensis* for the biological control of insect pests in Colombia. This country is severely affected by lepidopteran and coleopteran pests, such as larvae of the potato tuber moth, *T. solanivora*; the armyworm, *S. frugiperda*; the Andean weevil, *Premnotrypes vorax* and the coffee berry borer (CBB), *Hypothenemus hampei*.

5.3 Our experience with lepidopterans

We worked with the tobacco budworm, *Heliothis virescens* (Lepidoptera: Noctuidae), an important pest in the Americas. This insect is susceptible to the Cry1Aa, Cry1Ab, Cry1Ac, Cry1Ae, Cry1B, Cry1F, Cry1I, Cry1J, Cry2, Cry8 and Cry9 toxins. Cry1Ac is the most active toxin against this pest (van Frankenhuyzen, 2009). In collaborative work, we tested chimeric Cry1 proteins (Cry1Ba, Cry1Ca, Cry1Da, Cry1Ea, and Cry1Fb) containing domain III of Cry1Ac, which shows higher toxicity in the Cry1Ba, Cry1Ca and Cry1Fb proteins. In addition, we considered an analysis for toxicity against *H. virescens* with the Cry1Ac domain

III triple-mutant toxin, named Tmut (N506D, Q509E, Y513A), supplied by Dr. Ellar (Burton et al., 1999). The test was done by means of a competition-binding assay using an immunoblotting method on nitrocellulose paper. Brush border membrane vesicles (BBMVs) from the *H. virescens* midgut were incubated with biotin-labelled toxin and with increasing concentrations of homologous (identical) or heterologous (mutant) toxin (Figure 5). The Tmut toxin was not able to compete with the Cry1Ac protein for binding to BBMVs (Figure 5). Also the mutant toxicity was 7-fold lower than the toxicity of the reference Cry1Ac. It indicates that at least one of the three residues (N506, Q509 and Y513) has an important role in the biological activity of the toxin (Karlova et al., 2005).

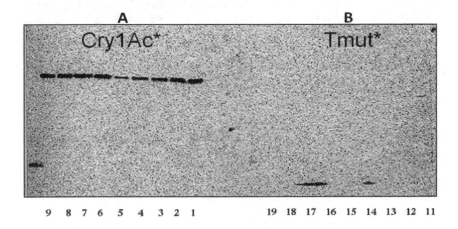

Fig. 5. The Cry1Ac binding reaction on *H. virescens* BBMVs. A. Lane 1, control with nothing added; lanes 2-5, homologous competition between parental Cry1Ac (10, 30, 90, 270 ng of the protein for each lane, respectively) and Cry1Ac labelled with biotin (10 ng); lanes 6-9, heterologous competition between the Cry1Ac domain III triple-mutant (named Tmut, which has the point mutations N506D, Q509E, and Y513A) toxin (10, 30, 90, 270 ng of the protein for each lane, respectively) and Cry1Ac labelled with biotin (10 ng). The first experiment (lanes 2-5) shows that the Cry1Ac wild-type protein (both the labelled and unlabelled proteins) binds to BBMVs (*i.e.*, competition was observed); the second experiment (lanes 6-9) indicates that the Cry1Ac domain III triple-mutant (Tmut) toxin was not able to bind to BBMVs and compete with the bound Cry1Ac wild type (labelled) protein (*i.e.*, competition was not visible). B was set up as follows: lane 11, a no-competitor control; lanes 12-15, heterologous competition between parental Cry1Ac (10, 30, 90, 270 ng of the protein for each lane, respectively) and the Cry1Ac domain III triple-mutant (Tmut) toxin labelled with biotin (10 ng); lanes 16-19, homologous competition between the Cry1Ac domain III triple-mutant (Tmut) toxin (10, 30, 90, 270 ng of protein for each line, respectively) and the Cry1Ac domain III triple-mutant (Tmut) toxin labelled with biotin (10 ng). However, the absence of bands in B confirmed that Tmut is unable to bind to BBMVs. The asterisk indicates the toxin labelled with biotin. Also see the text.

We collaborated in the genetic characterization of *S. frugiperda* (fall armyworm) strains from Brazil, Colombia and Mexico, all of which were correlated with vulnerability to the Latin American *B. thuringiensis* isolates and recombinant toxins (Monnerat et al., 2006). The recognition of genetic variability among insect strains is a decisive analysis for the development of improved pest control strategies, since the biological behaviour of Cry proteins on insect populations is dependent on the specific alleles (specially receptor related), the gene flow and fitness performance. Genetic analysis [molecular analysis for genetic variability was done with Random Amplification of polymorphic DNA (RAPD)] showed that these *S. frugiperda* populations had different levels of similarity among them (between 22% and 37%). *B. thuringiensis* isolates were found to have genes for Cry1 (Cry1Aa, Cry1Ab, Cry1Ac, Cry1B, Cry1C, Cry1D, Cry1E, Cry1G and Cry1I) and Cry2. The fall armyworm (*S. frugiperda*) groups differ in their susceptibilities to *B. thuringiensis*. The most toxic *B. thuringiensis* isolates for *S. frugiperda* had a mixture of genes for Cry1Aa, Cry1B and Cry1D. The Colombian population of this insect was the most susceptible to Latin American *B. thuringiensis* strains. The Mexican *S. frugiperda* was sensitive to recombinant Cry1Ca and Cry1Da. *S. frugiperda* from Brazil was highly susceptible to recombinant Cry1Ca, while the Colombian insects were susceptible to recombinant Cry1B, Cry1C and Cry1D proteins (Monnerat et al., 2006).

Recently we contributed to the determination of Cry1 toxicity against the first instar larvae of *T. solanivora*. We evaluated the products of the *cry1Aa*, *cry1Ab*, *cry1Ac*, *cry1Ca*, *cry1Da*, *cry1Ba*, *cry1Ea*, *cry1Fa* and *cry1Ia* genes and the gene for the hybrid protein SN1917 (encoding Cry1Ba and Cry1Ia in domain II) against the first instar larvae of this pest. We identified toxins with high activity relative to the Cry1Ba, Cry1Ac and SN1917 toxins (Martinez et al., 2003; López-Pazos et al., 2010).

5.4 Our experience with coleopterans

We researched the relationship between ecological niches of the Andean weevil, *P. vorax*, and the bacterium *B. thuringiensis*. We isolated and molecularly characterized *B. thuringiensis* native strains from potato areas (soil, store products and dead *P. vorax*). Bioassays were done using neonate larvae. In addition, the Cry3Aa recombinant toxin and its mutants (mutant 1: D354E; mutant 2: R345A, ΔY350, ΔY351; and mutant 3: Q482A, S484A, R485A) were constructed; and biological assays were performed. We found 300 strains (Bt index was 0.43, calculated as *B. thuringiensis* strains divided by the total amount of *Bacillus* strains) with 21 *cry* gene profiles. Unfortunately neither the isolates nor the recombinant Cry3Aa toxin were toxic against this coleopteran. However, a Cry3A triple mutant [R345A, ΔY350 (deletion), ΔY351 (deletion)] had a minor level of biological activity (mortality 21.87%), in contrast to wild-type Cry3Aa (<6%). This was probably due to site-directed modifications (López-Pazos et al., 2009b).

Coffee crops are severely affected by the CBB (coffee berry borer, *H. hampei*). Female insects drill fissures into the berry and lay their eggs, causing severe losses in production and quality. The entire metamorphosis takes place in the fruit (Damon 2000). This pest is currently present in more than 90% of the planted area (Bustillo 2006; Ramírez 2009). Recently, our research has been centred on the study of Cry toxins for the biological control of CBB, using recombinant proteins of Cry1B, Cry1I, Cry3A, Cry4, Cry9 and SN1917. Although the Cry1B and Cry3A proteins showed minor activity against the pest, the results support the hypothesis that toxicity could be indirect and due to physiological factors of the insect rather than directly from the toxicity of dedicated toxin molecules. Unfortunately the Cry1I, Cry4, Cry9 and SN1917 hybrids were not toxic to CBB (López-Pazos et al. 2010, 2009a). We wanted to learn about the possible interaction between Cry toxins and the receptors in midgut CBB. Brush border membrane vesicles (BBMVs) from the midgut of *H. hampei* were prepared according to Wolfersberger et al. (1987). We used the Cry1B, Cry1I, Cry3A (López-Pazos et al. 2009a; López-Pazos et al. 2010), Cry4 and Cry9 proteins (Figure 6). BBMVs divided by protein electrophoresis showed bands between 20–220 kDa (Figure 6). A blotting test was prepared to determine the weight of Cry-binding proteins in CBB-BBMVs. Cry1B recognized proteins of ~190, 140, 80, 75, 60, 50 and 40 kDa (Figure 6). A signal for Cry1I was also visible at 140 kDa (Figure 6). Cry3A binding proteins were detected at ~140 kDa, 120 kDa and 70 kDa (Figure 6). Cry4 and Cry9 were not detected by any protein on BBMVs (Figure 6). There appeared to be several Cry1B and Cry3A toxin binding sites and/or receptors in the midgut epithelia of CBB.

5.4.1 The modes of action of Cry toxins in coleopterans: the case of CBB

The specific conditions in CBB gut physiology (acidic pH, types of proteases or high proportions of insecticide resistance alleles) are not favourable to the modes of action of the Cry proteins (López-Pazos et al. 2009a). The presence of candidate receptors for Cry proteins in CBB offers evidence for the potential of Cry protein use for the control of this pest. Cadherin-like receptors (CADR) have been studied in lepidopteran and dipteran insects. CADRs were isolated from the coleopterans *Diabrotica virgifera virgifera* (191 kDa) and *Tenebrio molitor* (179 kDa) (Sayed 2007; Fabrick et al. 2009). The CADR receptors are highly variable, with molecular weights ranging from 175 to 210 kDa. An important Cry protein binding site was found to be contained in CADR repeat number 12 (Pigott & Ellar 2007; Hua et al. 2004). It was possible improve the toxicity of Cry3 proteins against coleopterans by adding a CADR fragment containing Cry protein binding site (Park et al. 2009).

Aminopeptidase N (APN) is an N-acetyl-D-galactosamine (GalNAc)-bearing glycoprotein. APN is a receptor for Cry toxins. Different APNs have molecular weights of 90-170-kDa. It was proposed that the Cry-APN interaction has two steps: carbohydrate recognition and irreversible protein-protein interaction (Pigott & Ellar 2007). More than 60 different APNs have been registered in databases. They are from 26% to 65% similar (Herrero et al. 2005, Nakanishi et al. 1999). The 140 kDa protein (from BBMV analysis) is consistent with its being an APN. We do not know if the multiple Cry-binding polypeptides detected in CBB are different proteins or if they are one APN glycosylated differently.

It is also known that CADRs are susceptible to proteolytic digestion and for producing a ~120 kDa fraction. For this reason, CADRs can be confused with APNs in protein-protein interaction blots (Martínez-Rámirez et al. 1994). Cry proteins have multiple binding determinants, possibly specified independently by domains II and III. Moreover, Cry toxins interact with other classes of proteins in the Coleoptera order, such as ALP (molecular weight ~65 kDa), V-ATPase and the Heat-Shock Cognate protein (~ 80 kDa) and the ADAM metalloprotease (~30 kDa) (Hua et al. 2001; Ochoa-Campuzano et al. 2007; Martins et al. 2010; Nakasu et al. 2010). Any signals in the ligand blot for Cry1B and Cry3A would be related with these proteic groups. However, we identified the minor biological activity of Cry1B and Cry3A proteins on CBB larvae (López- Pazos et al. 2009a); and none was seen

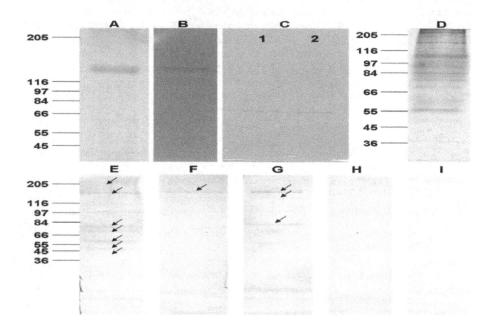

Fig. 6. SDS-polyacrylamide gel electrophoresis (SDS-PAGE) of recombinant toxins (A, B, and C) and ligand blots of Cry proteins on membrane vesicles from the midgut of the coffee berry borer (CBB-BBMVs) (E, F, G, H, and I). D shows SDS-PAGE of CBB-BBMV proteins. (A) Cry4 protoxin, (B) Cry9 protoxin, and (C) Cry4 (1) and Cry9 (2) protease-treated toxins; (D) brush-border-membrane-vesicle (BBMV) proteins from CBB. Cry-binding proteins (E-I) are indicated by the arrows. The biotin-labelled ligands (see below) are the following: (E) Cry1B, (F) Cry1I, (G) Cry3, (H) Cry4, and (I) Cry9. The numbers are molecular masses (kDa). Specifically, Cry4 and Cry9 were prepared for cloning by PCR amplification using the primers Cry4F (5'-ATGGGATCCTATCAAAATAAAAATGAATAT-3') with Cry4R (5'-TCACTCGTTCATGCCTGCAGATTCAAT GCT-3') and Cry9F (5'-ATGGGTACCAATAAACACGGAATTATTGGC-3') with Cry9R (5'-TTACTGCAGTGTTTCAACGAA TTCAATACT-3'), respectively. BamHI and KpnI restriction sites were added to the sequences of the Cry4 and Cry9 forward primers (underlined), respectively. PstI restriction sites were added to both the Cry4 and Cry9 reverse primers (underlined). The restriction sites were added to clone the amplified DNA fragment. The brush border membrane protein resolved on SDS-PAGE was transferred onto an Immobilon-P polyvinylidene difluoride (PVDF) membrane for blotting. The PVDF membrane was incubated with a biotin-labelled activated Cry toxin for binding, followed by washing with PBS/Tween (phosphate-buffered saline, pH7.4, containing 0.05% Tween-20) and incubation with streptavidin conjugated to peroxidise. The bands were visualized by peroxidase reacting with diaminobenzidine.

with Cry1I, Cry4, Cry9 and SN1917 hybrids. In this sense, there is a correlation between our data and ligand blot observations.

6. Conclusion

Insecticidal toxins are an important option for the biological control of lepidopteran and coleopteran insects. Their use in the genetic engineering of plants could provide a new generation of resistant crops. Such recombinant plants, thanks to their significant environmental and economic benefits, could help agricultural families in poor countries.

7. Acknowledgments

The authors are grateful to the Instituto de Biotecnología de la Universidad Nacional de Colombia. López-Pazos S.A. is grateful to Colciencias for a doctoral fellowship.

8. References

Abdelghany A. Y., Awadalla S. S., Abdel-Baky N. F., El-Syrafi H. A., Fields P. G. (2010). Effect of high and low temperatures on the drugstore beetle (Coleoptera: Anobiidae). *J. Econ. Entomol.* 103:1909-1914.

Adair J. R., Wallace T. P. (1998). Site-Directed Mutagenesis. *Molecular Biomethods Handbook,* Pages 347-360.

Adams M. E. (2004). Agatoxins: ion channel specific toxins from the American funnel web spider, *Agelenopsis aperta. Toxicon.* 43: 509–525.

Aravind L. (2000). Guilt by association: contextual information in genome analysis. *Genome Res.* 10: 1074–1077.

Arnold K., Bordoli L., Kopp J., Schwede T. (2006). The SWISS-MODEL Workspace: A web-based environment for protein structure homology modelling. *Bioinformatics.* 22:195-201.

Bloomquist J.R. (2003). Mode of action of atracotoxin at central and peripheral synapses of insects. *Invertebr. Neurosci.* 5:45-50.

Boonserm P., Davis P., Ellar D. J., Li J. (2005). Crystal structure of the mosquito larvicidal Toxin Cry4Ba and its biological implications. *J. Mol. Biol.* 348: 363-382.

Boonserm P., Mo M., Angsuthanasombat C., Lescar J. (2006). Structure of the functional form of the mosquito larvicidal Cry4Aa toxin from *Bacillus thuringiensis* at a 2.8-angstrom resolution. *J. Bacteriol.* 188: 3391-401.

Bosa C. F., Cotes A. M., Osorio P., Fukumoto T., Bengtsson M., Witzgall P. (2006). Disruption of pheromone communication in *Tecia solanivora* (Lepidoptera: Gelechiidae): flight tunnel and field studies. *J. Econ. Entomol.* 99:1245-1250.

Burton S.L., Ellar D.J., Li J., Derbyshire, D. J. (1999). N-Acetylgalactosamine on the putative insect receptor aminopeptidase N is recognised by a site on the domain III lectin-like fold of a *Bacillus thuringiensis* insecticidal toxin. *J. Mol. Biol.* 287:1011–1022.

Bustillo A. E. (2006). Una revisión sobre la broca del café, *Hypothenemus hampei* (Coleoptera: Curculionidae: Scolytinae), en Colombia. *Rev. Colomb. Entomol.* 32: 101-116.

Chagas Filho N. R., Boiça A. L. Jr., Alonso T. F. (2010). Biology of *Plutella xylostella* L. (Lepidoptera: Plutellidae) reared on cauliflower genotypes. *Neotrop. Entomol.* 39:253-259.

Chelius M. K., Triplett E. W. (2000). *Dyadobacter fermentans* gen. nov., sp. nov., a novel gram-negative bacterium isolated from surface-sterilized Zea mays stems. *Int. J. Syst. Evol. Microbiol.* 50 Pt 2: 751-758.

Chong Y., Hayes J. L., Sollod B., Wen S., Wilson D. T., Hains P. G., Hodgson W. C., Broady K. W., King G. F., Nicholson G. M. (2007). The ω-atracotoxins: selective blockers of insect M-LVA and HVA calcium channels. *Biochem. Pharmacol.* 74:623-638.

Cruz L. P., Gaitan A. L., Gongora C. E. (2006). Exploiting the genetic diversity of *Beauveria bassiana* for improving the biological control of the coffee berry borer through the use of strain mixtures. *Appl. Microbiol. Biotechnol.* 71: 918-926.

Damon A. (2000). A review of the biology and control of the coffee berry borer, *Hypothenemus hampei* (Coleoptera: Scolytidae). *Bull. Entomol.Res.* 90:453-465.

Deckert G., Warren P. V., Gaasterland T., Young W. G., Lenox A. L., Graham D. E., Overbeek R., Snead M. A., Keller M., Aujay M., Huber R., Feldman R. A., Short J. M., Olsen G. J., Swanson R. V. (1998). The complete genome of the hyperthermophilic bacterium *Aquifex aeolicus*. *Nature*.392: 353-358.

De Lima M. E., Figueiredo S. G., Pimenta AM. C., Santos D. M., Borges M. H., Cordeiro M. N., Richardson M., Oliveira L. C., Stankiewicz M., Pelhate M. (2007). Peptides of arachnid venoms with insecticidal activity targeting sodium channels. *Comp.Biochem.Physiol.* Part C. 146: 264–279.

Dereeper A., Guignon V., Blanc G., Audic S., Buffet S., Chevenet F., Dufayard J. F., Guindon S., Lefort V., Lescot M., Claverie J. M., Gascuel O. (2008). Phylogeny.fr: robust phylogenetic analysis for the non-specialist. *Nucleic. Acids. Res.* 36 (Web Server issue):W465-W469.

Du E., Ni X., Zhao H., Li X. (2011). Natural history and intragenomic dynamics of the Transib transposon Hztransib in the cotton bollworm *Helicoverpa zea.Insect. Mol. Biol.*20:291-301.

Ellegren H. (2008).Comparative genomics and the study of evolution by natural selection. *Mol. Ecol.* 17:4586-4596.

El-Sayed A., El-Sheikh Shizuo G. Kamita, Kiem Vu , Bruce D. Hammock. (2011). Improved insecticidal efficacy of a recombinant baculovirus expressing mutated JH esterase from *Manduca sexta*. *Biological control*. 58: 354-361.

Fabre C., Causse H., Mourey L., Koninkx J., Rivière M., Hendriks H., Puzo G., Samama J. P., Rougé P. (1998). Characterization and sugar-binding properties of arcelin-1, an insecticidal lectin-like protein isolated from kidney bean (*Phaseolus vulgaris* L. cv. RAZ-2) seeds. *Biochem. J.* 329: 551-560.

Fabrick J., Oppert C., Lorenzen M. D., Morris K., Oppert B., Jurat-Fuentes J. L. (2009). A novel *Tenebrio molitor* cadherin is a functional receptor for *Bacillus thuringiensis* Cry3Aa toxin. *J. Biol. Chem.* 284: 18401-18410.

Ferrat G., Bosmans F., Tytgat J., Pimentel C., Chagot B., Gilles N., Nakajima T., Darbon H., Corzo G. (2005). Solution structure of two insect-specific spider toxins and their pharmacological interaction with the insect voltage-gated Na+ channel. *Proteins* 59:368-379.

Fletcher J. I., Smith R., O'Donoghue S. I., Nilges M., Connor M., Howden M. E., Christie M. J., King G. F. (1997). The structure of a novel insecticidal neurotoxin, ω-atracotoxin-HV1, from the venom of an Australian funnel web spider. *Nat.Struct.Biol.* 4: 559-566.

Forster M. J. (2002).Molecular modeling in structural biology.*Micron*.33: 365–384.

Gabaldon T., Huynen M. A. (2004).Prediction of protein function and pathways in the genomeera.*Cell. Mol. Life Sci.* 61: 930–944.

Galagan J. E., Nusbaum C., Roy A., Endrizzi M. G., Macdonald P., FitzHugh W., Calvo S., Engels R., Smirnov S., Atnoor D., Brown A., Allen N., Naylor J., Stange-Thomann N., DeArellano K., Johnson R., Linton L., McEwan P., McKernan K., Talamas J., Tirrell A., Ye W., Zimmer A., Barber R. D., Cann I., Graham D. E, Grahame D. A., Guss A. M., Hedderich R., Ingram-Smith C., Kuettner H. C., Krzycki J. A., Leigh J. A., Li W., Liu J., Mukhopadhyay B., Reeve J. N., Smith K., Springer T. A., Umayam L. A., White O., White R. H., Conway de Macario E., Ferry J. G., Jarrell K. F., Jing H., Macario A. J., Paulsen I., Pritchett M., Sowers K. R., Swanson R. V., Zinder S. H., Lander E., Metcalf W. W., Birren B. (2002). The genome of *M. acetivorans* reveals extensive metabolic and physiological diversity. *Genome Res.* 12: 532-542.

Galitsky N., Cody V., Wojtczak A., Ghosh D., Luft J. R. (2001). Structure of the insecticidal bacterial δ-endotoxin Cry3Bb1 of *Bacillus thuringiensis. Act Crystallogr* 2001; D57:1101-1109.

Gentz M. C., Murdoch G., King G. F. (2010). Tandem use of selective insecticides and natural enemies for effective, reduced -risk pest management. *Biol. Control.* 52: 208-215.

Goldin, A.L. (2002). Evolution of voltage-gated Naþ channels.*J. Exp. Biol.* 205: 575-584.

Gordon D., Karbat I., Ilan N., Cohen L., Kahn R., Gilles N., Dong K., Stühmerd W., Tytgat J., Gurevitz M..(2007). The differential preference of scorpion α-toxins for insect or mammalian sodium channels: Implications for improved insect control. *Toxicon.* 49: 452–472.

Goudet C., Chi C.-W., Tytgat J. (2002). An overview of toxins and genes from the venom of the Asian scorpion *Buthus martensi* Karsch. *Toxicon.* 40: 1239–1258.

Gunning S. J, Maggio F., Windley M. J., Valenzuela S. M., King G. F., Nicholson G. M. (2008). The Janusfaced atracotoxins are specific blockers of invertebrate KCa channels. *FEBS J.* 275: 4045-4059.

Grochulski P., Masson L., Borisova S., Pusztai-Carey M., Schwartz J. L., Brousseau R., Cygler M. (1995). *Bacillus thuringiensis* CryIA(a) Insecticidal Toxin: Crystal Structure and Channel Formation. *J. Mol. Biol.* 254: 447-464.

Gruber C. W., Cemazar M., Anderson M. A., Craik D. J. (2007). Insecticidal plant cyclotides and related cystine knot toxins. *Toxicon.* 49: 561-575.

Guex, N., Peitsch, M. C. (1997). Swiss-model and the Swiss-PdbViewer: An environment for comparative protein modelling. *Electrophoresis.* 18: 2714-2723.

Guo M., Chong Y. E., Beebe K., Shapiro R., Yang X. L., Schimmel P. (2009). The C-Ala domain brings together editing and aminoacylation functions on one tRNA. *Science.* 325:744-747.

Gurevitz M., Karbat I., Cohen L., Ilan N., Kahn R., Turkov M., Stankiewicz M., Stuhmerc W., Dong K., Gordon D. (2007). The insecticidal potential of scorpion β-toxins. *Toxicon.* 49: 473-489.

Gurevitz M., Urbach D., Zlotkin E., Zilberberg N. (1991) Nucleotide sequence and structure analysis of a cDNA encoding anα-insect toxin from the scorpion *Leiurus quinquestriatus hebraeus. Toxicon* 29:1270-1272.

Hammock B.D., Bonning B.C., Possee R.D., Hanzlik T.N., Maeda S. (1990). Expression and effects of the juvenile hormone esterase in a baculovirus vector. *Nature*.344: 458–461.

Han S., Craig J. A., Putnam C. D., Carozzi N. B., Tainer J. A. (1999) Evolution and mechanism from structures of an ADP ribosylating toxin and NAD complex. *Nat. Struct.Biol.* 6:932-936.

Hao C. J., Xu C. G., Wang W., Chai B. F., Liang A. H. (2005). Expression of an insect excitatory toxin, BmK IT, from the scorpion, *Buthus martensii* Karsch, and its biological activity. *Biotechnol. Lett.*27:1929-1934.

Herrero S., Gechev T., Bakker P. L., Moar W. J., de Maagd R. A. (2005).*Bacillus thuringiensis* Cry1Ca-resistant *Spodoptera exigua* lacks expression of one of four aminopeptidase N genes. *BMC Genomics*.6: 96.

Howard A. F., N'guessan R., Koenraadt C. J., Asidi A., Farenhorst M., Akogbéto M., Thomas M. B., Knols B. G., Takken W. (2010). The entomopathogenic fungus *Beauveria bassiana* reduces instantaneous blood feeding in wild multi-insecticide-resistant *Culex quinquefasciatus* mosquitoes in Benin, West Africa. *Parasit. Vectors.* 15: 87.

Hua G., Jurat-Fuentes J. L., Adang M. J. (2004). Bt-R1a extracellular cadherin repeat 12 mediates *Bacillus thuringiensis* Cry1Ab binding and toxicity. *J. Biol. Chem.* 279: 28051–28056.

Hua G., Masson L., Jurat-Fuentes J. L., Schwab G., Adang M. J. (2001). Binding analyses of *Bacillus thuringiensis* Cry δ-endotoxins using brush border membrane vesicles of *Ostrinia nubilalis*. *Appl. Environ. Microbiol.*67: 872-879.

James R. M., Dickinson P. (1998). Site-Directed Mutagenesis. *Molecular Biomethods Handbook.* Pages 361-381.

Ji Y. H., Mansuelle P., Terakawa S., Kopeyan C., Yanaihara N., Hsu K., Rochat H.(1996). Two neurotoxins (Bmk I and Bmk II) from the venom of the scorpion *Buthus martensi* Karsch: purification, amino acid sequences and assessment of specific activity. *Toxicon*.34: 987-1001.

Jablonsky M. J., Jackson P. L., Krishna N. R. (2001). Solution structure of an insect-specific neurotoxin from the new world scorpion*Centruroides sculpturatus* Ewing. *Biochemistry.* 40: 8273-8282.

Johnson J. H., Bloomquist J. R., Krapcho K. J., Kral R. M. Jr, Trovato R., Eppler K. G., Morgan T. K., DelMar E. G. (1998). Novel insecticidal peptides from *Tegenaria agrestis* spider venom may have a direct effect on the insect central nervous system. *Arch Insect Biochem Physiol.* 38: 19-31.

Kamita S. G., Hinton A. C., Wheelock C. G., Wogulis M. D., Wilson D. K., Wolf N. M., Stok J. E., Hock B., Hammock B. D. (2003). Juvenile hormone (JH) esterase: why are you so JH specific? *Insect Biochemistry and Molecular Biology.* 33: 1261–1273.

Kanchiswamy C. N., Takahashi H., Quadro S., Maffei M. E., Bossi S., Bertea C., Zebelo S. A., Muroi A., Ishihama N., Yoshioka H., Boland W., Takabayashi J., Endo Y., Sawasaki T., Arimura G. (2010). Regulation of *Arabidopsis* defense responses against Spodoptera littoralis by CPK-mediated calcium signaling.*BMC Plant.Biol.* 10: 97.

Karbat I., Frolow F., Froy O., Gilles N., Cohen L., Turkov M., Gordon D., Gurevitz M. (2004). Molecular Basis of the High Insecticidal Potency of Scorpion α toxins. *The Journal of Biological Chemistry*.279: 31679–31686.

Karbat I., Turkov M., Cohen L., Kahn R., Gordon D., Gurevitz M., Frolow F. (2007). X-ray structure and mutagenesis of the scorpion depressant toxin LqhIT2 reveals key determinants crucial for activity and anti-insect selectivity. *J. Mol. Biol.* 366: 586-601.

Karlova R., Weemen-Hendriks M., Naimov S., Ceron J., Dukiandjiev S., de Maagd R. (2005).*Bacillus thuringiensis*δ-endotoxin Cry1Ac domain III enhances activity against *Heliothis virescens* in some, but not all Cry1-Cry1Ac hybrids. *J. Invertebr. Pathol.* 88: 169–172.

Keszthelyi S., Pál-Fám F., Kerepesi I. (2011). Effect of cotton bollworm (*Helicoverpa armigera* Hübner) caused injury on maize grain content, especially regarding to the protein alteration. *Acta Biol. Hung.* 62: 57-64.

King G. F., Gentz M. C., Escoubas P., Nicholson G. M. (2008). A rational nomenclature for naming peptide toxins from spiders and other venomous animals. *Toxicon.* 52: 264-276.

King G. F., Tedford H. W., Maggio F. (2002). Structure and function of insecticidal neurotoxins from Australian funnel web spiders. *Toxin Reviews.* 21: 361-389.

Krimm I., Gilles N., Sautière P., Stankiewicz M., Pelhate M., Gordon D., Lancelin J. M. (1999). NMR structures and activity of a novel α-like toxin from the scorpion *Leiurus quinquestriatus* hebraeus. *J. Mol. Biol.* 285:1749-1763.

Lechner S., Mayr R., Francis K. P., Prüss BM, Kaplan T., Wiessner-Gunkel E., Stewart G. S., Scherer S. (1998).*Bacillus weihenstephanensis* sp. *nov.* is a new psychrotolerant species of the Bacillus cereus group. *Int. J. Syst. Bacteriol.*48: 1373-1 382.

Lee W., Moore C. H., Watt D. D., Krishna N. R. (1994). Solution structure of the variant-3 neurotoxin from *Centruroides sculpturatus* Ewing. *Eur. J. Biochem.*218: 89-95.

Lee D., Redfern O., Orengo C. (2007). Predicting protein function from sequence and structure. *Nat. Rev. Mol. Cell. Biol.* 8: 995-1005.

Li C., Guan R. J., Xiang Y., Zhang Y., Wang D. C. (2005). Structure of an excitatory insect-specific toxin with an analgesic effect on mammals from the scorpion *Buthus martensii* Karsch. *Acta Crystallogr. D Biol. Crystallogr.* 61(Pt 1):14-21.

Li J., Carroll J., Ellar D. J. (1991).Crystal Structure of Insecticidal δ-endotoxin from *Bacillus thuringiensis* at 2.5 Å Resolution.*Nature.*353:815-821.

Liang S. (2004). An overview of peptide toxins from the venom of the Chinese bird spider *Selenocosmia huwena* Wang [=*Ornithoctonus huwena* (Wang)] *Toxicon.*43: 575-585.

Likitvivatanavong S., Katzenmeier G., Angsuthanasombat C. (2006). Asn183 in α5 is essential for oligomerisation and toxicity of the *Bacillus thuringiensis* Cry4Ba toxin. *Archives of Biochemistry and Biophysics.*445: 46-55.

Liu Y. J., Cheng C. S., Lai S. M., Hsu M. P., Chen C. S., Lyu P. C. (2006). Solution structure of the plant defensin VrD1 from mung bean and its possible role in insecticidal activity against bruchids. *Proteins.*63:777-786.

López-Pazos S. A., Cerón J. A. (2007). Three-dimensional structure of *Bacillus thuringiensis* toxins: a review *Acta. Biol. Colomb.*12: 19-32.

López-Pazos S. A., Cortázar J. E., Cerón J. A. (2009a). Cry1B and Cry3A are active against *Hypothenemus hampei* Ferrari (Coleoptera: Scolytidae). *J. Invertebr. Pathol.*101: 242-245.

López-Pazos S. A., Martínez J. W., Castillo A. X., Cerón Salamanca J. A.(2009b). Presence and significance of *Bacillus thuringiensis* Cry proteins associated with the Andean weevil *Premnotrypes vorax* (Coleoptera: Curculionidae). *Rev. Biol. Trop.* 57: 1235-1243.

López-Pazos S. A., Rojas Arias A. C., Ospina S. A., Cerón J. (2010). Activity of *Bacillus thuringiensis* hybrid protein against a lepidopteran and a coleopteran pest. *FEMS Microbiol. Lett.* 302: 93-98.

Loris R., Hamelryck T., Bouckaert J., Wyns L. (1998). Legume lectin structure.*Biochim. Biophys.* Acta.1383: 9–36.

Lykidis A., Pérez-Pantoja D., Ledger T., Mavromatis K., Anderson I. J., Ivanova N. N., Hooper S. D., Lapidus A., Lucas S., González B., Kyrpides N. C. (2010). The complete multipartite genome sequence of *Cupriavidus necator* JMP134, a versatile pollutant degrader. *PLoS One.* 5: e9729.

Martínez-Ramírez A. C., González-Nebauer S., Escriche B., Real M. D. (1994). Ligand blot identification of a *Manduca sexta* midgut binding protein specific to three *Bacillus thuringiensis* CryIA-type ICPs. *Biochem. Biophys. Res. Commun.*201: 782–787.

Martínez W., Uribe D., Cerón J. (2003). Efecto tóxico de proteínas Cry1 de *Bacillus thuringiensis* sobre larvas de *Tecia solanivora* (Lepidoptera: Gelechiidae). *Rev. Colomb. Entomol.*29: 89–93.

Martins E. S., Monnerat R. G., Queiroz P. R., Dumas V. F., Braz S. V., de Souza Aguiar R. W., Gomes A. C., Sánchez J., Bravo A., Ribeiro B. M. (2010). Midgut GPI-anchored proteins with alkaline phosphatase activity from the cotton boll weevil (Anthonomus grandis) are putative receptors for the Cry1B protein of *Bacillus thuringiensis. Insect. Biochem. Mol. Biol.* 40: 138-145.

Miller E. A., Lee M. C. S., Atkinson A. H. O., Anderson M. A. (2000). Identification of a novel four-domain member of the proteinase inhibitor II family from the stigmas of *Nicotiana alata. Plant. Mol. Biol.* 42: 329-333.

Monnerat R., Martins E., Queiroz P., Ordúz S., Jaramillo G., Benintende G., Cozzi J., Real M. D., Martinez-Ramirez A., Rausell C., Cerón J., Ibarra J. E., Del Rincon-Castro M. C., Espinoza A. M., Meza-Basso L., Cabrera L., Sánchez J., Soberon M., Bravo A. (2006). Genetic variability of *Spodoptera frugiperda* Smith (Lepidoptera: Noctuidae) populations from Latin America is associated with variations in susceptibility to *Bacillus thuringiensis* Cry toxins. *Appl. Environ. Microbiol.* 72:7029-7035.

Moran Y., Cohen L., Kahn R., Karbat I., Gordon D., Gurevitz M. (2006). Expression and Mutagenesis of the Sea Anemone Toxin Av2 Reveals Key Amino Acid Residues Important for Activity on Voltage-Gated Sodium Channels.*Biochemistry.* 45: 8864-8873.

Moran Y., Gordon D., Gurevitz M. (2009). Sea anemone toxins affecting voltage-gated sodium channels-molecular and evolutionary features. *Toxicon.* 54: 1089-1101.

Morse R. J., Yamamoto T., Stroud R. M. (2001). Structure of Cry2Aa suggests an unexpected receptor binding epitope. *Structure.*9:409-417.

Mourey L., Pédelacq J. D., Birck C., Fabre C., Rougé P., Samama J. P. (1998). Crystal structure of the arcelin-1 dimer from *Phaseolus vulgaris* at 1.9-A resolution.*J. Biol. Chem.* 273:12914-12922.

Naganuma M., Sekine S., Fukunaga R., Yokoyama S. (2009). Unique protein architecture of alanyl-tRNA synthetase for aminoacylation, editing, and dimerization. *Proc. Natl. Acad. Sci.* USA.106: 8489-8494.

Nakanishi K., Yaoi K., Shimada N., Kadotani T., Sato R. (1999).*Bacillus thuringiensis* insecticidal Cry1Aa toxin binds to a highly conserved region of aminopeptidase N in the host insect leading to its evolutionary success. *Biochim. Biophy. Acta.*1432: 57-63.

Nakasu E. Y., Firmino A. A., Campos Dias S., Lima Rocha T., Batista Ramos H., Ramos de Oliveira G., Lucena W., Ribeiro da Silva Carlini C. R., Grossi de Sá M. F. (2010). Analysis of Cry8Ka5-binding proteins from *Anthonomus grandis* (Coleoptera: Curculionidae) midgut.*J Invertebr. Pathol.* 104: 227-230.

Ochoa-Campuzano C., Real M. D., Martinez-Ramirez A. C., Bravo A., Rausell C. (2007). An ADAM metalloprotease is a Cry3Aa *Bacillus thuringiensis* toxin receptor.*Biochem. Biophys. Res. Comm.* 362: 437-442.

Oerke E.-C., Dehne H.-W. (2004). Safeguarding production-losses in major crops and the role of crop protection. *Crop Protection.* 23: 275-285.

Ohashi S., Miyashita H., Okada N., Lemura T., Watanabe T., Kobayashi M. (2008). Unique photosystems in *Acaryochloris marina*. Photosynth.Res. 98:141-149.

Omecinsky D. O, Holub K. E., Adams M. E., Reily M. D. (1996). Three-dimensional structure analysis of μ-agatoxins: further evidence for common motifs among neurotoxins with diverse ion channel specificities. *Biochemistry.*35: 2836-2844.

Oren D. A., Froy O., Amit E., Kleinberger-Doron N., Gurevitz M., Shaanan B. (1998). An excitatory scorpion toxin with a distinctive feature: an additional α helix at the C terminus and its implications for interaction with insect sodium channels. *Structure.* 6:1095-1103.

Pai A., Bernasconi G. (2008). Polyandry and female control: the red flour beetle *Tribolium castaneum* as a case study. *J. Exp. Zool. B. Mol. Dev. Evol.* 310:148-159.

Paramasivan R., Sivaperumal R., Dhananjeyan K. J., Thenmozhi V., Tyagi B. K.(2006). Prediction of 3-dimensional structure of salivary odorant-binding protein-2 of the mosquito Culex quinquefasciatus, the vector of human lymphatic filariasis. *In Silico Biol.* 7: 1-6.

Park Y., Abdullah M. A., Taylor M. D., Rahman K., Adang M. J. (2009). Enhancement of *Bacillus thuringiensis* Cry3Aa and Cry3Bb Toxicities to Coleopteran Larvae by a Toxin-Binding Fragment of an Insect Cadherin.*Appl. Environ. Microbiol.*75: 3086-3092.

Pava-Ripoll M., Posada F. J., Momen B., Wang C., St Leger R. (2008). Increased pathogenicity against coffee berry borer, *Hypothenemus hampei* (Coleoptera: Curculionidae) by *Metarhizium anisopliae* expressing the scorpion toxin (AaIT) gene.*J. Invertebr. Pathol.*99: 220-226.

Pechan T., Cohen A., Williams W. P., Luthe D. S. (2002). Insect feeding mobilizes a unique plant defense protease that disrupts the peritrophic matrix of caterpillars. *Proc. Natl. Acad. Sci. U S A.* 99: 13319-13323.

Petrey D., Honig B. (2009). Is protein classification necessary? Toward alternative approaches to function annotation.*Curr Opin Struct Biol* 19:363-368.

Peumans W. J., Van Damme E. J. (1995a). Lectins as plant defence proteins. *Plant.Physiol.* 109: 347-352.

Peumans W. J., Van Damme E. J. (1995b).Role of lectins in plant defense.*Histochem.J.* 27: 253-271.

Philpott, M.L., Hammock, B.D. (1990). Juvenile hormone esterase is a biochemical anti-juvenile hormone agent. *Insect Biochemistry* 20: 451–459.

Pigott C.R., Ellar D.J. (2007). Role of Receptors in *Bacillus thuringiensis* Crystal Toxin Activity.*Microbiol. Mol. Biol. Rev.* 71: 255–281.

Possani L. D., Becerril B., Delepierre M., Tytgat J. (1999).Scorpion toxins specific for Na+-channels.*Eur. J. Biochem.* 264: 287–300.

Qin Y., Ying S. H., Chen Y., Shen Z. C., Feng M. G. (2010). Integration of insecticidal protein Vip3Aa1 into *Beauveria bassiana* enhances fungal virulence to *Spodoptera litura* larvae by cuticle and per Os infection. *Appl. Environ. Microbiol.*76:4611-4618.

Ralph S. G., Yueh H., Friedmann M., Aeschliman D., Zeznik J. A., Nelson C. C., Butterfield Y. S., Kirkpatrick R., Liu J., Jones S. J., Marra M. A., Douglas C. J., Ritland K., Bohlmann J. (2006). Conifer defence against insects: microarray gene expression profiling of Sitka spruce (*Picea sitchensis*) induced by mechanical wounding or feeding by spruce budworms (*Choristoneura occidentalis*) or white pine weevils (*Pissodes strobi*) reveals large-scale changes of the host transcriptome. *Plant. Cell. Environ.* 29: 1545-1570.

Ramírez R. (2009). La broca del café en Líbano. Impacto socioproductivo y cultural en los años 90.*Revista de Estudios Sociales.*32: 158-171.

Rayapuram C., Baldwin I. T. (2008). Host-plant-mediated effects of Nadefensin on herbivore and pathogen resistance in *Nicotiana attenuata*. *BMC Plant.Biol.* 8: 109.

Ribas de Pouplana L., Schimmel P. (2000). A view into the origin of life: aminoacyl-tRNA synthetases *Cell. Mol. Life Sci.* 57: 865–870.

Rodríguez de la Vega R. C., Schwartz E. F., Possani L. D. (2010). Mining on scorpion venom biodiversity. *Toxicon.* 56: 1155–1161.

Roh J. Y., Choi J. Y., Li M. S., Jin B. R., Je Y. H.(2007). *Bacillus thuringiensis* as a Specific, Safe, and Effective Tool for Insect Pest Control. *J. Microbiol. Biotechnol.* 17: 547–559.

Rosengren K. J., Daly N. L., Plan M. R., Waine C., Craik D. J. (2003). Twists, knots, and rings in proteins. Structural definition of the cyclotide framework. *J. Biol. Chem.* 278:8606-8616.

Sayed A., Nekl E. R., Siqueira H. A., Wang H. C., Ffrench-Constant R. H., Bagley M., Siegfried B. D. (2007). A novel cadherin-like gene from western corn rootworm, *Diabrotica virgifera virgifera* (Coleoptera: Chrysomelidae), larval midgut tissue. *Insect.Mol. Biol.* 16: 591–600.

Schirra H. J., Anderson M. A., Craik D. J. (2008). Structural refinement of insecticidal plant proteinase inhibitors from *Nicotiana alata*. *Protein. Pept. Lett.*15:903-909.

Schirra H. J., Scanlon M. J., Lee M. C., Anderson M. A., Craik D. J. (2001). The solution structure of C1-T1, a two-domain proteinase inhibitor derived from a circular precursor protein from *Nicotiana alata*. *J. Mol. Biol.* 306:69-79.

Schrank A., Vainstein M. H. (2010). *Metarhizium anisopliae* enzymes and toxins.*Toxicon.* 56: 1267-1274.

Schnepf E., Crickmore N., Van Rie J., Lereclus D., Baum J., Feitelson J., Zeigler D.R., Dean D.H. (1998). *Bacillus thuringiensis* and Its Pesticidal Crystal Proteins. *Microbiol. Mol. Biol. Rev.* 62: 775-806.

Shah P. A., Pell J. K. (2003). Entomopathogenic fungi as biological control agents.*Appl. Microbiol. Biotechnol.* 61: 413-423.

Shu Q., Lu S. Y., Gu X. C., Liang S. P. (2002). The structure of spider toxin huwentoxin-II with unique disulfide linkage: evidence for structural evolution. *Protein Sci.* 11:245-252.

Skinner W. S., Dennis P. A., Li J.P., Summerfelt R.M., Carney R. L., Quistad G. B. (1991). Isolation and Identification of Paralytic Peptides from Hemolymph of the Lepidopteran Insects Manduca sexta, Spodoptera exigua, and *Heliothis virescens. J. Biol. Chem.* 266: 12873-12877.

St. Leger R. J., Wang C. (2010). Genetic engineering of fungal biocontrol agents to achieve greater efficacy against insect pests.*Appl. Microbiol. Biotechnol.* 85:901-907.

Suckling D. M., Brockerhoff E. G. (2010). Invasion biology, ecology, and management of the light brown apple moth (Tortricidae). *Annu. Rev. Entomol.* 55:285-306.

Szolajska E., Poznanski J., López M., Michalik J., Gout E., Fender P., Bailly I., Dublet B., Chroboczek J. (2004). Poneratoxin, a neurotoxin from ant venom. Structure and expression in insect cells and construction of a bio-insecticide.*Eur. J. Biochem.*271: 2127-2136.

Tedford H. W., Sollod B. L., Maggio F., King G. F.(2004). Australian funnel-web spiders: master insecticide chemists. *Toxicon.* 43:601-618.

Tindall K. V., Stewart S., Musser F., Lorenz G., Bailey W., House J., Henry R., Hastings D., Wallace M., Fothergill K. (2010). Distribution of the long-horned beetle, *Dectes texanus*, in soybeans of Missouri, Western Tennessee, Mississippi, and Arkansas.*J. Insect.Sci.* 10:178.

Tugarinov V., Kustanovich I., Zilberberg N., Gurevitz M., Anglister J. (1997). Solution structures of a highly insecticidal recombinant scorpion α-toxin and a mutant with increased activity. *Biochemistry.*36: 2414-2424.

van Frankenhuyzen K. (2009).Insecticidal activity of *Bacillus thuringiensis* crystal proteins. *J. Invertebr. Pathol.*101: 1-16.

Venter, J.C., et.al.(2004). Environmental Genome Shotgun Sequencing of the Sargasso Sea.*Science.*304: 66-74.

Vaishnavi C. (2010). Clinical spectrum & pathogenesis of *Clostridium difficile* associated diseases. *Indian. J. Med. Res.* 131: 487-499.

Vassilevski A., Kozlov S. A., Grishin E. V. (2009). Molecular Diversity of Spider Venom. *Biochemistry (Moscow).*74: 1505-1534.

Wang X. H., Connor M., Wilson D., Wilson H. I., Nicholson G. M., Smith R., Shaw D., Mackay J. P., Alewood P. F., Christie M. J., King G. F. (2001).Discovery and structure of a potent and highly specific blocker of insect calcium channels.*J. Biol. Chem.* 276: 40306-40312.

Wang L., Wang S., Li Y., Paradesi M. S. R., Brown S. J. (2007). BeetleBase: the model organism database for *Tribolium castaneum*. Nucleic. Acids. Res. 35: D476-D479.

Warmke, J. W., Reenan, A. G. R., Wang, P., Qian, S., Arena, J. P., Wang, J., Wunderler, D., Liu, K., Kaczorowski, G. J., Van der Ploeg, L. H. T., Ganetzky, B., and Cohen, C. J. (1997). Functional expression of Drosophila para sodium channels.Modulation by membrane protein TipE and toxin pharmacology.*J. Gen. Physiol.* 110: 119-133.

Whetstone P. A., Hammock B. D. (2007).Delivery methods for peptide and protein toxins in insect control.*Toxicon.* 49: 576-596.

Wolfersberger M. G., Luethy P., Maurer A., Parenti P., Sacchi F. V., Giordana B., Hanozet G. M. (1987). Preparation and partial characterization of amino acid transporting brush border membrane vesicles from the larval midgut of the cabbage butterfly (*Pieris brassicae*). *Comp. Biochem.Physiol.* 86:301–308.

Yamaji N., Sugase K., Nakajima T., Miki T., Wakamori M., Mori Y., Iwashita T. (2007). Solution structure of agelenin, an insecticidal peptide isolated from the spider *Agelena opulenta*, and its structural similarities to insect-specific calcium channel inhibitors. *FEBS Lett.*581: 3789-3794.

Yu X. Q., Prakash O., Kanost M. R. (1999).Structure of a paralytic peptide from an insect, *Manduca sexta.J. Pept. Res.* 54:256-261.

Zilberberg N., Froy O., Loret E., Cestele S., Arad D., Gordon D., Gurevitz M. (1997). Identification of structural elements of a scorpion α-neurotoxin important for receptor site recognition.*J. Biol. Chem.* 272: 14810–14816.

Zlotkin E., Fishman Y., Elazar M. (2000).AaIT: from neurotoxin to insecticide.*Biochimie.*82: 869-881.

New Tools or Approaches for Molecular Genetics

7

Studying Cell Signal Transduction with Biomimetic Point Mutations

Nathan A. Sieracki and Yulia A. Komarova
University of Illinois – Chicago
USA

1. Introduction

Post-translational modification (PTM), the chemical modification of a protein after its translation, represents an evolutionarily conserved mechanism of regulation of protein function. (Deribe et al., 2010) Through modulation of conformational change, enzymatic activity or interaction with other proteins, it often provides a transient switch between various intracellular signals to orchestrate an integrated response of the cell to environmental cues.

Site-directed mutagenesis has proven to be an essential tool in studying the molecular and signaling mechanisms of complex cellular functions through the ability to mutate a specific residue in the sequence of native protein in order to change its function. In combination with proteomics and database analysis of primary sequences, it offers a tool for exploring the transient nature of protein-protein interactions along with activity of enzymes, the primary assemblies of signaling nodes.

This chapter will highlight the use of site-directed mutagenesis for studying signal transduction networks. It will first describe how the diversity of the natural 22 amino acid 'toolbelt' results in fortuitous isostructural and isoelectronic similarity to several key post-translationally modifications (e.g. phosphorylation, sulfoxidation and nitrosylation). This similarity has allowed researchers to introduce gain- or loss-of-function substitutions into primary sequences and thus produce a form of the protein that mimicks its active or inactive state. The substitution of serine or threonine for phosphomimetic aspartic or glutamic acid or non-phosphorylatable residues, such as alanine, will be discussed, along with other permutations, including natural analogs of oxidized methionine and cysteine.

While discussing how site-directed mutagenesis offers to researchers a tool to engineer a constitutively active OFF and ON forms of a protein, we will also provide examples of how PTM translates into a chemical phenotype through the regulation of enzyme activity or modulation of binding affinity to partners. Also discussed in this chapter are limitations and considerations to be taken when interpreting results obtained with biomimetic tools.

With advanced computing analysis and algorithms operating on primary sequences, the predictability of PTM sites is rapidly becoming an automated process. Complex networks can be accessed and verified. Even in the age of high-resolution mass spectrometry, site-directed mutagenesis will continue to be a powerful tool for studying and validation of signaling networks.

2. Harnessing nature's toolbelt with biomimetic point mutations

Site-directed mutagenesis, a molecular biology technique used to introduce a point mutation into primary DNA sequence at a defined site, was first described by Michael Smith in 1978. (Hutchison et al, 1978) His group induced specific mutations into the viral DNA strand of the bacteriophage phiX174 *am3* using complementary oligodeoxyribonucleotides and specific mutagens. "It appears to us that general methods for the construction of any desired mutant sequence would greatly increase the ability of sequencing methods to yield biologically relevant information. Such methods should also make it possible to construct DNAs with specific modified biological functions" M. Smith and his colleagues proposed in the paper. This was proven to be correct by vast number of studies, which exploited site-directed mutagenesis as a tool for testing structure-function relationships over the course of three decades. Providing a constitutively activated or deactivated form of the protein or removing entirely the potential for protein modification, it has also been instrumental in dissecting signal transduction networks. It is instructive to understand the precise strengths and limitations of biomimetic approaches in order to extract the most meaningful data from experiments.

2.1 Layers of diversity

The quest to understand and control structure-function relationship has revealed that Nature is an exquisite chemist. The already significant chemical diversity of the amino acid pool is amplified profoundly when individual amino acids are condensed into peptide chains. Individual residues can act in a seemingly modular fashion, or residues can coordinate to bestow more complex and macroscopic functional and catalytic characteristics to the proteins and peptides on which they reside. This diversity, however, is even more dramatic than the genetic message may initially predict.

As summarized in an excellent review by Wash and co-authors, (Walsh et al., 2005), Nature has devised two mechanisms of 'proteome diversification' over the course of evolution. The amount of unique proteins per unit RNA transcript has been dramatically amplified to be far more diverse than Central Dogma of protein expression would have suggested. The first method lies at the level of the transcript. Alternate splicing of the same gene product can result in a variety of unique proteins with unique functions. The second is through targeted post-translational modification of existing proteins and peptides to chemically 'tag' them. These "tags" can directly influence the structure and function of the protein, or it may be interpreted by another docking small molecule, protein, or assembly of proteins as a biological ON/OFF switch. This largely reversible labelling offers the ability to tune and coordinate signalling networks on a timescale faster than is required for protein synthesis and proteolysis. A necessity for life, it is not surprising that many aspects of signalling cascades predate species divergence from unicellular organisms. (Tan, 2011)

While Nature has found a way to amplify the diversity of the pool of amino acid building blocks (in direct analogy to the diversification of DNA messages through arrangement of the nucleotide code), the selection pressures of evolution have still adhered to chemical intuition and energy conservation. In fact, evolution exerts less pressure on similar chemically structured residues. It seems the idea of a structural mimic is, ironically, bio-inspired.

2.2 Post-translational modification offers fast and dynamic response to stimulus

Post-translational modification of proteins offers another layer of complexity in the regulation of cellular processes. It provides a powerful tool for fast and reversible modification of protein structure, which effects dynamic changes in protein function, and ultimately translates into a cellular phenotype. The most easily deciphered systems are those in which there is a large chemical difference between the modified and non-modified group and systems that are controlled by a unimodal 'adding' or 'removing' of an effector moiety, as in the prototypical case of phosphorylation (*i.e.*, addition or removal of a phosphate group with a kinase or phosphatase, respectively). Chemical modification of an amino acid residue on an existing polypeptide is an energy-efficient means of reversibly and quickly altering the charge or steric properties of a peptide to result in a new molecular phenotype.

For example, a mere phosphorylation event induces a large structural change in the bacterial receiver domain of the nitrogen regulatory protein C (NtrC) (Kern et al., 1999) (Fig. 1). Phosphorylation of NtrC at this position Asp_{54} causes the movement of three alpha-helices, resulting in the formation of a hydrophobic patch thought to be necessary for downstream signal transduction.

Asp 54
Phospho-Asp 54

Fig. 1. Phosphorylation induces conformational changes. Crystal structure of NtrC before (blue) and after (orange) phosphorylation of Asp_{54}. A single residue modification from Asp_{54} (blue sticks) to the phosphorylated form (red sticks) results in major protein restructuring, including movement of α-helix 4 (black arrows) away from the active site to generate a hydrophobic patch. Structural coordinates were obtained from the Protein Databank PDB code: (1DC7, 1DC8).

To date, over 200 post-translation modifications have been identified; and they have been explored to varying extents, based on a mixture of propensity and technical accessibility. Phosphorylation, although overshadowed in propensity by glycosylation, is by far the most largely studied and characterized modification in the context of a signal relay in cellular systems. This is owed in large part to the availability of tools to efficiently mimic the isostructural property of phosphorylated residues, not the least of which is site-directed mutagenesis and the use of natural amino acids as constitutive substitutes.

2.3 Natural amino acids as substitutes for modified amino acids

Invention of site-directed mutagenesis offered scientists, for the first time, access to the entire amino acid regimen for a given residue location. Chemical similarity between certain post-translationally modified residues and some natural amino acids became central to the field of genetic-based manipulation of protein properties. Shown in Fig. 2 are examples of natural mimics of some well-studied post-translational modifications, highlighting structural similarities.

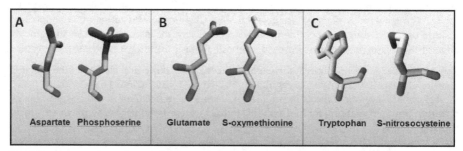

Fig. 2. Isostructural similarities between natural amino acids and post-translational modifications. Structural coordinates for natural amino acids and products of (A) phosphorylation (B), sulfoxidation, and (C) nitrosylation. Carbon atoms are shown in cyan; oxygen, in red; nitrogen, in blue; and phosphorous in gold. Hydrogen atoms have been omitted for clarity.

While site-directed mutagenesis was first described in 1978 (Hutchison et al., 1978), the first examples of a PTM mimicking appeared in 1988 and 1989; the latter was described four years before the Nobel Prize in Chemistry was awarded to Michael Smith and Kary B. Mullis, for the development of site-directed mutagenesis and the polymerase chain reaction, respectively. When investigating the effect of phosphorylation of the large T antigen portion of Simian Virus 40 (SV40) on DNA-binding and replication, it was found that substitution of a known phosphorylatable Thr_{124} residue by glutamic acid, as opposed to alanine or cysteine, resulted in partial retention of native DNA replication activity. (Schneider et al., 1988) It was suggested that a negative charge, and not a phospho-group *per-se*, was required to promote DNA replication. Shortly thereafter, two papers revealed a remarkable structural and functional similarity between the phosphorylated form of serine and glutamic acid. In agreement with similar structural deviations of phosphorylated serine and glutamic acid, the substitution of the regulatory Ser_{46} for glutamic acid in HPr (Wittekind et al., 1989), a protein component of the bacterial PEP-dependent sugar-transport system, resulted in enzyme inactivation indistinguishable from the native phosphorylation. (Reizer et al., 1989) The crystal structure of unmodified and phospho-Ser_{46} would later be determined (Jia et al, 1994) (Audette et al., 2000), and an image of the electrostatic potential shows the large surface change that accompanies phosphorylation (Figure 3).

This provided the first direct evidence for the structural mimicking of a phosphate group using site-directed mutagenesis. It showed that an easily genetically encoded glutamic acid residue was a strong phosphoserine mimic. The result was an explosion in the usage of point mutations as surrogates for phosphorylated or phospho-defective forms of potential regulatory sites.

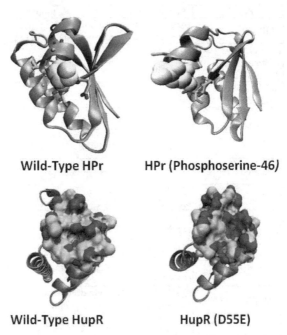

Wild-Type HPr   **HPr (Phosphoserine-46)**

Wild-Type HupR   **HupR (D55E)**

Fig. 3. Phosphorylation induces marked structural and surface-contour changes. (Top) Ribbon diagrams of crystal structures of native and phosophorylated forms of HPr; Ser$_{46}$ is highlighted in yellow, and nearby residues are shown. (Bottom) Surface representation of the wild-type and the phosphomimetic Asp$_{55}$E variants of Hydrogen Uptake Protein Regulator (*HupR*). A phosphomimetic mutation induced marked surface-contour changes that affected protein dimerization and function. Structural coordinates were obtained from the Protein Databank (PDB code: 1FU0, 1PTF, 2JK1, 2VUH). Highlighted residues and the surface layer proximal to residue 55 in HupR were color-coded by atom type: carbon atoms are cyan, nitrogen atoms are blue, oxygen atoms are red, and phosphorous atoms are green (inhibitor compound).

Another example of how substitution with phospho-mimetic glutamic acid can lead to a conformational change is shown in Fig. 3 (bottom). A surface-contour map of the receiver domain of nitrogen regulatory protein HupR (in which phosphorylation is actually inhibitory to transcription) before and after mutation of Asp$_{55}$ to glutamic acid (Davies et al., 2009) illustrates a marked surface change due to mutation of a single residue, which leads to deactivation of the receiver domain in a manner similar to that observed for authentic phosphorylation.

Phosphorylation is not the only form of post-translation regulation that has been investigated with natural amino acid mimics. Biomimetic point mutations have been fundamental for studying some other means of protein modification including sulfoxidation, S-nitrosylation, tyrosine nitration, acetylation, methylation, etc. Several examples are illustrated in Table 1.

Modification	Protein	Substitution	Readout	Ref
Phosphorylation	CLIP-170	Ser→Ala Ser→Glu	Co-immunoprecipitation, protein kinetics at MT tip	(Lee et al., 2010)
	Inositol 1,4,5- Trisphosphate Receptor (InsP3R)	Ser→Ala Ser→Glu	Ca^{+2} release from ER	(Wagner et al., 2004)
	AMPA (GluA2 subunit)	Ser→Ala Ser→Glu	Conductance	(Kristensen et al., 2011)
	Itch (E3-Ubiquitin Ligase)	Ser→Ala Ser→Glu Thr→Ala Thr→Glu	Co-immunoprecipitation /Proteolysis/ Ubiquitinylation activity	(Gallagher et al., 2006)
Sulfinic acid	Peroxiredoxin	Cys→Glu	Crystallography	(Jonsson et al., 2009)
Sulfoxidation	Slo-1	Met→Leu Met→Glu	Ionic current (Inside-out patch clamp)	(Santarelli et al., 2006)
Acetylation	Heat shock protein- (HSP-90)	Lys→Arg Lys→Gln Cys→Ala	Chaperon activity	(Scroggins et al., 2007)
S-nitrosylation	Syntaxin-1	Cys→Ser Cys→Trp Cys→Met	Exocytosis (amperometry)	(Palmer et al., 2008)
	Heat shock protein- (HSP-90)	Cys→Ala Cys→Asn Cys→Asp	Protein dimerization (FRET) ATP-ase activity Reverse Transcriptase activity	(Retzlaff et al., 2009)
Methylation	Flap endonuclease 1 (FEN1)	Arg→Ala Arg→Phe Arg→Lys	Endonuclease activity	(Guo et al., 2010)

Table 1. Representative post-translational modifications mimicked by natural point mutations using site-directed mutagenesis.

Amino acid oxidation is a common mechanism for diversification of certain residues; and, like phosphorylation, there are both addition and removal effector proteins associated with the modification. The reactive sulfur group of the methionine residue can be oxidatively modified with a number of reagents to methionine sulfoxide or methionine sulfone, dramatically increasing the polarity and hydrophobicity of the residue. (Black and Mould, 1991) (Hoshi and Heinemann, 2001) A family of enzymes known as methionine sulfoxide reductases have been found competent to reduce the methionine sulfoxide back to a thioether, a necessary requirement for a signalling relay. Substitution of a leucine residue for methionine offers a suitably isostructural null mimic, removing the oxidizeable sulphur atom while maintaining similar geometric and hydrophobic characteristics. (Ciorba et al., 1997) (Chen et al., 2000) In one example, this modification, which adds electron density onto the long methionine residue, has been mimicked *in vitro* by glutamic acid substitution. Interestingly, it was recently found that calcium and voltage gated BK (Big Potassium) ion channels, responsible for regulating K^+ ion flux, are sensitive to oxidation. This has been accomplished through lining of ion-selective pore with the hydrophobic Met_{536}, Met_{712}, and Met_{739} residues, which can be oxidatively modified to the more polar methionine sulfoxide, offering tuning of the hydrophobicity of channel. Oxidation of these methionine residues results in a decrease in half-activation voltage, increasing channel activity. Interestingly, one must mutate all three methionine residues within the channel to leucine to render the channel insensitive to oxidant treatment. If, however, one replaces any one of these three residues with glutamic acid, the channel behaves just like it does in the wild-type protein after treatment with the oxidant. (Santarelli et al., 2006) Thus, the technique of site-directed mutagenesis revealed that any of the three separate methionine residues can confer the oxidant-sensitivity phenotype to the protein, highlighting a robust selection pressure for that function.

Oxidized forms of cysteine, including nitrosyl, sulfenic acid, and sulfinic acid variants, have been implicated in reversible signalling arrays (Reddie and Carroll, 2008), and some have been mimicked with the use of natural amino acids. For example, cysteine sulfinic acid differs from aspartic acid by a single-atom replacement of sulphur atom with a carbon atom. This phosphomimicking strategy has been used in the investigation of the mechanism of peroxiredoxin (Pxr) (Jonsson et al., 2009), an enzyme involved in the ATP-dependent reduction of cysteine sulfinic acid back to a thiol residue. Reduction of an active site cysteine-sulfinic acid is first accomplished through ATP-dependent phospho-deficient of the sulfinic acid residue, followed by attack of a neighboring cysteine from sulfiredoxin (Srx) enzyme on the sulfinic phosphory ester intermediate. Remarkably, replacement of the catalytic cysteine in Pxr with an aspartic acid residue results in phosphorylation of the aspartic acid, but does not allow subsequent reactivity with the docked Srx. Thus, the single-atom replacement offered by native site-directed mutagenesis allowed for detailed mechanistic study of a trapped intermediate complex, inaccessible with other techniques. Substitution of a cysteine residue with a serine residue offers an effective single-atom replacement of a sulphur atom with an oxygen atom and eliminates sulphur-based chemistry from the site, while minimizing steric differences.

In several cases, tryptophan was reported to be an effective mimic of nitroso-cysteine (Palmer et al., 2008) (Wang et al., 2009) as computer simulations of the cysteine to tryptophan mutant suggest it contains steric bulk not offered by the next closest steric match, methionine, despite the absence of a sulphur atom in the former. In another case, nitroso-cysteine modification was mimicked by substitution with asparagine in studies investigating the effect of nitrosylation on ATPase activity of the 90 kDa heat shock protein (γHSP90). (Retzlaff et al., 2009) (Scroggins and Neckers, 2009) In this study an NO-insensitive γHSP90 was rendered sensitive to inactivation by NO treatment due to replacement of an alanine residue in the C-terminal domain with a cysteine, as this corresponding site is known to be a cysteine in all other HSP variants. Mutation of the engineered site to asparagine (γHSP90 A577N) recapitulates NO treatment of the engineered cysteine mutant, in contrast to a control isoleucine mutant (γHSP90 A577I), which serves as an unmodifiable mimic (shown in Fig 4). Although no citation or rationale was used for the mutant choices, these mimicking residues indeed recapitulated ATPase activity and chaperone activity of the NO-free and NO-bound forms of engineered γHSP90 protein.

As virtually every intracellular process is under dynamic regulation, it is no surprise that PTM-mimicking mutations have been used in conjunction with a myriad of analytical techniques. Shown in Table 1 are some of examples relevant to native site-directed mutagenesis - giving a flavour of the breadth of the strategy. In a manuscript that would bring the strategies of the last nearly 35 years full circle, it has recently been suggested that throughout evolution, phosphoserine residues on proteins have been mutated, through genetic drift. In one study they were more frequently converted into glutamic acid and

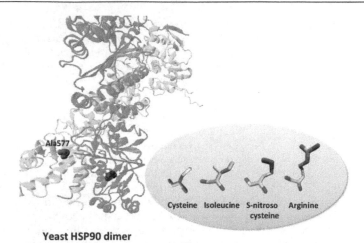

Yeast HSP90 dimer

Fig. 4. Mimicking S-nitrosylation in an engineered regulatory site. Ribbon representation of X-ray crystal structure of the Yeast HSP90 dimer; native Ala 577 (left) and engineered variants at position 577 (right) are highlighted in red. Note that there are structural similarities between cysteine and the non-modifiable isoleucine and between S-nitrosocysteine and arginine. Structural coordinates were obtained from the Protein Databank (PDB code: 2CG9) Stick-models of residues were color-coded by atom type: carbon atoms are cyan, nitrogen atoms are blue, and oxygen atoms are red.

aspartic acid than into any other residue, demonstrating evolution away from the switchable into permanent ON or OFF functions. (Kurmangaliyev et al., 2011) Although this work focused on non-structured regions and solely on phosphoserine evolution (the only post-translational modification for which there were sufficient data sets for statistical interpretation), these results might be viewed as an acknowledgement from biology of a mere requirement for a well-placed charge to induce the biological phenotype for which the switch evolved.

2.4 Site-directed mutagenesis in animal models

The ultimate verification of the experimental models obtained in *in vitro* assays or even in cultured cells comes with experiments in living organisms. Site-directed mutagenesis, in combination with other technologies for delivering a tailored plasmid to more complex genomes, has allowed the knock-in of mutations for individual engineered proteins in entire organisms, including mammals. (Yang et al., 1997) This technology has been used to investigate the role of individual amino acids in many biological processes and disease states, such as ASD (Autism Spectrum Disorder) (Bader et al., 2011), malignant hyperthermia (associated with heat stroke susceptibility) (Durham et al., 2008), and depression (Talbot et al., 2010), in the context of a living mammals. Phosphomimetic and phospho-deficient mutations have been knocked-in as well, leading to new understanding of downstream events to protein modification in the context of a living organism. In one prominent example, phosphorylation of ribosomal protein S6 was linked to control of cell size and glucose homeostasis in certain cell types through the use of a phospho-deficient knock-in mouse model. (Ruvinsky et al., 2005) In this model, all five potentially phosphorylatable serine residues in the putative region of rS6 were mutated to alanine (rpS6$^{P-/-}$) to abolish the potential for their phosphorylation. Smaller cell size, increased protein synthesis, and a severe decrease of circulating insulin were among the phenotypic signatures of the phospho-deficient variant. The phenotype was later shown to include decreased muscle mass and energy availability.

In another salient example, both phosphomimetic and phospho-deficient mutations were generated in a transgenic mouse model of Huntington's disease - a neurodegenerative disorder characterized by motor impairments and various psychiatric symptoms. (Gu et al., 2009) In this model, two phosphorylatable serine residues (13 and 18) in the N-terminus of a disease-promoting mutant of the huntingtin protein (*mhtt*) were mutated to either glutamate (phosphomimetic) or alanine (phospho-deficient) residues to investigate effect of phosphorylation on modification of the *htt* protein with a polyglutamine (PolyQ) extension, and on the onset of Huntington's Disease. Both mutants were viable and successfully rescued the embryonic lethality of an *htt* knockout mutant. A remarkable difference was observed between the Ala and Glu variants. The mice that were phosphodeficient (Ala) in the two serine residues showed onset of disease pathogenesis from the mutant *mhtt* - as measured by neuropathological examination of brain tissue sections, as well as by various motor and psychological tests. Remarkably, the phosphomimetic mutant mice (Glu) showed resistance to *mhtt*-induced disease pathogenesis. This technique clearly demonstrated that loss-of-phosphorylation of these two serine sites is a key signature of the pathogenesis *in vivo*. These results further suggested that the N-terminus of *htt* may be a target for future drug design to treat disease, even after the initializing mutation is present.

Indeed these examples show the power of site-directed mutagenesis in studying the signaling events associated with the disease state. However, caution must always be taken in the interpretation of data obtained from studies using biomimetic constructs.

2.5 Limitations of site-directed mutagenesis for targeting signaling nodes

Despite vast precedent for the utilization of amino acid surrogates and mimics of post-translational modifications in the literature, it must be remembered that the data and interpretation are only as robust as the chemical match of the surrogate moiety to the target. For example, the carboxylic acid functionality of glutamic acid has differing fundamental chemical properties from a serine-phosphate group, including, but not limited to, access to a (-2) charge state (in the case of phosphate) and different acid dissociation constants. Despite this, the single mutation has proven a valuable tool in the mimicking of phosphorylated residues. In addition it has become a common and quite inexpensive laboratory procedure, which is qualified as a 'quick-check' protocol to produce constitutively active, inactive or null mutants of potential phosphorylation sites.

As mentioned, PTMs that result in a significant size or charge difference from the unmodified states have been among the earliest developed. For example, while phosphorylation is the most thoroughly studied form of modification, it has been predicted that protein methylation will be among the slowest developed signalling pathways. This is due to the subtle spatial and charge density differences between methylated and non-methylated forms of an amino acid residue. With mass spectrometry virtually alone in the struggle to address such modifications, the decoding of such signalling networks awaits the development of sufficient chemical tools to recognize such subtle nodes and expedite interpretation and clarification of mass spectral data. (Huttlin et al., 2010)

It is probable that certain interactions, or classes of interactions, may be more stringent in which the electronic features of post-translationally modified residues are recognized and interpreted as signals. The importance of a cautious eye when utilizing signalling mimics was recently made salient when a phosphometic serine to glutamic acid mutant recapitulated one aspect of a signalling cascade, but not another. (Paleologou et al., 2010) While investigating potential phosphorylation sites in α-synuclein - a major component of the Alzheimer's disease amyloid - the authors generated serine to glutamate and serine to alanine mutants and compared them to wild-type and authentic phosphorylated material in ability to localize at the membrane and participate in fibril formation. In a model wherein phosphorylation of Ser_{87} blocks membrane recruitment and fibril formation, it was shown that substitution of native Ser_{87} with alanine resulted in wild-type-like membrane association and fibril formation, while authentic phospho-Ser_{87}, along with the $Ser_{87}E$ mutant, showed blunted formation of fibrils and poor membrane recruitment. Circular dichroism studies showed that, in a micellar environment, the authentic phospho-Ser_{87} modification resulted in an unstructuring of the random coiled protein, whereas the phosphomimic mutant could not recapitulate the membrane-associate phenotype. This unstructuring has been shown to be important in the interaction with membrane contents, and may further alter protein specificity. Thus, the phosphomimetic mutation introduced with site-directed mutagenesis could only reproduce one aspect of the biological phenotype. In this case, the authors were able to view the limits of the phosphomimic approach because they had access to constructs containing the authentic phosphoprotein for comparison. Nevertheless, it serves as an instructive case in which it is acknowledged that a given PTM can have multiple roles or be required in multiple steps of a signalling cascade or pathway.

3. Conclusion

We have demonstrated that site-directed mutagenesis has played an indispensable role in the unravelling of signal transduction pathways by offering access to constitutively active or inactive forms of several common PTMs. A critical eye must be used when utilizing surrogate residues that lack the capacity for a switchable context, and it is important to understand the limits of any biomimetic approach before results can be properly interpreted. Even with these limitations, it is no wonder the technique of site-directed mutagenesis has penetrated the field of signal transduction since the very conception of the technology. Indeed the utility of site-directed mutagenesis for the development and verification of models for biological processes and disease states will remain beneficial for years to come.

4. Acknowledgment

This work is supported by NIH Training Grant T32 HL007829 to Sieracki N. and RO1 HL103922 to Komarova Y.

5. References

Audette G.F., Engelmann R., Hengstenberg W., Deutscher J., Hayakawa K., Quail J.W., Delbaere L.T.J., 2000. The 1.9 Å Resolution Structure of Phospho-Serine 46 HPr from Enterococcus Faecalis. J. Mol. Biol. 303, 545-553.

Bader P.L., Faizi M., Kim L.H., Owen S.F., Tadross M.R., Alfa R.W., Bett G.C.L., Tsien R.W., Rasmusson R.L., Shamloo M. 2011. Mouse model of Timothy syndrome recapitulates triad of autistic traits. Proc. Natl Acad. Sci. USA 108, 15432-15437

Black S.D., Mould D.R., 1991. Development of Hydrophobicity Parameters to Analyze Proteins Which Bear Post- or Cotranslational Modifications. Anal. Biochem. 193, 72-82.

Chen J., Avdonin V., Ciorba M.A., Heinemann S.H., Hoshi T., 2000. Acceleration of P/C-Type Inactivation in Voltage-Gated K+ Channels by Methionine Oxidation. Biophys. J. 78, 174-187.

Ciorba M.A., Heinemann S.H., Weissbach H., Brot N., Hoshi T., 1997. Modulation of Potassium Channel Function by Methionine Oxidation and Reduction. Proc. Natl Acad. Sci. USA 94, 9932-9937.

Davies K.M., Lowe E.D., Vénien-Bryan C., Johnson L.N., 2009. The HupR Receiver Domain Crystal Structure in its Nonphospho and Inhibitory Phospho States. J. Mol. Biol. 385, 51-64.

Deribe Y.L., Pawson T., Dikic I., 2010. Post-Translational Modifications in Signal Integration. Nat. Struct. Mol. Biol. 17, 666-672.

Durham W.J., Aracena-Parks P., Long C., Rossi A.E., Goonasekera S.A., Boncompagni S., Galvan D.L., Gilman C.P., Baker M.R., Shirokova N., Protasi F., Dirksen R., Hamilton S.L. 2008. RyR1 S-Nitrosylation Underlies Environmental Heat Stroke and Sudden Death in Y522S RyR1 Knockin Mice. Cell 133, 53-65.

Gallagher E., Gao M., Liu Y.-C., Karin M., 2006. Activation of the E3 Ubiquitin Ligase Itch Through a Phosphorylation-Induced Conformational Change. Proc. Natl Acad. Sci. USA 103, 1717-1722.

Guo Z., Zheng L., Xu H., Dai H., Zhou M., Pascua M.R., Chen Q.M., Shen B., 2010. Methylation of FEN1 Suppresses Nearby Phosphorylation and Facilitates PCNA Binding. Nat. Chem. Biol. 6, 766-773.

Hoshi T., Heinemann S.H., 2001. Regulation of Cell Function by Methionine Oxidation and Reduction. J. Physiol. 531, 1-11.

Hutchison C.A., Phillips S., Edgell M.H., Gillam S., Jahnke P., Smith M., 1978. Mutagenesis at a Specific Position in a DNA Sequence. J. Biol. Chem. 253, 6551-6560.

Huttlin E.L., Jedrychowski M.P., Elias J.E., Goswami T., Rad R., Beausoleil S.A., Villén J., Haas W., Sowa M.E., Gygi S.P., 2010. A Tissue-Specific Atlas of Mouse Protein Phosphorylation and Expression. Cell 143, 1174-1189.

Jia Z., Vandonselaar M., Hengstenberg W., Wilson Quail J., Delbaere L.T.J., 1994. The 1.6 Å Structure of histidine-Containing Phosphotransfer Protein HPr from Streptococcus Faecalis. J. Mol. Biol. 236, 1341-1355.

Jonsson T.J., Johnson L.C., Lowther W.T., 2009. Protein Engineering of the Quaternary Sulfiredoxin-Peroxiredoxin Enzyme-Substrate Complex Reveals the Molecular Basis for Cysteine Sulfinic Acid Phosphorylation. J. Biol. Chem. 284, 33305-33310.

Kern D., Volkman B.F., Luginbuhl P., Nohaile M.J., Kustu S., Wemmer D.E., 1999. Structure of a Transiently Phosphorylated Switch in Bacterial Signal Transduction. Nature 402, 894-898.

Kristensen A.S., Jenkins M.A., Banke T.G., Schousboe A., Makino Y., Johnson R.C., Huganir R., Traynelis S.F., 2011. Mechanism of Ca2+/Calmodulin-Dependent Kinase II Regulation of AMPA Receptor Gating. Nature 14, 727-735.

Kurmangaliyev Y., Goland A., Gelfand M., 2011. Evolutionary Patterns of Phosphorylated Serines. Biol. Direct 6, 8.

Lee H-S., Komarova Y.A., Nadezhdina E.S., Anjum R., Peloquin J.G., Schober J.M., Danciu O., van Haren J., Galjart N., Gygi S.P., Akhmanova A, Borisy G.G., 2010.

Phosphorylation Controls Autoinhibition of Cytoplasmic Linker Protein-170. Mol. Biol. Cell 21, 2661-2673.

Paleologou K.E., Oueslati A., Shakked G., Rospigliosi C.C., Kim H.-Y., Lamberto G.R., Fernandez C.O., Schmid A., Chegini F., Gai W.P., Chiappe D., Moniatte M., Schneider B.L., Aebischer P., Eliezer D., Zweckstetter M., Masliah E., Lashuel H.A., 2010. Phosphorylation at S87 Is Enhanced in Synucleinopathies, Inhibits a-Synuclein Oligomerization, and Influences Synuclein-Membrane Interactions. J. Neurosci. 30, 3184-3198.

Palmer Z.J., Duncan R.R., Johnson J.R., Lian L.-Y., Mello L.V., Booth D., Barclay J.W., Graham M.E., Burgoyne R.D., Prior I.A., Morgan A., 2008. S-Nitrosylation of Syntaxin 1 at Cys145 is a Regulatory Switch Controlling Munc18-1 Binding. Biochem. J. 413, 479-491.

Reddie K.G., Carroll K.S., 2008. Expanding the Functional Diversity of Proteins Through Cysteine Oxidation. Curr. Opin. Chem. Biol. 12, 746-754.

Reizer J., Sutrina S.L., Saier M.H., Stewart G.C., Peterkofsky A., Reddy P., 1989. Mechanistic and Physiological Consequences of HPr(ser) Phosphorylation on the Activities of the Phosphoenolpyruvate:Sugar Phosphotransferase System in Gram-Positive Bacteria: Studies with Site-Specific Mutants of HPr. EMBO J. 8, 2111-2120.

Retzlaff M., Stahl M., Eberl H.C., Lagleder S., Beck J., Kessler H., Buchner J., 2009. Hsp90 is Regulated by a Switch Point in the C-terminal Domain. EMBO Rep. 10, 1147-1153.

Santarelli L.C., Wassef R., Heinemann S.H., Hoshi T., 2006. Three Methionine Residues Located Within the Regulator of Conductance for K+ (RCK) Domains Confer Oxidative Sensitivity to Large-Conductance Ca2+-Activated K+ Channels. J. Physiol. 571, 329-348.

Schneider J., Fanning E., 1988. Mutations in the Phosphorylation Sites of Simian Virus 40 (SV40) T Antigen Alter its Origin DNA-Binding Specificity for Sites I or II and Affect SV40 DNA Replication Activity. J. Virol. 62, 1598-1605.

Scroggins B.T., Neckers L., 2009. Just Say NO: Nitric Oxide Regulation of Hsp90. EMBO Rep. 10, 1093-1094.

Scroggins B.T., Robzyk K., Wang D., Marcu M.G., Tsutsumi S., Beebe K., Cotter R.J., Felts S., Toft D., Karnitz L., Rosen N., Neckers L., 2007. An Acetylation Site in the Middle Domain of Hsp90 Regulates Chaperone Function. Mol. Cell 25, 151-159.

Tan C.S.H., 2011. Sequence, Structure, and Network Evolution of Protein Phosphorylation. Sci. Signal. 4, 6.

Wagner L.E., Li W.-H., Joseph S.K., Yule D.I., 2004. Functional Consequences of Phosphomimetic Mutations at Key cAMP-Dependent Protein Kinase Phosphorylation Sites in the Type 1 Inositol 1,4,5-Trisphosphate Receptor. J. Biol. Chem. 279, 46242-46252.

Walsh C.T., Garneau-Tsodikova S., Gatto G.J., 2005. Protein Posttranslational Modifications: The Chemistry of Proteome Diversifications. Angew. Chem. Int. Edit. 44, 7342-7372.

Wang P., Liu G.-H., Wu K., Qu J., Huang B., Zhang X., Zhou X., Gerace L., Chen C., 2009. Repression of Classical Nuclear Export by S-nitrosylation of CRM1. J. Cell Sci. 122, 3772-3779.

Wittekind M., Reizer J., Deutscher J., Saier M.H., Klevit R.E., 1989. Common Structural Changes Accompany the Functional Inactivation of HPr by Seryl Phosphorylation or by Serine to Aspartate Substitution. Biochemistry 28, 9908-9912.

Yang X.W., Model P., Heintz N., 1997. Homologous recombination based modification in Escherichia coli and germline transmission in transgenic mice of a bacterial artificial chromosome. Nat Biotechnol. 15(9):859-65.

8

Using Genetic Reporters to Assess Stability and Mutation of the Yeast Mitochondrial Genome

Shona A. Mookerjee[1] and Elaine A. Sia[2]
[1]*The Buck Institute for Research on Aging*
[2]*Department of Biology, The University of Rochester*
USA

1. Introduction

The mitochondrion has been an identified subcellular organelle since the late 1800s, though its function and importance remained relatively cryptic until almost a century later. With the molecular renaissance of the 1950s and 1960s came the tools and perspectives, including peptide sequencing, microscopy, and the chemiosmotic hypothesis, with which to formulate and test crucial questions about what mitochondria do and how they function. In addition, shortly after the structure of DNA was elucidated and its role as the genetic material of the cell established, the presence of extranuclear DNA was revealed by electron microscopy (Nass and Nass, 1963) and by density gradient fractionation (Haslbrunner et al., 1964), identifying this DNA as part of highly purified mitochondria.

The goal of targeted genetic manipulation within the mitochondrial genome is rapidly driving a second renaissance in mitochondrial biology. Current approaches to the mitochondrial uptake of DNA have recently been reviewed in detail (Mileshina et al., 2011a). Several model organisms are currently amenable to this technique although no vertebrate species are among them. In our view, successful mitochondrial transformation is hindered by our limited knowledge of the fundamental processes of mitochondrial DNA maintenance and repair. We are hopeful that a greater understanding of the replication, repair, and maintenance of the mitochondrial genome, afforded by the tools currently in existence and presented here, will soon allow the construction of mammalian mitochondrial disease models (Dunn et al., 2011), or lead directly to gene therapy for the treatment of mitochondrial diseases in patients (Schon and Gilkerson, 2010). In this review, we will discuss the current state of mitochondrial genetic reporters and their application toward understanding and manipulating mitochondrial genome maintenance.

In cells, mitochondrial DNA is organized into protein-associated structures called nucleoids. Quantitative PCR coupled with immunofluorescence microscopy revealed an estimated 2-8 mitochondrial genomes per nucleoid in human immortalized cell culture (Legros et al., 2004); higher-resolution microscopy identified more nucleoids per cell, bringing this estimate down to only 1-2 copies per nucleoid (Kukat et al., 2011). A growing number of the proteins associated with the mitochondrial nucleoid have been identified from eukaryotic model systems (Bogenhagen et al., 2003; Garrido et al., 2003; Kienhöfer et al., 2009). The conserved and abundant HMG protein, Abf2p in yeast, and mtTFA (or TFAM) in vertebrates, is thought to organize and compact the mitochondrial genome, and is proposed to play a histone-like role in mitochondrial DNA organization (Kaufman et al., 2007; Pohjoismaki et al., 2006). In addition to these properties, TFAM has recently been reported to contribute to mitochondrial DNA replication (Pohjoismaki et al., 2006) and repair (Canugovi et al., 2010).

The mitochondrial genome is vitally important to the organelle's proper function. Its encoded proteins are almost all subunits of the mitochondrial respiratory complexes; mutations to these genes can disrupt bioenergetic function and control. There is also growing evidence that proper maintenance of the mitochondrial genome is tightly associated with normal mitochondrial behavior, including fusion, fission, and intracellular migration (Baker and Haynes, 2011; Chen et al., 2010; Gilkerson, 2009). The reasons for this are currently unknown, but we can speculate that nucleoids may help to define a minimal mitochondrial "unit" and that disruption of mitochondrial DNA resolution affects mitochondrial dynamics and transmission in dividing cells (Margineantu et al., 2002). Alternately, nucleoids may help to catalyze organization of cristae, respiratory complexes, or other submitochondrial structures, as has been suggested for the Complex V ATP synthase (Strauss et al., 2008), and faulty mitochondrial DNA maintenance disrupts this patterning.

Fission and fusion, coupled with autophagic degradation of mitochondrial material ("mitophagy"), are increasingly appreciated as a primary means of mitochondrial quality control. Instability of the mitochondrial genome is therefore not just a problem affecting its encoded products, but potentially the structural integrity of the entire organelle. Recent proposals suggest that the cellular dysfunction underlying neurodegeneration in Alzheimer's and Parkinson's diseases is caused or at least exacerbated by defective mitophagy and increased mitochondrial DNA instability (Chang, 2000; Corral-Debrinski et al., 1994; Coskun et al., 2011; Narendra et al., 2010; Narendra et al., 2008; Sasaki et al., 1998; Suen et al., 2010).

1.1 Yeast gene nomenclature

A comprehensive guide to yeast gene nomenclature is both published (*Trends in Genetics, Volume 14, Issue 11, Supplement 1, 1998, Pages S.10-S.11*) and available online (http://www.yeastgenome.org/help/yeastGeneNomenclature.shtml). We will summarize here the points that are most relevant to the following sections.

There are two nomenclature systems, reflecting the advent of genomic analysis. The ORF naming system, instituted in the post-genomic era, gives each predicted open reading frame a systematic designation reflecting its chromosomal position and strand orientation. This system will not be discussed here. We will focus on gene symbol nomenclature, which applies only to ORFs that express a known gene product and is in common use for most proteins.

In gene symbol nomenclature, yeast genes are given a three-letter "name" followed by a number. Gene names are italicized, with dominant (usually wild-type) alleles in capital and recessive (usually mutant) alleles in lowercase type. For example, *ABC1* denotes a wild-type, dominant gene; *abc1* a generic mutant. The names themselves correspond to characterized features, either phenotypic, biochemical, enzymatic, or genetic. Mutant alleles can carry a second number to specify a particular mutation, e.g., *abc1-1*, or a descriptive name. For example, *abc1-G43V* describes a point mutant that bears a glycine-to-valine amino acid substitution at position 43. Gene knockouts are denoted with a delta symbol, e.g., *abc1-Δ*. Gene disruptions by insertion of another gene are denoted with a double colon, e.g., *abc1::URA3*. If a gene is fully replaced with the insertion, this is indicated as *abc1Δ::URA3*. In yeast, gene products may also carry the gene name, e.g., Abc1p or abc1p-G43V. Gene product names are interchangeable with a functional name, for example, Arg8p = acetylornithone transaminase.

Mitochondrial dysfunction can occur through single gene mutations in both nuclear and mitochondrial genomes, forming two classes of mutants. *Pet* mutants are those with nuclear DNA mutations, while *mit* mutants are those with mitochondrial DNA mutations. Additionally, mitochondrial genome status is also categorized. Respiring yeast strains carrying wild-type mitochondrial DNA are referred to as ρ+. When yeast cells are grown on a fermentable carbon source such as glucose, spontaneous variants arise that cannot respire and form smaller colonies, termed *"petite"* mutants. The vast majority of spontaneous *petite* strains contain large-scale mitochondrial DNA deletions and rearrangements. Often these strains carry only a small fraction of their original mitochondrial DNA; retention of only 1% of the genome is commonly observed. The total mitochondrial DNA content of ρ- and ρ+ strains are equivalent, however, as the remaining fragments in ρ- strains are amplified accordingly (Dujon, 1981).

Generation of mitochondrial DNA-free derivatives of any strain is achieved by culturing cells with ethidium bromide, which blocks mitochondrial DNA replication without affecting nuclear DNA (Meyer and Simpson, 1969). These ρ0 strains can be studied directly or used as a tool for mitochondrial DNA manipulation.

1.2 The presence of mitochondrial DNA repair mechanisms

The assertion that mitochondrial DNA did not undergo repair was made as late as 1990 (Singh and Maniccia-Bozzo, 1990), in agreement with early reports (Clayton et al., 1974). Part of the appeal of this idea was the observation that mitochondrial DNA depletion can be induced by various insults, including oxidative stress (Shokolenko et al., 2009), ethanol (Ibeas and Jimenez, 1997) and zidovudine (AZT) (Arnaudo et al., 1991) treatment. This was interpreted as evidence that, when damaged, mitochondrial DNA was simply eliminated rather than repaired. However, in 1992, the first evidence for photolyase repair of UV-induced mitochondrial DNA damage in yeast was provided (Yasui et al., 1992), followed quickly by work from the Bohr and Campbell groups demonstrating uncharacterized repair activities and homologous recombination, respectively, in two mammalian mitochondrial systems (LeDoux et al., 1992; Thyagarajan et al., 1996). The fifteen years since have revealed an extensive set of mitochondrial DNA repair pathways, including base excision repair (BER), homologous recombination (HR), and non-homologous end joining (NHEJ). A recent review summarizes our current understanding of these pathways in mitochondria and other organellar genomes (Boesch et al., 2010).

1.3 Nuclear and mitochondrial DNA repair pathways share protein components

Many of the known mitochondrial DNA repair pathway proteins are mitochondrially-localized proteins initially characterized in nuclear repair. The first mitochondrial DNA repair protein identified, photolyase, was demonstrated to be one such dual-localized protein (Green and MacQuillan, 1980). Subsequent studies indicated localization of base excision repair proteins to both subcellular compartments, including the yeast glycosylases Ntg1p (You et al., 1999), Ung1p (Chatterjee and Singh, 2001) and Ogg1p (Singh et al., 2001), the mammalian glycosylases UNG1 (Nilsen et al., 1997), MTH1 (Kang et al., 1995), OGG1 (Nishioka et al., 1999), and MYH (De Souza-Pinto et al., 2009; Nakabeppu et al., 2006), and the yeast AP endonuclease Apn1p (Ramotar et al., 1993). In human lymphoblasts, BER proteins were associated with the mitochondrial inner membrane fraction, where mitochondrial nucleoids are also found (Stuart and Brown, 2006). These findings illustrate the high evolutionary conservation of mitochondrial BER. Factors that regulate the subcellular localization of these proteins are not well understood; however, changes to localization in response to stress has recently been demonstrated (Griffiths et al., 2009; Swartzlander et al., 2010). This apparent recruitment of DNA repair proteins to the mitochondria may represent a DNA-specific communication pathway between the intramitochondrial and extramitochondrial environments.

Other DNA repair proteins have been shown to affect mitochondrial DNA maintenance in mammalian cells, including the BER flap endonuclease FEN1 (Liu et al., 2008), DNA double-strand break repair proteins, Rad51p (Sage et al., 2010), Mre11 (Dmitrieva et al., 2011) and Ku80 (Coffey et al., 1999), and the nucleotide excision repair protein CSA (Kamenisch et al., 2010). In addition, in yeast, DNA damage tolerance pathways that utilize the translesion polymerase complexes encoded by Rev1p, Rev3p, and Rev7p also impact mitochondrial mutagenesis (Kalifa and Sia, 2007; Zhang et al., 2006).

2. Manipulation of the yeast mitochondrial genome

2.1 Basic features of the mitochondrial genome

The yeast mitochondrial genome consists of 75-85 kb of double-stranded DNA, encoding seven protein products (Cox I, II, III, Atpase 6, 9, cyt b, Var1), 2 rRNAs and 24 tRNAs, while the human mitochondrial genome is a much smaller 16.5 kb and encodes 13 protein products (cox I, II, III, ND1-6, 4L, Atpase 6, 8, cyt b), 2 rRNAs and 22 tRNAs. Aside from the presence of complex I (NADH:ubiquinone oxidoreductase) subunit genes in the human mitochondrial genome and not yeast, and the presence of non-coding regions in the yeast genome, the two are remarkably similar in structure and encoded products, giving yeast mitochondrial genome manipulation great power to inform our understanding of mammalian mitochondrial DNA defects.

Mitochondrial and nuclear DNA in yeast are compositionally different; mitochondrial DNA is relatively AT-rich and highly repetitive, with G and C bases further segregated in coding regions. This repetition made initial sequencing of the entire yeast mitochondrial genome difficult (Foury et al., 1998) and is a continued challenge in targeted gene manipulation, particularly in the intergenic AT-rich regions. Yeast mitochondrial DNA has multiple regions of non-coding DNA, which are the primary contributors to the 83% AT bias of the genome. The size difference between human and yeast mitchondrial DNA is almost entirely due to the absence of these intergenic regions in the human mitochondrial genome.

2.2 Organisms with tractable mitochondrial genomes

To date, the number of organisms that have successfully undergone mitochondrial transformation remains small and is restricted to unicellular eukaryotes. The most widely used model remains the budding yeast *Saccharomyces cerevisiae* (Johnston et al., 1988), which presents multiple biological advantages beyond the accessibility of its mitochondrial genome. Budding yeast are facultative anaerobes. This organism can survive by fermentation when oxygen is unavailable or when respiratory mechanisms are disrupted, as occurs during biolistic transformation (Section 2.3). Moreover, our comprehensive understanding and molecular tool base for the nuclear genome enables efficient analysis and powerful interpretation of mitochondrial phenotypes. Finally, homologous recombination in yeast is highly efficient in both nuclear and mitochondrial genomes, facilitating the introduction of targeted sequences to either genome.

The mitochondria of hyphal yeast *Candida glabrata* have also recently been reported to take up biolistically delivered DNA. Unlike wild-type *S. cerevisiae*, *C. glabrata* can maintain relatively stable heteroplasmy, maintaining mixed populations of transformed and endogenous DNA, which the authors propose as an ideal characteristic for studying the regulation of mitochondrial DNA transmission. Under selective pressure the exogenous genotype can be fixed to homoplasmy (Zhou et al., 2010). Heteroplasmy is the state of harboring multiple different mitochondrial genomes, while in homoplasmy only one type is present.

An important requirement of successful mitochondrial manipulation is the ability to select and purify a rare transformation event; in *S. cerevisiae*, the efficiency of transformation using microprojectile bombardment is on the order of one in 10^7 (Bonnefoy and Fox, 2007). As described in Section 2.3, this is the sole method currently known for successful mitochondrial incorporation of exogenous DNA into whole cells.

The only known algal species to date that can be biolistically transformed is the green alga *Chlamydomonas reinhardtii*. This species can incorporate exogenous DNA into both mitochondrial and chloroplast genomes, providing a unique model for studying the genetic and functional interactions of these two organelles (Randolph-Anderson et al., 1993). Here, the development of a selectable mutant facilitated the initial isolation of mitochondrial transformants. *C. reinhardtii* are normally able to grow in the dark if supplemented with acetate as a carbon source; due to a deletion disruption of the mitochondrial CYB gene encoding apocytochrome b, the *dum-1* mutant cannot. Transforming *dum-1* cells with mitochondrial DNA isolated from a *DUM-1* strain and growing cells in darkness plus acetate selects for restoration of a growth phenotype, indicating mitochondrial uptake of the wild-type CYB gene.

Biolistic transformation in multicellular organisms is hampered by an inability to select and amplify individual transformants. A promising recent attempt to transform mitochondria in a mouse embryonic fibroblast line with a "universal" neomycin marker via a bacterial conjugation-like mechanism was not successful in generating mitochondrial transformants (Yoon and Koob, 2011). However, mitochondria isolated from both mammalian and plant sources have been successfully transformed *in vitro* (Mileshina et al., 2011b; Yoon, 2005). Multiple methods, including DNA targeting with a protein localization signal, electroporation, and spontaneous mitochondrial uptake of linear DNA have all given rise to mitochondrial uptake of exogenous DNA, but these constructs could not be propagated in the mitochondrial genome of a viable and dividing cell (Mileshina et al., 2011a).

2.3 Biolistic transformation of yeast and selection of transformed clones

Microprojectile bombardment of DNA on a carrier is an effective method for delivering DNA past the plasma membrane and two mitochondrial membranes into the mitochondrial matrix. This method was pioneered in plants by John Sanford (Sanford et al., 1987), and first demonstrated in yeast by Sanford, Butow and colleagues (Johnston et al., 1988). Described below is the general transformation procedure used for *S. cerevisiae* (Bonnefoy and Fox, 2007).

The linear or circular plasmid DNA to be transformed is alcohol-precipitated onto a carrier substrate, usually tungsten or gold particles <1μm. The bombardment itself occurs in a biolistic gun chamber (Sanford, 1988), where rising vacuum pressure ruptures a pressure sensitive disk holding the DNA-precipitated particles, driving the particles onto a freshly plated lawn of haploid yeast cells. The plate medium then selects for uptake of either the plasmid of interest or of a co-transformed marker if the target plasmid does not confer a selectable phenotype.

The target cells for mitochondrial transformation are typically ρ^0 (non-mitochondrial DNA-containing) derivatives of a chosen strain, to ensure that the only mitochondrial DNA present is transformation-derived. After selection for the co-transformed nuclear marker, transformants must be screened for the presence of the transforming mitochondrial DNA. Following mitochondrial uptake of the desired DNA, positive haploid clones are mated to a strain containing wild-type mitochondrial genomes, allowing mixing of the transformed DNA with the target mitchondrial DNA. Generally, the desired outcome is integration of the synthetic DNA construct into the mitochondrial genome, although mitochondrial plasmid maintenance can also occur. The specific example of *ARG8^m* integration is provided below in Section 3.1.

3. The *ARG8^m* auxotrophic mitochondrial reporter gene

3.1 Building an auxotrophic mitochondrial reporter: the *ARG8^m* gene

The phenotypic output of a genetic reporter system determines its strengths and weaknesses as an analytic tool. In yeast, multiple auxotrophic (factor-requiring) mutants have historically been used with great success as both selective markers and phenotypic reporters. The defined requirements of yeast grown in culture allow for synthetic reconstitution of growth media lacking a specific amino acid. Commonly used auxotrophic markers in yeast include growth status on media lacking uracil, histidine, leucine, arginine, methionine, and lysine. Many laboratory strains, including our wild-type strain, DFS188 and its derivatives, lack the ability to make these amino acids due to specific nuclear mutations. These strains are typically maintained in rich media, allowing unrestricted growth. Withdrawal of the amino acid in question results in cell death. Rescue of a growth phenotype occurs when the gene that complements the nuclear mutation for that amino acid's synthesis is supplied on a plasmid or as part of a conditional reversion construct, allowing cells to regain prototrophic (factor-independent) growth.

While an ideal system to assess mechanisms associated with nuclear gene expression, the mitochondrial genome has long been inaccessible to auxotrophic reporter manipulation because it does not encode any amino acid biosynthetic enzymes. Direct insertion of a nuclear gene is impossible, as the codon usage of the mitochondrial genome differs from the nuclear genome both in the preferred codon frequency and in some codon products. Multiple nuclear leucine codons encode threonine in mitochondria, and a nuclear UGA stop codon encodes tryptophan in the mitochondria of yeast (Bonitz et al., 1980). Generating a mitochondrial auxotrophic reporter thus requires mutating a nuclear gene to enable its expression from the mitochondrial genome. Once constructed, this gene must be introduced into the mitochondrial genome with the appropriate transcriptional and translational cues. To date, the only such auxotrophic marker gene to be engineered in this way is the synthetic *ARG8m* gene, made by Tom Fox and colleagues (Steele et al., 1996).

The nuclear *ARG8* gene encodes acetylornithine transaminase, which catalyzes an early step in the biosynthesis of ornithine, a precursor to both arginine and proline. The Arg8 protein is normally localized to the mitochondrial matrix and yields the active mitochondrial transaminase following cleavage of its N-terminal targeting sequence (Steele et al., 1996). It is therefore an ideal candidate for reconfiguration as a mitochondrial gene, as its product functions within the matrix and does not require mitochondrial export for phenotypic expression.

Fox and group began by synthetically generating a 1.3 kilobase fragment encoding the entire 423 amino acid acetylornithine transaminase enzyme. Substitutions were made at 12 CUN codons (n: Leu; mt: Thr) and 6 AUA codons (n: Ile; mt: Met) to maintain the Leu and Ile residues. In addition, each of the two Trp codons was changed to UGA (n: STOP; mt: Trp) ensuring *ARG8m* expression from the mitochondrial genome only. This construct was introduced into a plasmid containing mitochondrial DNA sequence flanking the *COX3* gene, providing sequences for recombination-dependent integration and ensuring the presence of correct transcriptional and translational processing signals.

Several steps were then required to generate the desired end product of the *cox3::ARG8m* sequence incorporated into the mitochondrial genome while maintaining isogenicity of both mitochondrial and nuclear genomes. These steps are a "shell game" of genetic manipulation, designed to shield various DNA pools from one another until the desired product is achieved (Fig. 1).

The plasmid containing *cox3::ARG8m* was biolistically transformed into a ρ^0 haploid yeast strain, ensuring that plasmid DNA was the only DNA present in the mitochondrial compartment (Fig. 1). In addition, these yeast cells lack the nuclear *ARG8* gene and carry a mutation in the *KAR1* gene that prevents nuclear fusion during mating. A second plasmid carrying a functional *LEU2* allele was co-transformed to allow selection of successful DNA uptake by a Leu+ phenotype. Note that the mitochondrial DNA targeted construct, though present, is inactive; screening for its presence therefore requires selection of a DNA-dependent effect. Transformants were screened for the presence of the *ARG8m* gene in the mitochondria by mating with a strain carrying a mitochondrial genome deletion in the 5′ untranslated region upstream of the *COX3* gene. Only Leu+ transformants that also carry the mitochondrial plasmid with the wild-type *COX3* 5′ sequence will complement the deletion to give rise to respiring recombinants. Positive mitochondrial transformants identified by this test mating were purified and used to generate the integrated reporters.

To allow mixing of the mitochondrial plasmid DNA with intact mitochondrial genomes, the biolistically transformed cells were mated with a second haploid strain bearing normal mitochondrial DNA (Fig. 1, Fig. 2A). Since one strain is karyogamy-deficient, the nuclear envelopes do not fuse. Cell division gives rise to haploid cells, and one haploid genome can be selected in subsequent divisions. The mitochondria, however, undergo rapid fusion, allowing interaction between plasmid and mitochondrial DNA. This process is known as

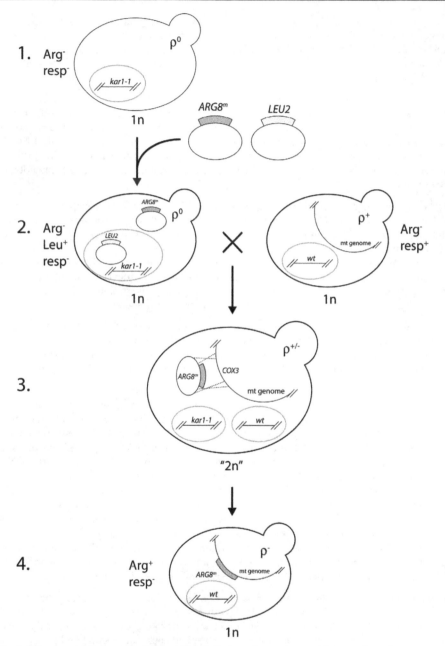

Fig. 1. Diagrammatic representation of biolistic transformation and integration of exogenous DNA into the yeast mitochondrial genome. Mitochondrial genome status is given in ρ

nomenclature. Slash marks indicate endogenous genomic DNA; ovals with bars for relevant genes represent plasmid DNA. Gray line denotes the nucleus. Selectable phenotypes of the genotype shown are written outside the cell. Ploidy is written below the cell. **Step 1.** A haploid, ρ^0, nuclear fusion-deficient mutant (*kar1-1*) is biolistically transformed with two plasmids, one bearing the *LEU2* allele and one the *ARG8^m* allele with *COX3* flanking sequence. **Step 2.** Transformants are selected by Leu+ growth, reflecting nuclear uptake of the *LEU2*-bearing plasmid (and possible uptake of the *ARG8^m*-bearing plasmid; confirmation of uptake is described in Sections 2.3 and 3.1). Leu+ transformants are crossed with wild-type cells. **Step 3.** Mixing of the *ARG8^m* plasmid with endogenous mitochondrial genomes; homologous regions of *COX3* non-coding sequence flanking *ARG8^m* mediate recombination at the mitochondrial *COX3* locus. **Step 4.** Replacement of endogenous *COX3* with *ARG8^m* in the mitochondrial genome, maintained by Arg+ selection. Loss of *COX3* confers respiration failure.

cytoduction. Homologous recombination between the *COX3* flanking sequences mediated replacement of the endogenous *COX3* gene with plasmid-derived *cox3::ARG8^m* sequence. By simultaneously selecting for both mitochondrial (Arg+) and nuclear markers, a haploid strain of the desired nuclear and mitochondrial backgrounds results that is phenotypically Arg+ and respiration-deficient, requiring a fermentable carbon source.

Mitochondrial genome incorporation of *cox3::ARG8^m* was confirmed by Southern blotting, while its requirement for the Arg+ phenotype was shown by curing yeast of their mitochondrial DNA and observing a reversion to an Arg- phenotype. The Arg+ phenotype was also dependent on the *COX3* translational activation complex, as deletion of the genes encoding the complex components result in Arg- cells. Finally, immunoblot analysis demonstrated that the protein product of the *ARG8^m* gene is identical in size to that of the nuclear *ARG8*, suggesting correct N-terminal processing. These controls elegantly demonstrate the correct location, expression and control of the *cox3::ARG8^m* reporter gene (Steele et al., 1996).

The *ARG8^m* reporter was initially used by Fox and group to examine mechanisms of mitochondrial translation (Bonnefoy and Fox, 2000; Dunstan et al., 1997), but its utility as a reporter extends to any assay in which reporter expression can be made a meaningful indicator of the function of interest. Fox and colleagues have also used the *ARG8^m* and other reporters (including a recoded mitochondrial GFP (Cohen and Fox, 2001)) to measure peptide import and export from the mitochondria (He and Fox, 1997; Torello et al., 1997), and to generate mutations in respiratory complex subunits (Ding et al., 2008). In addition, we and others have used *ARG8^m* expression to measure various aspects of mitochondrial DNA instability. The following sections will describe the construction, insertion, and use of these reporters in detail.

3.2 *ARG8^m* as a reporter of mitochondrial translation, DNA repair, recombination, and heteroplasmy

Prior to the construction of the *ARG8^m* mitochondrial reporter and the advent of qPCR and high-throughput sequencing, only two methods were available to easily assess mitochondrial DNA stability. First, ρ- yeast cells with grossly defective mitochondrial DNA are respiration deficient and display slower growth on fermentable media, forming "petite"

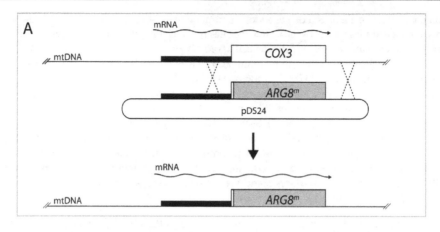

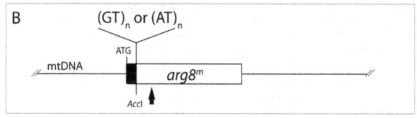

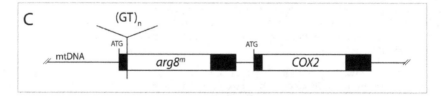

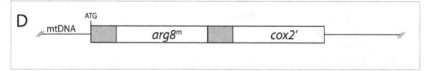

Fig. 2. *ARG8^m* construct diagrams. Slash marks indicate endogenous genomic DNA; ovals with bars for relevant genes represent plasmid DNA. **A.** Endogenous *COX3* locus (top) and replacement by *ARG8^m* borne on plasmid pDS24 to generate the mitochondrial genome-integrated *ARG8^m* allele (bottom). Thin black bar represents non-translated sequence; labeled white and gray bars denote the *COX3* and *ARG8^m* ORFs, respectively. Adapted from (Steele et al., 1996). **B.** Insertion of GT or AT dinucleotide repeats into a *cox3::ARG8^m* reporter fusion using an internal *Acc*I restriction endonuclease site. Black bar represents *COX3* coding sequence, ATG denotes *COX3* translational start, arrow represents post-translational cleavage site to yield functional Arg8p. Adapted from (Sia et al., 2000). **C.** Respiring

microsatellite reporter insertion upstream of the endogenous *COX2* locus, flanked by *COX2* non-coding sequence (black bars) including the translational start (ATG). Adapted from (Mookerjee and Sia, 2006). **D.** *Rep96::arg8^m^::cox2'* direct repeat-mediated deletion (DRMD) reporter. Gray bars represent the first 96 base pairs of *COX2* translated sequence, either containing (ATG, left of *arg8^m^* locus) or missing (right of *arg8^m^* locus) the ATG translational start codon. Adapted from (Phadnis et al., 2005).

colonies. Though petite formation is a somewhat useful indicator of gross mitochondrial DNA abnormality, its pleiotropic nature (the petite phenotype may result from nuclear *pet* as well as mitochondrial *mit* mutations), its origin via multiple types of mitochondrial DNA mutation, and the lack of a mechanistic explanation for the generation of ρ⁻ genomes all limit its utility. Second, because mitochondria retain many prokaryotic properties, they are selectively sensitive to antibiotic drugs, including erythromycin. Erythromycin binds to the 21S mitochondrial ribosomal RNA, disrupting mitochondrial protein synthesis. Specific point mutations to the mitochondrial 21S rRNA gene prevent erythromycin from binding to the gene product, conferring an erythromycin resistant (Ery^R^) phenotype (Cui and Mason, 1989; Kalifa and Sia, 2007). In this way, the appearance of Ery^R^ colonies can be used to estimate mitochondrial DNA point mutation accumulation rates using various calculation methods (Lea and Coulson, 1949; Luria and Delbruck, 1943). Ery^R^ acquisition is useful because it is largely restricted to one mutation type. However, the spectrum of mutations that can be obtained is biased in that mutants must maintain mitochondrial ribosome function to preserve respiration competence. Since erythromycin does not affect yeast viability, only respiration, yeast strains are resistant to erythromycin in fermentable media. This makes it impossible to select point mutations under non-respiring conditions using Ery^R^.

A spontaneous *cox3::arg8^m^* mutant, isolated by Fox and group and reported by Strand and Copeland, is an Arg⁻ revertant that was shown to contain two nucleotide substitutions and a +1 frameshift insertion mutation (Strand and Copeland, 2002). Deletion of a single nucleotide restores the proper reading frame and allows reversion to an Arg⁺ phenotype (Zhang et al., 2006). This mutant was proposed as a replacement for the Ery^R^ mutation assay, which measures true point mutations. While this reporter has some advantages over Ery^R^ acquisition, this approach fails to distinguish between different types of mitochondrial DNA mutation and is therefore less helpful at elucidating the mechanisms responsible for their generation. Our subsequent work has demonstrated that the proteins that impact mitochondrial nucleotide substitutions and mitochondrial insertion/deletion mutations are not always the same, allowing the dissection of pathways that differentially affect point and frameshift mutation as described below.

4. Measuring mitochondrial microsatellite instability

Short, repetitive sequences consisting of di- and tri-nucleotide repeats are abundant in the nuclear genome, in both coding and non-coding regions. Their appearance and inherent instability in the coding regions of several proteins is the underlying cause of the polyglutamine diseases, including Huntington's disease and multiple types of spinocerebellar ataxia. These microsatellites are an important source of mutation in nuclear DNA through the internal repetition that facilitates repeat length changes by polymerase slippage, and through the ability of separate repeated regions to undergo homologous recombination in *trans*.

Yeast mitochondrial DNA is highly microsatellite-enriched, partially due to the abundance (83%) of A and T. Mammalian mitochondrial DNA lacks most of the non-coding AT-rich DNA (~44% A/T) that gives rise to this bias, but still contains AT-rich repetitive regions (Anderson et al., 1980). If this repetition confers higher mitochondrial DNA instability, similar to its effects on nuclear DNA (Wierdl et al., 1997), it could play an important role in mitochondrial dysfunction as it relates to aging and aging-related diseases. These findings provided the impetus for making a mitochondrial microsatellite reporter system.

In 2000, Petes, Fox, and colleagues published an analysis of mitochondrial DNA microsatellite instability using the *ARG8m* reporter as a marker of frameshift mutation within a microsatellite repeat (Sia et al., 2000). This approach built on a previous reporter of yeast nuclear microsatellite instability (Henderson and Petes, 1992). They generated *cox3Δ::arg8m*-bearing plasmids that contained poly-GT and poly-AT tracts 15-17 repeats in length 5' to the *ARG8m* sequence, shifting the reading frame of the gene either 1 or 2 nucleotides out of frame (Fig 2B). Once incorporated and expressed, these *cox3Δ::arg8m*(G/A-T) constructs express a nonsense transcript. However, a frameshift mutation in the microsatellite tract that restores the correct reading frame restores a functional gene product, conferring an Arg+ phenotype. As a control, plasmids were also made bearing microsatellite inserts that did not result in an *ARG8m* frameshift, conferring a constitutive Arg+ phenotype, confirming that insertion of the repetitive sequence did not disrupt *ARG8m* function.

The experiments carried out with these strains were among the first to demonstrate fundamental differences between nuclear and mitochondrial DNA processing and maintenance. Unlike nuclear DNA, in which poly(AT) and poly(GT) tracts have similar levels of instability, with repeat addition favored, mitochondrial poly(AT) tracts are much more stable than poly(GT) tracts, and repeat deletion predominates. These and other differences suggest that assumptions of mitochondrial DNA behavior based on nuclear DNA may be inherently flawed, preventing a clear understanding of how to predictably manipulate mitochondrial DNA.

To allow microsatellite instability measurement in respiring cells, a state that imposes a respiration requirement on mitochondrial DNA maintenance, we developed and characterized a version of the microsatellite reporter that is respiration competent (Kalifa and Sia, 2007; Mookerjee and Sia, 2006). This ensures that the measured microsatellite instability occurs in an otherwise functional background. This new reporter serves two useful purposes, allowing us to determine the effect of an active respiratory chain on mitochondrial mutagenesis, and to assess microsatellite instability in mutant strains that only maintain mitochondrial DNA under constant respiratory selection. For this reporter, instead of replacing *COX3*, *ARG8m* was inserted with *COX2* flanking sequence upstream of the endogenous *COX2* locus (Fig. 2C). These untranslated flanking sequences ensure correct expression while avoiding disruption of cytochrome oxidase subunits. In this new genetic and functional context, poly(GT)$_{16}$ repeats in the +1 frame were approximately 10-fold more unstable than the original *cox3Δ::arg8m*(G/T)$_{16}$ (Kalifa and Sia, 2007). Further experiments will be required to determine whether this is due to the altered flanking sequence or respiration status of the cell.

5. Measuring direct repeat-mediated deletions

Accumulation of mitochondrial deletions is associated with multiple pathologies and with aging. These deletions are commonly flanked by direct repeats, raising speculation that they are recombination-mediated. A detailed understanding of mitochondrial DNA recombination has lagged far behind equivalent nuclear recombination processes, which in turn limits our efforts to use recombination both as a molecular tool for genome manipulation and as a biological correlate of mitochondrial dysfunction. By manipulating the sequence context of the *ARG8m* gene, we developed reporters to measure and characterize mitochondrial recombination in yeast. These studies have revealed detailed information about the requirements for and mechanisms of recombination in the yeast mitochondrial genome.

5.1 DFS188 *Rep96::ARG8^m::cox2'* reporter and variants

We generated a synthetic deletion substrate with the *ARG8m* gene fused in frame to the first 99 bp of the mitochondrial *COX2* gene. This construct was followed by the entire *COX2* coding sequence, lacking the 5' start codon, giving rise to *ARG8m* flanked by 96 bp of directly repeated *COX2* sequence (Fig. 2D). This deletion substrate expresses functional acetylornithine transaminase and confers an Arg⁺ phenotype. *COX2* is not expressed, as translation terminates with Arg8p. These cells are therefore non-respiring. However, deletion between the flanking 96-bp repeats excises *ARG8m* and restores a functional *COX2* sequence with the appropriate initiation and termination signals. Cells that have undergone either sufficient recombination events, or homoplasmic fixation of one or a few recombination events, display a phenotypic shift to respiration competence and arginine auxotrophy. The work that followed development of the *Rep96::ARG8m::cox2'* reporter used either this form or modified versions with changes to the flanking repeats.

Yeast bearing the *Rep96::ARG8m::cox2'* reporter can be assayed by imposing different nutrient conditions that select for either the original construct or the deleted one. Cells are initially grown on medium lacking arginine to maintain the original reporter. Individual colonies are then separately diluted and plated on glycerol medium (YPG) to select for deletion events, which are scored after 3 days. The median number of colonies appearing on YPG plates is used to estimate the mutation rate using the method of the median (Lea and Coulson, 1949), which incorporates statistical assumptions about the number of mutational events vs. colony number.

5.2 Characterizing direct-repeat mediated deletion (DRMD) in yeast

Work by Phadnis *et al.* (2005) demonstrates the utility of this reporter in characterizing mitochondrial recombination. First, generation of deletion reporters containing different repeat lengths revealed that the rate of *ARG8m* deletion is linearly dependent on repeat lengths between 33 and 96 bp; a 21-bp repeat did not facilitate a significant deletion rate. This establishes the minimal efficient processed sequence (MEPS) between 21 and 33 bp long. This MEPS is longer than the direct repeats flanking mitochondrial DNA deletions in mammalian systems, including the 13 bp flanking the 5-kb "common deletion" (Schon et al., 1989) and a 7-bp recombination-associated sequence (Myers et al., 2008). It should be noted that in humans, direct repeats mediate only some of the total detectable mitochondrial DNA deletions (Guo et al., 2010; Srivastava and Moraes, 2005).

Second, the effects of heterology on deletion efficiency were tested by introducing silent mutations into either the leading or following repeat (relative to the direction of transcription), giving rise to ~2% heterology between the repeat sequences. This design was meant to allow comparison to similar work in the yeast nuclear genome, where a 3% heterology between 205 bp repeats decreased the rate of deletion formation 6-fold (Sugawara et al., 2004) These experiments revealed similar behavior in mitochondrial DNA, where a 3- to 4-fold reduction in deletion formation rate was observed. Interestingly, this effect was dependent on mutation placement in the leading repeat; the same mutations in the following repeat had no effect on deletion rate. One explanation for this is that the heteroduplex rejection mechanisms may act more stringently on particular mispair orientations. In successful deletion events, the final sequence was almost always that of the repeat closest to the remaining COX2 sequence. These findings are likely to be related to the specific types of mechanisms that mediate DRMD, but at present are unexplained.

The possible mechanisms mediating repeat-dependent deletion can be partially distinguished based on the DNA products they generate, namely, whether detectable products are reciprocal or non-reciprocal. Unlike qPCR detection methods that are commonly used to measure mitochondrial DNA deletions *in vivo*, we used Southern blot analysis to examine the products of direct repeat dependent deletion. In theory, the reciprocal products, either a circular molecule or tandem duplication on another mitochondrial genome would both be detectable. We did not observe such species using this technique. However, we detected reciprocal products through PCR amplification and electrophoresis, indicating that reciprocal events do occur. This method cannot distinguish between circular "pop-outs" or tandem duplications. This analysis revealed that the majority of deletion events were non-reciprocal, suggesting a replication- or single strand annealing-based mechanism.

Third, genes believed to be involved in repeat-mediated deletion were tested to determine their effects on the rate of mutation. Mutations to the proof-reading domain of the mitochondrial DNA polymerase were shown to affect *Rep96::ARG8^m::cox2'* deletion. The *mip1-D347A* exonuclease proofreading mutation, which confers a approximately 500-fold increase in point mutation rates as measured by Ery^R, actually decreases direct repeat-mediated deletion nearly 5-fold (Phadnis et al., 2005). Interestingly, several years later Vermulst *et al.* (2008) determined that the analogous Poly mutation shifts deletion formation from a repeat-mediated to a non-repeat-mediated mechanism (Vermulst et al., 2008). That the analogous mutation appears to limit DRMD in both yeast and vertebrates suggests that at least some of the mechanisms for mitochondrial deletion may be conserved between these organisms.

5.3 Assaying heteroplasmy in yeast

The *Rep96::ARG8^m::cox2'* reporter, by virtue of its construction, can also be used as a phenotypic marker of mitochondrial DNA heteroplasmy. Unlike mammalian cells, yeast containing multiple different mitochondrial DNA types will purify this population to fix one type within 6-10 budding cycles (Dujon, 1981). Several factors are thought to be involved in this process of homoplasmic sorting. First, compaction of multiple, possibly clonal genome copies into nucleoids reduces the number of heritable units. Second, a limited number of nucleoids are transmitted to daughter cells, and may undergo regulated sorting, further restricting mitochondrial DNA inheritance (Spelbrink, 2009). In mammaliam systems, a third factor may be that nucleoids do not appear to readily mix, even when given the opportunity to do so by cytoduction (Schon and Gilkerson, 2010).

As stated earlier, the *Rep96::ARG8ᵐ::cox2'* reporter confers respiration incompetence and prototrophy. This phenotype is mutually exclusive with that of the deleted reporter, which confers respiration competence and arginine auxotrophy. Therefore, growing yeast under conditions that select for both respiration competence and arginine prototrophy, representing one phenotype of each possible state of the reporter, should select for yeast that maintain heteroplasmy. This was shown successfully in an analysis of point mutants of *MSH1* a yeast homolog of the MutS mismatch repair initiation protein (Mookerjee and Sia, 2006). In contrast to wild-type yeast, in which heteroplasmic maintenance occurred at a frequency on the order of 10^{-5}, *msh1* alleles permitted frequencies of heteroplasmy up to 0.25, 4 to 5 orders of magnitude greater. The role of Msh1p in homoplasmic sorting remains undetermined.

6. Elucidation of mitochondrial DNA repair pathways

With specific reporters of microsatellite instability, point mutation, and direct repeat instability, coupled with direct sequencing and Southern blotting, the pathways of mitochondrial DNA repair become more readily accessible to quantitative analysis. This section will discuss some of the research findings resulting from use of the mitochondrial reporters described above.

6.1 Mismatch recognition combines with recombination and base excision repair pathways

Use of the *Rep96::ARG8ᵐ::cox2'* reporter allowed the initial characterization of mitochondrial DNA recombination requirements, including repeat size, degree of sequence identity, and directional/positional repair bias as described in Section 5.2 and in Phadnis et al. (2005). We have further applied the reporters to determining the total complement of mitochondrial DNA repair mechanisms and their interactions

Mismatch repair has been a predicted pathway of mitochondrial DNA repair since the identification of the mitochondrially-localized MutS homolog, Msh1p. However, there is currently no direct evidence in yeast for mismatch repair activity. Human mitochondria do have a putative mismatch repair mechanism (De Souza-Pinto et al., 2009), but do not possess a MutS homolog. Point mutation accumulation rates increase with Msh1p disruption, but evidence from multiple groups suggests that this is due to base excision repair (BER), rather than mismatch repair (MMR), defects. Further, no other characterized mismatch repair proteins are known to localize to the mitochondria.

Haploid yeast strains with deletions of the *MSH1* gene cannot maintain wild-type mitochondrial DNA, generating ρ⁻ petites at a high frequency even in the presence of selection on a non-fermentable carbon source (Chi and Kolodner, 1994). To explore this problem in detail, we characterized the mutagenic consequences of three point mutations to *MSH1*. These mutations are analogous to well-studied mutations in *E. coli* MutS and yeast nuclear MutS homologs and were chosen based on the biochemical functions of the mutant proteins and their ability to maintain mitochondrial function when under selection. The *msh1-F105A* substitution lies in the conserved DNA binding domain, and is predicted by its homology to MutS and Msh6p to impair DNA binding and mismatch recognition (Bowers et al., 1999; Schofield et al., 2001). The *msh1-G776D* and *msh1-R813W* substitutions both lie within the highly conserved ATPase domain, although they are predicted to have different phenotypic and biochemical consequences. The yeast msh6p-G987D, analogous to our msh1p-G776D mutant, is significantly impaired in ATP-binding and displays an ability to bind mismatches, but is defective in further processing. The msh2p-R730W mutation, analogous to our msh1p-R813W, is able to bind ATP, but is defective in hydrolysis. While complexes containing this mutant form of Msh2p cannot perform mismatch repair, they remain at least partially functional for promoting deletions at directly repeated sequences (Kijas et al., 2003; Studamire et al., 1998).

By comparison with the known mutations, all three *msh1* mutations were predicted to result in loss of any mismatch repair activity. Consistent with this hypothesis, all three *msh1* mutants displayed increased point mutation rates. However, this increase was insufficient to explain the catastrophic loss of wild-type mitochondrial DNA and respiratory function (Mookerjee et al., 2005; Mookerjee and Sia, 2006), as mutations in the proof-reading domain of the mitochondrial replicative polymerase display more than ten-fold higher rates of point mutation, but can maintain ρ+ DNA (Foury and Vanderstraeten, 1992).

We then characterized the effects of *MSH1* mutation on microsatellite instability, hypothesizing that since nuclear mismatch repair disruption greatly increases nuclear microsatellite instability, a similar observation in mitochondria would support a bona fide mismatch repair function. Examination of *msh1* alleles with disruptions in the DNA-binding (*msh1-F105A*) and ATPase (*msh1-R813W*) domains revealed no significant changes in GT microsatellite instability, suggesting that Msh1p initiates non-mismatch repair mechanisms. It is formally possible that mitochondrial mismatch repair is not equivalent with respect to microsatellites between mitochondria and the nucleus due to other differences between the two compartments. Still, this finding prompted us to search for other repair activities involving Msh1p.

Mismatch repair proteins have been shown to function in other DNA repair pathways, including BER, nucleotide excision repair (NER), and homologous recombination (Goldfarb and Alani, 2005; Polosina and Cupples, 2010). Using the mitochondrial reporters, we were able to examine the genetic interaction of *MSH1* with putative recombination (Mookerjee and Sia, 2006) and BER components (Pogorzala et al., 2009).

Though widely accepted as a functional mechanism in mitochondrial DNA, the proteins that carry out recombination, and the specific mechanisms themselves, are largely unknown. Due to differences in the available proteins, in the substrate, or in the presumably constant availability of a homologous template, mitochondria may combine existing repair components is ways not seen in the nucleus (Masuda et al., 2009). We speculated that Msh1p, like its nuclear homologs, may play a role in the generation of deletions at directly-repeated sequences, and therefore would be predicted to result in reduced DRMD. Unexpectedly, we found that all three *msh1* mutations increase deletion rate approximately 100-fold, revealing a novel role for Msh1p function in mitochondrial recombination suppression that is not mirrored by any of its nuclear homologs.

Previously, examination of the deletion junctions of ρ- genomes in spontaneous *petite* strains had suggested that recombination utilizing repeated sequences may be the initiating event in their generation (Dujon, 1981). If so, all three of the *msh1* mutants would be predicted to give rise to similar, high levels of non-respiring cells. However, while the *msh1-*

F105A, msh1-G776D, and the *msh1-R813W* mutant strains all display high rates of point mutation and DRMD, the *msh1-R813W* mutant strain displays significantly lower rates of respiration loss than strains expressing the other two *MSH1* mutant alleles. This result calls into question several assumptions about how cells become ρ- *petite*. In these cells, respiration capability is lost with deletion of the majority of the mitochondrial genome, but it is clear that gene disruptions leading to increased rates of both point mutation accumulation and repeat-mediated deletion do not necessarily lead to loss of ρ+ mitochondrial DNA.

If Msh1p mutation allows it to bind but not release mismatched DNA, it is tempting to speculate that DNA molecules might become aberrantly linked. While not direct confirmation of this, we did find that the *msh1-G776D* and *msh1-R813W* alleles, and to a lesser extent, the *msh1-F105A* allele, conferred increases in the frequency of heteroplasmic maintenance of at least three or more orders of magnitude.

In addition to suppressing DRMD, Msh1p is also an important component of mitochondrial base excision repair. Through conventional epistasis analysis of *MSH1* and the BER genes *OGG1, NTG1,* and *APN1,* we found that Msh1p defects gave rise to different mutator phenotypes depending on the form of BER, and proposed that mismatch or lesion recognition contributes to short-patch BER, while DNA binding plays a stabilizing role during long-patch BER (Pogorzala et al., 2009).

6.2 Translesion polymerases facilitate frameshifts but suppress mitochondrial point mutation

The high fidelity of replicative DNA polymerases arises through extremely stringent requirements for nucleotide binding in the active site. While normally desirable, the inflexibility of these domains to accept alternate substrates causes replication fork stalling or collapse in the presence of cyclobutane-pyrimidine dimers, a common UV-induced product, and other damage resulting in large adducts.

To remedy this, a second class of polymerases exists that have much less stringent binding requirements for nucleotide incorporation. These translesion polymerases sacrifice fidelity for greater flexibility in template usage, favoring processivity but leading to mutagenic DNA synthesis that introduces both point and frameshift mutations. Consequently, their disruption gives rise to higher sensitivity to DNA damaging agents (e.g., UV) but lower nuclear DNA mutation rates in surviving cells.

Polγ is the only known mitochondrial replicative polymerase and, until recently, the only polymerase with known mitochondrial DNA activity. In yeast, *REV3*, encoding the catalytic subunit of the Polζ translesion polymerase, was previously implicated in mitochondrial DNA maintenance (Smolińska, 1987). Rev1p and Rev7p interact with Rev3p and are required for Rev3p-dependent synthesis opposite damaged templates (Lawrence et al., 2000). Subsequently, the Singh group observed that Rev1p, Rev3p, and Rev7p have N-terminal mitochondrial localization signals and demonstrated the ability of these N-terminal sequences to deliver GFP to mitochondria. They used the *cox3::arg8m* variant isolated by T. Fox to analyze Arg+ reversion as a measure of mutation, and found that single deletion of Rev1p, Rev3p, and Rev7p all result in a decrease in mutation frequency (Zhang et al., 2006). This reporter reverts to Arg+ via loss of a single base pair, indicating that Polζ is important in the generation of mitochondrial insertion/deletion (indel) mutations.

Using the respiring ($GT_{16}+1$) and ($GT_{16}+2$) reporters (Fig 2C), (Mookerjee and Sia, 2006), we found that single deletions of *REV1*, *REV3*, and *REV7* all led to decreases in both spontaneous and UV-induced frameshift consistent with previously published work and supporting a role for all three gene products in the generation of mitochondrial indels. However, unlike nuclear DNA, which displays decreases in both frameshifts and point mutations, the *rev3*-Δ and *rev7*-Δ strains also showed unexpected increases in spontaneous mitochondrial DNA point mutation rate, and all three deletions gave rise to increases in UV-induced point mutation rates. Sequence analysis of these mutants revealed a UV-dependent shift favoring A to T transversion in the absence of Rev1p, consistent with repair of a thymine dimer by a mechanism biased towards adenine insertion (Kalifa and Sia, 2007).

These observations emphasize the importance of examining the effect of disrupting repair or damage tolerance pathways on multiple types of mutations, as they are often generated and repaired via distinct mechanisms. The employment of multiple mutagenesis reporters allows the proper dissection of these pathways.

7. Conclusion

Increasingly, proteins previously considered to be nuclear DNA repair factors are found to also display mitochondrial localization (Section 1.3), suggesting significant overlap between nuclear and mitochondrial DNA repair pathways with respect to their protein components. However, our analysis of the phenotypic consequences of mutating the relevant genes reveals significant differences in the contribution of these proteins to mitochondrial mutagenesis and repair (Sections 5 and 6). These differences likely result from compositional DNA differences, the different packaging and DNA-binding proteins, the different regulatory control of mitochondrial DNA replication and transmission, the exposure to certain kinds of damage, and the availability of other repair proteins, between nuclear and mitochondrial DNA. We should not expect that studies of nuclear DNA repair can simply be extrapolated to generate correct mitochondrial models.

Careful analysis of these pathways within the mitochondrion will require tools like those we have described here. These reporters provide us with the ability to differentiate between point substitutions, frameshift mutations, and deletion events and will be critical to elucidating specific pathways. While *Saccharomyces cerevisiae* is currently the only model system available for these studies, the conservation of DNA repair pathways among eukaryotes supports the hypothesis that some, if not most of these repair proteins will have conserved mitochondrial roles. Finally, understanding mitochondrial DNA repair serves two practical purposes. First, it allows us to understand an important genetic system whose failure is heavily implicated in aging. Second, understanding the mechanisms of repair may facilitate their exploitation for mitochondrial genome manipulation in other systems.

8. Acknowledgements

Work described in this chapter was supported by the National Institutes of Health grant GM63626 and the National Science Foundation grants MCB0543084 and MCB0841857 to E. A. S.

9. References

Anderson, S., Bankier, A. T., Barrell, B. G., de Bruijn, M. H., Coulson, A. R., Drouin, J., Eperon, I. C., Nierlich, D. P., Roe, B. A., Sanger, F., Schreier, P. H., Smith, A. J., Staden, R., Young, I. G., 1980. Sequence and organization of the human mitochondrial genome. Nature. 290, 457-465.

Arnaudo, E., Dalakas, M., Shanske, S., Moraes, C. T., DiMauro, S., Schon, E. A., 1991. Depletion of muscle mitochondrial DNA in AIDS patients with zidovudine-induced myopathy Lancet. 337, 508–510.

Baker, B. M., Haynes, C. M., 2011. Mitochondrial protein quality control during biogenesis and aging. Trends Biochem Sci. 1-8.

Boesch, P., Weber-Lotfi, F., Ibrahim, N., Tarasenko, V., Cosset, A., Paulus, F., Lightowlers, R. N., Dietrich, A., 2010. DNA repair in organelles: Pathways, organization, regulation, relevance in disease and aging. Biochim Biophysica Acta. 1813, 186-200.

Bogenhagen, D. F., Wang, Y., Shen, E. L., Kobayashi, R., 2003. Protein components of mitochondrial DNA nucleoids in higher eukaryotes. Mol Cell Proteomics. 2, 1205-16.

Bonitz, S. G., Berlani, R., Coruzzi, G., Li, M., Macino, G., Nobrega, F. G., Nobrega, M. P., Thalenfeld, B. E., Tzagoloff, A., 1980. Codon recognition rules in yeast mitochondria. Proc Natl Acad Sci USA. 77, 3167-70.

Bonnefoy, N., Fox, T. D., 2000. *In vivo* analysis of mutated initiation codons in the mitochondrial *COX2* gene of *Saccharomyces cerevisiae* fused to the reporter gene *ARG8^m* reveals lack of downstream reinitiation. Mol Gen Genet. 262, 1036-46.

Bonnefoy, N., Fox, T. D., 2007. Directed alteration of *Saccharomyces cerevisiae* mitochondrial DNA by biolistic transformation and homologous recombination. Methods Mol Biol. 372, 153-66.

Bowers, J., Sokolsky, T., Quach, T., Alani, E., 1999. A mutation in the MSH6 subunit of the *Saccharomyces cerevisiae* MSH2-MSH6 complex disrupts mismatch recognition. J Biol Chem. 274, 16115-16125.

Canugovi, C., Maynard, S., Bayne, A.-C. V., Sykora, P., Tian, J., Souza-Pinto, N. C. d., Croteau, D. L., Bohr, V. A., 2010. The mitochondrial transcription factor A functions in mitochondrial base excision repair. DNA Repair. 9, 1080-1089.

Chang, S., 2000. The frequency of point mutations in mitochondrial DNA is elevated in the Alzheimer's brain. Biochem Biophys Res Comm. 273, 203-208.

Chatterjee, A., Singh, K. K., 2001. Uracil-DNA glycosylase-deficient yeast exhibit a mitochondrial mutator phenotype. Nucleic Acids Res. 29, 4935-4940.

Chen, H., Vermulst, M., Wang, Y. E., Chomyn, A., Prolla, T. A., Mccaffery, J. M., Chan, D. C., 2010. Mitochondrial fusion is required for mtDNA stability in skeletal muscle and tolerance of mtDNA mutations. Cell. 141, 280-9.

Chi, N. W., Kolodner, R. D., 1994. Purification and characterization of MSH1, a yeast mitochondrial protein that binds to DNA mismatches. J Biol Chem. 269, 29984-92.

Clayton, D. A., Doda, J. N., Friedberg, E. C., 1974. The absence of a pyrimidine dimer repair mechanism in mammalian mitochondria. Proc Natl Acad Sci USA 71, 2777-2781.

Coffey, G., Lakshmipathy, U., Campbell, C., 1999. Mammalian mitochondrial extracts possess DNA end-binding activity. Nucleic Acids Res. 27, 3348-3354.

Cohen, J. S., Fox, T. D., 2001. Expression of green fluorescent protein from a recoded gene inserted into *Saccharomyces cerevisiae* mitochondrial DNA. Mitochondrion. 1, 181-9.

Corral-Debrinski, M., Horton, T., Lott, M. T., Shoffner, J. M., McKee, A. C., Beal, M. F., Graham, B. H., Wallace, D. C., 1994. Marked changes in mitochondrial DNA deletion levels in Alzheimer brains. Genomics. 23, 471-6.

Coskun, P., Wyrembak, J., Schriner, S., Chen, H.-W., Marciniack, C., Laferla, F., Wallace, D. C., 2011. A mitochondrial etiology of Alzheimer and Parkinson disease. Biochim Biophys Acta. 1-33.

Cui, Z., Mason, T. L., 1989. A single nucleotide substitution at the *rib2* locus of the yeast mitochondrial gene for 21S rRNA confers resistance to erythromycin and cold-sensitive ribosome assembly. Curr Genet. 16, 273-279.

De Souza-Pinto, N. C., Mason, P. A., Hashiguchi, K., Weissman, L., Tian, J., Guay, D., Lebel, M., Stevnsner, T. V., Rasmussen, L. J., Bohr, V. A., 2009. Novel DNA mismatch-repair activity involving YB-1 in human mitochondria. DNA Repair. 1-16.

Ding, M. G., Butler, C. A., Saracco, S. A., Fox, T. D., Godard, F., di Rago, J.-P., Trumpower, B. L., 2008. Introduction of cytochrome b mutations in *Saccharomyces cerevisiae* by a method that allows selection for both functional and non-functional cytochrome b proteins. Biochim Biophys Acta. 1777, 1147-56.

Dmitrieva, N. I., Malide, D., Burg, M. B., 2011. Mre11 is expressed in mammalian mitochondria where it binds to mitochondrial DNA. Am J Phys. 301, R632-40.

Dujon, B., 1981. Mitochondrial genetics and functions. In: J. N. Strathern, et al., Eds.), Molecular biology of the yeast *Saccharomyces*: Life Cycle and Inheritance. Cold Spring Harbor Laboratory Press, Cold Spring Harbor, New York.

Dunn, D. A., Cannon, M. V., Irwin, M. H., Pinkert, C. A., 2011. Animal models of human mitochondrial DNA mutations. Biochim Biophys Acta. 1-30.

Dunstan, H. M., Green-Willms, N. S., Fox, T. D., 1997. In vivo analysis of *Saccharomyces cerevisiae COX2* mRNA 5'-untranslated leader functions in mitochondrial translation initiation and translational activation. Genetics. 147, 87-100.

Foury, F., Roganti, T., Lecrenier, N., Purnelle, B., 1998. The complete sequence of the mitochondrial genome of *Saccharomyces cerevisiae*. FEBS Lett. 440, 325-31.

Foury, F., Vanderstraeten, S., 1992. Yeast mitochondrial DNA mutators with deficient proofreading exonucleolytic activity. EMBO J. 11, 2717-26.

Garrido, N., Griparic, L., Jokitalo, E., Wartiovaara, J., van der Bliek, A. M., Spelbrink, J. N., 2003. Composition and dynamics of human mitochondrial nucleoids. Mol Bio Cell. 14, 1583-1596.

Gilkerson, R. W., 2009. Mitochondrial DNA nucleoids determine mitochondrial genetics and dysfunction. Int J Biochem Cell Bio. 41, 1899-906.

Goldfarb, T., Alani, E., 2005. Distinct roles for the *Saccharomyces cerevisiae* mismatch repair proteins in heteroduplex rejection, mismatch repair and nonhomologous tail removal. Genetics. 169, 563-74.

Green, G., MacQuillan, A., 1980. Photorepair of ultraviolet-induced petite mutational damage in *Saccharomyces cerevisiae* requires the product of the *PHR1* gene. J Bact. 144, 826-829.

Griffiths, L. M., Swartzlander, D., Meadows, K. L., Wilkinson, K. D., Corbett, A. H., Doetsch, P. W., 2009. Dynamic compartmentalization of base excision repair proteins in response to nuclear and mitochondrial oxidative stress. Mol Cell Biol. 29, 794-807.

Guo, X., Kudryavtseva, E., Bodyak, N., Nicholas, A., Dombrovsky, I., Yang, D., Kraytsberg, Y., Simon, D. K., Khrapko, K., 2010. Mitochondrial DNA deletions in mice in men: substantia nigra is much less affected in the mouse. Biochim Biophys Acta. 1-11.

Haslbrunner, E., Tuppy, H., Schatz, G., 1964. Deoxyribonucleic acid associated with yeast mitochondria. Biochem Biophys Res Comm. 15, 127-132.

He, S., Fox, T. D., 1997. Membrane translocation of mitochondrially coded Cox2p: distinct requirements for export of N and C termini and dependence on the conserved protein Oxa1p. Mol Bio Cell 8, 1449-1460.

Henderson, S. T., Petes, T. D., 1992. Instability of simple sequence DNA in *Saccharomyces cerevisiae*. Mol Cell Biol. 12, 2749-57.

Ibeas, J. I., Jimenez, J., 1997. Mitochondrial DNA loss caused by ethanol in *Saccharomyces* flor yeasts. Appl Envt Microbiol. 63, 7-12.

Johnston, S. A., Anziano, P. Q., Shark, K., Sanford, J. C., Butow, R. A., 1988. Mitochondrial transformation in yeast by bombardment with microprojectiles. Science. 240, 1538-1541.

Kalifa, L., Sia, E. A., 2007. Analysis of Rev1p and Pol zeta in mitochondrial mutagenesis suggests an alternative pathway of damage tolerance. DNA Repair 6, 1732-9.

Kamenisch, Y., Fousteri, M., Knoch, J., von Thaler, A.-K., Fehrenbacher, B., Kato, H., Becker, T., Dollé, M. E., Kuiper, R., Majora, M., Schaller, M., van der Horst, G. T., van Steeg, H., Röcken, M., Rapaport, D., Krutmann, J., Mullenders, L. H., Berneburg, M., 2010. Proteins of nucleotide and base excision repair pathways interact in mitochondria to protect from loss of subcutaneous fat, a hallmark of aging. J Exp Med. 207, 379-90.

Kang, D., Nishida, J., Iyama, A., Nakabeppu, Y., Furuichi, M., Fujiwara, T., Sekiguchi, M., Takeshige, K., 1995. Intracellular localization of 8-oxo-dGTPase in human cells, with special reference to the role of the enzyme in mitochondria. J Biol Chem. 270, 14659-14665.

Kaufman, B. A., Durisic, N., Mativetsky, J. M., Costantino, S., Hancock, M. A., Grutter, P., Shoubridge, E. A., 2007. The mitochondrial transcription factor TFAM coordinates the assembly of multiple DNA molecules into nucleoid-like structures. Mol Biol Cell. 18, 3225-36.

Kienhöfer, J., Häussler, D. J. F., Ruckelshausen, F., Muessig, E., Weber, K., Pimentel, D., Ullrich, V., Bürkle, A., Bachschmid, M. M., 2009. Association of mitochondrial antioxidant enzymes with mitochondrial DNA as integral nucleoid constituents. FASEB J. 23, 2034-44.

Kijas, A. W., Studamire, B., Alani, E. A., 2003. Msh2 separation of function mutations confer defects in the initiation steps of mismatch repair. J Mol Bio. 331, 128-138.

Kukat, C., Wurm, C. A., Spåhr, H., Falkenberg, M., Larsson, N.-G., Jakobs, S., 2011. Super-resolution microscopy reveals that mammalian mitochondrial nucleoids have a uniform size and frequently contain a single copy of mtDNA. Proc Natl Acad Sci USA. 108, 13534-9.

Lawrence, C. W., Gibbs, P. E., Murante, R. S., Wang, X. D., Li, Z., McManus, T. P., McGregor, W. G., Nelson, J. R., Hinkle, D. C., Maher, V. M., 2000. Roles of DNA polymerase zeta and Rev1 protein in eukaryotic mutagenesis and translesion replication Cold Spring Harb Symp Quant Biol. 65, 61-69.

Lea, D. E., Coulson, C. A., 1949. The distribution of the number of mutants in bacterial populations. J Genet. 49, 264-285.

LeDoux, S. P., Wilson, G. L., Beecham, E. J., Stevnsner, T., Wassermann, K., Bohr, V. A., 1992. Repair of mitochondrial DNA after various types of DNA damage in Chinese hamster ovary cells. Carcinogenesis. 13, 1967-73.

Legros, F., Malka, F., Frachon, P., Lombès, A., Rojo, M., 2004. Organization and dynamics of human mitochondrial DNA. J Cell Sci. 117, 2653-62.

Liu, P., Qian, L., Sung, J. S., De Souza-Pinto, N. C., Zheng, L., Bogenhagen, D. F., Bohr, V. A., Wilson, D. M., Shen, B., Demple, B., 2008. Removal of oxidative DNA damage via FEN1-dependent long-patch base excision repair in human cell mitochondria. Mol Cell Biol. 44.

Luria, S. E., Delbruck, M., 1943. Mutations of bacteria from virus sensitivity to virus resistance. Genetics 28, 491-511.

Margineantu, D. H., Gregory Cox, W., Sundell, L., Sherwood, S. W., Beechem, J. M., Capaldi, R. A., 2002. Cell cycle dependent morphology changes and associated mitochondrial DNA redistribution in mitochondria of human cell lines. Mitochondrion. 1, 425-435.

Masuda, T., Ito, Y., Terada, T., Shibata, T., Mikawa, T., 2009. A non-canonical DNA structure enables homologous recombination in various genetic systems. J Biological Chem. 284, 30230-9.

Meyer, R. R., Simpson, M. V., 1969 DNA biosynthesis in mitochondria. Differential inhibition of mitochondrial and nuclear DNA polymerases by the mutagenic dyes ethidium bromide and acriflavin. Biochem Biophys Res Comm 34, 238-244.

Mileshina, D., Ibrahim, N., Boesch, P., Lightowlers, R. N., Dietrich, A., Weber-Lotfi, F., 2011a. Mitochondrial transfection for studying organellar DNA repair, genome maintenance and aging. Mech Ageing Dev. 1-12.

Mileshina, D., Koulintchenko, M., Konstantinov, Y., Dietrich, A., 2011b. Transfection of plant mitochondria and *in organello* gene integration. Nucleic Acids Res. 10.1093/nar/gkr517.

Mookerjee, S. A., Lyon, H. D., Sia, E. A., 2005. Analysis of the functional domains of the mismatch repair homologue Msh1p and its role in mitochondrial genome maintenance. Curr Genet. 47, 84-99.

Mookerjee, S. A., Sia, E. A., 2006. Overlapping contributions of Msh1p and putative recombination proteins Cce1p, Din7p, and Mhr1p in large-scale recombination and genome sorting events in the mitochondrial genome of Saccharomyces cerevisiae. Mut Res. 595, 91-106.

Myers, S., Freeman, C., Auton, A., Donnelly, P., Mcvean, G., 2008. A common sequence motif associated with recombination hot spots and genome instability in humans. Nat Genetics. 40, 1124-1129.

Nakabeppu, Y., Kajitani, K., Sakamoto, K., Yamaguchi, H., Tsuchimoto, D., 2006. MTH1, an oxidized purine nucleoside triphosphatase, prevents the cytotoxicity and neurotoxicity of oxidized purine nucleotides. DNA Repair. 5, 761-772.

Narendra, D. P., Jin, S. M., Tanaka, A., Suen, D.-F., Gautier, C. A., Shen, J., Cookson, M. R., Youle, R. J., 2010. PINK1 is selectively stabilized on impaired mitochondria to activate Parkin. PLoS Bio. 8, e1000298.

Narendra, D. P., Tanaka, A., Suen, D.-F., Youle, R. J., 2008. Parkin is recruited selectively to impaired mitochondria and promotes their autophagy. J Cell Bio. 183, 795-803.

Nass, M. M., Nass, S., 1963. Intramitochondrial fibers with DNA characteriztics. I. Fixation and electron staining reactions. J Cell Bio. 19, 593-611.

Nilsen, H., Otterlei, M., Haug, T., Solum, K., Nagelhus, T. A., Skorpen, F., Krokan, H. E., 1997. Nuclear and mitochondrial uracil-DNA glycosylases are generated by alternative splicing and transcription from different positions in the UNG gene. Nucleic Acids Res. 25, 750-755.

Nishioka, K., Ohtsubo, T., Oda, H., Fujiwara, T., Kang, D., Sugimachi, K., Nakabeppu, Y., 1999. Expression and differential intracellular localization of two major forms of human 8-oxoguanine DNA glycosylase encoded by alternatively spliced OGG1 mRNAs. Mol Bio Cell. 10, 1637-1652.

Phadnis, N., Sia, R. A., Sia, E. A., 2005. Analysis of repeat-mediated deletions in the mitochondrial genome of *Saccharomyces cerevisiae*. Genetics. 171, 1549-59.

Pogorzala, L., Mookerjee, S., Sia, E. A., 2009. Evidence that msh1p plays multiple roles in mitochondrial base excision repair. Genetics. 182, 699-709.

Pohjoismaki, J. L. O., Wanrooij, S., Hyvarinen, A. K., Goffart, S., Holt, I. J., Spelbrink, J. N., Jacobs, H. T., 2006. Alterations to the expression level of mitochondrial transcription factor A, TFAM, modify the mode of mitochondrial DNA replication in cultured human cells. Nucleic Acids Res. 34, 5815-5828.

Polosina, Y. Y., Cupples, C. G., 2010. Wot the 'L-Does MutL do? Mut Res. 705, 228-38.

Ramotar, D., Kim, C., Lillis, R., Demple, B., 1993. Intracellular localization of the Apn1 DNA repair enzyme of Saccharomyces cerevisiae: Nuclear transport signals and biological role. J Biol Chem. 268, 20533-9.

Randolph-Anderson, B. L., Boynton, J. E., Gillham, N. W., Harris, E. H., Johnson, A. M., Dorthu, M. P., Matagne, R. F., 1993. Further characterization of the respiratory deficient dum-1 mutation of Chlamydomonas reinhardtii and its use as a recipient for mitochondrial transformation. Mol Gen Genet. 236, 235-44.

Sage, J. M., Gildemeister, O. S., Knight, K. L., 2010. Discovery of a novel function for human Rad51: maintenance of the mitochondrial genome J Biol Chem. 285, 18984-18990.

Sanford, J. C., 1988. The biolistic process. Trends Biotech. 6, 299-302.

Sanford, J. C., Klein, T. M., Wolf, E. D., Allen, N., 1987. Delivery of substances into cells and tissues using a particle bombardment process. J Particulate Sci Tech. 5, 27-37.

Sasaki, N., Fukatsu, R., Tsuzuki, K., Hayashi, Y., Yoshida, T., Fujii, N., Koike, T., Wakayama, I., Yanagihara, R., Garruto, R., Amano, N., Makita, Z., 1998. Advanced glycation end products in Alzheimer's disease and other neurodegenerative diseases. Am J Path. 153, 1149-55.

Schofield, M. J., Brownewell, F. J., Nayak, S., Du, C., Kool, E. T., Hsieh, P., 2001. The Phe-X-Glu DNA binding motif of MutS. J Biol Chem. 276, 45505-45508.

Schon, E. A., Gilkerson, R. W., 2010. Functional complementation of mitochondrial DNAs: Mobilizing mitochondrial genetics against dysfunction. Biochim Biophys Acta. 1800, 245-249.

Schon, E. A., Rizzuto, R., Moraes, C. T., Nakase, H., Zeviani, M., DiMauro, S., 1989. A direct repeat is a hotspot for large-scale deletion of human mitochondrial DNA. Science. 244, 346-349.

Shokolenko, I., Venediktova, N., Bochkareva, A., Wilson, G. L., Alexeyev, M. F., 2009. Oxidative stress induces degradation of mitochondrial DNA. Nucleic Acids Res. 37, 2539-2548.

Sia, E. A., Butler, C. A., Dominska, M., Greenwell, P., Fox, T. D., Petes, T. D., 2000. Analysis of microsatellite mutations in the mitochondrial DNA of Saccharomyces cerevisiae. Proc Natl Acad Sci USA. 97, 250-5.

Singh, G., Maniccia-Bozzo, E., 1990. Evidence for lack of mitochondrial DNA repair following cis-dichlorodiammineplatinum treatment. Cancer Chemother Pharmacol. 26, 97-100.

Singh, K. K., Sigala, B., Sikder, H. A., Schwimmer, C., 2001. Inactivation of Saccharomyces cerevisiae OGG1 DNA repair gene leads to an increased frequenncy of mitochondrial mutants. Nucleic Acids Res. 29, 1381-1388.

Smolińska, U., 1987. Mitochondrial mutagenesis in yeast: mutagenic specificity of EMS and the effects of RAD9 and REV3 gene products. Mut Res. 179, 167-74.

Spelbrink, J. N., 2009. Functional organization of mammalian mitochondrial DNA in nucleoids: History, recent developments, and future challenges. IUBMB Life. 62, 19-32.

Srivastava, S., Moraes, C. T., 2005. Double-strand breaks of mouse muscle mtDNA promote large deletions similar to multiple mtDNA deletions in humans. Hum Mol Gen. 14, 893-902.

Steele, D. F., Butler, C. A., Fox, T. D., 1996. Expression of a recoded nuclear gene inserted into yeast mitochondrial DNA is limited by mRNA-specific translational activation. Proc Natl Acad Sci USA. 93, 5253-7.

Strand, M. K., Copeland, W. C., 2002. Measuring mtDNA mutation rates in *Saccharomyces cerevisiae* using the mtArg8 assay. Methods Mol Biol. 197, 151-7.

Strauss, M., Hofhaus, G., Schröder, R. R., Kühlbrandt, W., 2008. Dimer ribbons of ATP synthase shape the inner mitochondrial membrane. EMBO J. 27, 1154-60.

Stuart, J., Brown, M., 2006. Mitochondrial DNA maintenance and bioenergetics. Biochimica et Biophysica Acta 1757, 79-89.

Studamire, B., Quach, T., Alani, E., 1998. *Saccharomyces cerevisiae* Msh2p and Msh6p ATPase activities are both required during mismatch repair. Mol Cell Bio. 18, 7590-7601.

Suen, D.-F., Narendra, D. P., Tanaka, A., Manfredi, G., Youle, R. J., 2010. Parkin overexpression selects against a deleterious mtDNA mutation in heteroplasmic cybrid cells. Proc Natl Acad Sci USA. 107, 11835-40.

Sugawara, N., Goldfarb, T., Studamire, B., Alani, E. A., Haber, J. E., 2004. Heteroduplex rejection during single-strand annealing requires Sgs1 helicase and mismatch repair proteins Msh2 and Msh6 but not Pms1. Proc Natl Acad Sci USA. 101, 9315-9320.

Swartzlander, D. B., Griffiths, L. M., Lee, J., Degtyareva, N. P., Doetsch, P. W., Corbett, A. H., 2010. Regulation of base excision repair: Ntg1 nuclear and mitochondrial dynamic localization in response to genotoxic stress. Nucleic Acids Res. 38, 3963-3974.

Thyagarajan, B., Padua, R. A., Campbell, C., 1996. Mammalian mitochondria possess homologous DNA recombination activity. J Biol Chem. 271, 27536-43.

Torello, A. T., Overholtzer, M. H., Cameron, V. L., Bonnefoy, N., Fox, T. D., 1997. Deletion of the leader peptide of the mitochondrially encoded precursor of *Saccharomyces cerevisiae* cytochrome c oxidase subunit II. Genetics. 145, 903-10.

Vermulst, M., Wanagat, J., Kujoth, G. C., Bielas, J. H., Rabinovitch, P. S., Prolla, T. A., Loeb, L. A., 2008. DNA deletions and clonal mutations drive premature aging in mitochondrial mutator mice. Nat Gen. 40, 392-4.

Wierdl, M., Dominska, M., Petes, T. D., 1997. Microsatellite instability in yeast: dependence on the length of the microsatellite. Genetics. 146, 769-79.

Yasui, A., Yajima, H., Kobayashi, T., Eker, A. P., Oikawa, A., 1992. Mitochondrial DNA repair by photolyase. Mut Res. 273, 231-236.

Yoon, Y. G., 2005. Transformation of isolated mammalian mitochondria by bacterial conjugation. Nucleic Acids Res. 33, e139-e139.

Yoon, Y. G., Koob, M. D., 2011. Toward genetic transformation of mitochondria in mammalian cells using a recoded drug-resistant selection marker. J Genet Genomics. 38, 173-179.

You, H. J., Swanson, R. L., Harrington, C., Corbett, A. H., Jinks-Robertson, S., Senturker, S., Wallace, S. S., Boiteux, S., Dizdaroglu, M., Doetsch, P. W., 1999. *Saccharomyces cerevisiae* Ntg1p and Ntg2p: broad specificity N-glycosylases for the repair of oxidative DNA damage in the nucleus and mitochondria. Biochemistry. 38, 11298-11306.

Zhang, H., Chatterjee, A., Singh, K. K., 2006. *Saccharomyces cerevisiae* polymerase zeta functions in mitochondria. Genetics. 172, 2683-8.

Zhou, J., Liu, L., Chen, J., 2010. Mitochondrial DNA Heteroplasmy in *Candida glabrata* after Mitochondrial Transformation. Eukaryotic Cell. 9, 806-814.

9

Site-Directed and Random Insertional Mutagenesis in Medically Important Fungi

Joy Sturtevant
LSUHSC School of Medicine
USA

1. Introduction

Site-directed and random mutagenesis have been useful tools in molecular biology. The application of directed mutagenesis in medically important fungi has been limited by the availability of molecular genetic techniques. Even species in which efficient genetic transformation methodologies exist, mutagenesis approaches were sparsely used due to diploidism. Lack of genetic tools hindered understanding of virulence mechanisms of medically important fungi. With the arrival of whole-genome sequencing, as well as improved techniques of genetic manipulation, the ability to address these questions is improving. A comprehensive review of mutagenesis in pathogenic fungi is outside the scope of this review, so not all studies were included. The intent of this review is to educate the reader on applications of site-directed and random insertional mutagenesis in medically important fungi in order to provide ideas for novel approaches to address major issues in pathogenic fungal research.

2. Site-directed mutagenesis

Site-directed mutagenesis has been exploited to understand signaling pathways, mechanisms of drug resistance, and identification of promoter DNA binding sites. Applications used less frequently have included protein localization and function of specific genes. In most instances, the site of the mutation was selected due to homology to model species or mammalian genes.

2.1 Signaling pathways

The most commonly reported application of site-directed mutants is the construction of dominant-negative and dominant-active alleles. The ability to make dominant-active alleles is particularly useful in diploid strains, since both endogenous alleles do not have to be disrupted. The amino acids chosen for mutation are often based on homology to *Saccharomyces cerevisiae* or *Aspergillus nidulans*. Although the roles of genes in signaling pathways were identified in model fungi, the regulation and downstream effects of these pathways are often very different in medically important fungi.

2.1.1 Phosphomimetics

The introduction of an amino acid substitution so the residue acts as constitutively phosphorylated or non-phosphorylated is a common technique to study cellular processes. Phosphomimetics have been used to study MAPK, cAMP-PKA, calcineurin, and two-component signaling, as well as cytokinesis and the heat shock response, in medically important fungi (Bockmühl and Ernst, 2001; Fox and Heitman, 2005; Hicks et al., 2005; Li et al., 2008; Menon et al., 2006; Nicholls et al., 2011).

The cAMP-PKA pathway regulates multiple cellular processes in eukaryotes. cAMP levels are regulated by phosphodiesterases (*PDE1, PDE2*), which in turn are regulated by protein kinase A (*PKA*) in some species. In *Cryptococcus neoformans*, the cAMP pathway is involved in multiple cellular processes, including virulence factor expression (melanin and capsule formation) (Hicks et al., 2005). Since the role in cAMP degradation and regulation by *PKA* of *PDE1* and *PDE2* differ among species, the goal of this study was to learn the functions of the PDEs in *C. neoformans*. In order to identify if *PDE1* was regulated by *PKA* through phosphorylation, a site-directed mutation in the *PDE1* at a putative *PKA* phosphorylation site was introduced based on work in *Saccharomyces*. Site-directed mutagenesis was performed by overlap PCR (Section 2.7), and the product was ligated into a *C. neoformans* transformation vector. The predicted outcome was an inactive *PDE1* and, thus, increased activation of the PKA pathway. In this way, the use of site-directed mutagenesis validated that PKA directly regulated the activation of *PDE1* in *C. neoformans* (Hicks et al., 2005).

Putative phosphorylated residues have not always been identified previously in model fungi. Consequently, *in silico* analysis can be utilized to identify putative phosphorylation sites (Bockmühl and Ernst, 2001; Li et al., 2008). *In silico* analysis predicted certain threonine residues as phosphorylation sites in the *Candida albicans* APSES protein Efg1p. These sites were mutated. Phenotypic analysis demonstrated that the mutations differentially affected morphogenesis, an important virulence attribute of *C. albicans* (Bockmühl and Ernst, 2001).

Unlike previous studies, target residues in the two-component response regulator, Ssk1p, were identified by sequence comparison to a bacterial response regulator (Menon et al., 2006). Invariant aspartic acid residues were substituted using site-directed mutagenesis. This study demonstrated that phosphorylation of two different residues affects regulation of different cellular processes involved in virulence (Menon et al., 2006).

2.1.2 G protein signaling

Another common application for site-directed mutagenesis has been G protein signaling. In *Aspergillus fumigatus,* asexual sporulation results in release of spores that are inhaled by man, which can lead to serious manifestations. In the non-pathogen, *Aspergillus nidulans*, it was known that G protein signaling pathways were responsible for both vegetative growth and conidiation. Activation of *flbA* is required for conidiation, and this was probably through the activation of the GTPase activity of the G alpha protein FadA. Mah et al. (2006) used this framework to determine if similar regulation occurred in the pathogen *A. fumigatus* (Mah and Yu, 2006). They were able to confirm that Af*flb* regulated the G protein signaling through GpaA (homolog of FadA). Gene disruption and random chemical mutagenesis confirmed the role of Af*flb* in conidiation. Dominant-active and dominant-negative mutant alleles of *gpaA* (made by overlap PCR) demonstrated that it is a downstream target in this pathway. Interestingly conidiation appeared even in the absence of *Afflb* (Mah and Yu, 2006), which may aid in dissemination.

As in *Aspergillus*, G protein signaling is also responsible for cellular differentiation in *C. albicans*. Much of the initial molecular dissection of the signaling pathways involved in morphogenesis was deciphered by constructing dominant-active and dominant-negative alleles of the G signaling proteins. Site-specific mutations were introduced in *CDC42, RAC1, GPA2, RAS1* and *RAS2* based on homology to *Saccharomyces* and mammalian G proteins (Bassilana and Arkowitz, 2006; Feng et al., 1999; Sanchez-Martinez and Perez-Martin, 2002; vandenBerg et al., 2004). The mutant alleles were introduced into exogenous loci under the expression of constitutive or regulatable promoters (Bassilana and Arkowitz, 2006; Feng et al., 1999; Sanchez-Martinez and Perez-Martin, 2002). However, since *CDC42* is essential, the mutated alleles were introduced at the endogenous locus in a *CDC42/cdc42* heterozygote (VandenBerg et al., 2004). These studies demonstrated the existence of different hyphal induction pathways, cross-talk between the MAPK and cAMP pathways, and distinction between growth and morphogenesis.

2.2 Mechanisms of drug resistance

Many fungal species present antifungal drug resistance *in vivo*. Studies have enhanced our understanding of this resistance. In most species, drug resistance is due to increased expression of export channels and/or mutations in target genes of the antifungal agent. Further studies confirmed the importance of the mutations.

The azoles interfere with ergosterol biosynthesis by targeting lanosterol 14 alpha-demethylase (*ERG11, CYP51*). Mutations in the gene resulted in reduced binding by the azole compound. Mutational hotspots were identified by sequencing the gene of interest from fungal strains isolated from patients or strains that have been passaged in the presence of the drug *in vitro*. It was then necessary to confirm that the mutation correlated with reduced susceptibility to the drug. Site-directed mutagenesis is an ideal method for validation. Due to homology of the *ERG11* gene among fungal species, many studies were performed in the more genetically malleable yeasts, *S. cerevisiae* or *Pichea pastoris* (Alvarez-Rueda et al., 2011). These studies confirmed that the mutated expressed protein is more or less susceptible to drug, but they do not definitively prove that the mutation was the reason for the clinical resistance. With the advent of genetic transformation techniques in medically important fungi, it is now possible to perform these experiments in the appropriate fungal host.

The most studied gene is *ERG11* in *C. albicans*, reviewed in Morio et al. (2010). Over 144 amino acid substitutions have been identified. It is less clear how many of these contribute to *in vivo* resistance. Additionally, some mutations may result from *in vitro* manipulations. A recent screen of azole-susceptible and resistant clinical isolates demonstrated that mutations are associated with both susceptibility and reduced susceptibility. Only 18% of isolates had no polymorphisms (Morio et al., 2010). These results highlighted the need to confirm that a specific mutation correlated with acquisition of resistance. In *C. albicans*, this has been approached by site-directed mutagenesis. Initial studies cloned the *ERG11* open reading frame (ORF) in a plasmid and then used PCR to introduce site-directed mutations. The PCR products were ligated into a *S. cerevisiae* expression vector. *S. cerevisiae* (azole susceptible strain) was transformed with plasmids containing the mutated *C. albicans ERG11* gene and tested in a series of assays for reduced azole susceptibility. In this manner, azole resistance was correlated with specific amino acid substitutions (Kakeya et al., 2000; Lamb et al., 2000; Lamb et al., 1997; Sanglard et al., 1998; Sheng et al., 2010). Direct mutagenesis of the *ERG11* gene in *C. albicans* has not been reported. However, direct mutagenesis of the azole-target gene *cyp51A*, was performed in *A. fumigatus* (Mellado et al., 2007; Snelders et al., 2011). Two approaches were used. In the first study, mutated sequences were amplified by PCR from clinical isolates that demonstrated reduced susceptibility to itraconazole (Mellado et al., 2007). Previously it was known that itraconazole resistance correlated with specific amino acid mutations at G54 and M220 (See Table 1 in the chapter by Figurski et al. for the amino acid codes). However, this study identified a new mutation site (L98H) in conjunction with a duplication in the promoter sequence. In order to confirm the importance of these mutations, an azole-susceptible strain was transformed with the mutated allele; and transformants were plated on itraconazole (Mellado et al., 2007). Although the importance of the mutation sites were confirmed, the transformation selection criteria were not efficient. A second study expanded upon this approach and used 3-D modeling to determine a mechanistic reason for the azole resistance conferred by the mutations (Snelders et al., 2011). Specific amino acids were substituted in the *cyp51A* using the QuickChange XL Site-Directed Mutagenesis Kit (Stratagene) (Section 2.7). The appropriate PCR products were cloned into a vector that contained a hygromycin resistance marker and flanking sequences for introduction into the endogenous *cyp51A* site. Therefore, positive transformants were selected on hygromycin and further tested for azole resistance/susceptibility phenotypes. Consequently, the inclusion of a dominant selective marker improved the efficiency of screening transformants (Snelders et al., 2011). To further expand identification of potential mutation sites, the same experimental approach could be used by subjecting the *cyp51A* gene to random PCR-directed mutagenesis (Palmer and Sturtevant, 2004) and thereby identify new mutation sites that confer altered susceptibility.

2.3 Promoter response elements

Mutagenesis is a common approach to identify DNA binding sites in promoters. Nested deletions are probably the most commonly reported method used in the medically important fungi. Site-directed mutagenesis has been used to introduce point mutations in the *hapB* promoter in *A. nidulans*; so this approach may be used in *A. fumigatus* in the future (Brakhage and Langfelder, 2002). Site-directed mutagenesis has been used to identify putative promoter elements in chitin synthases (*CHS2, CHS8*) in *C. albicans* (Lenardon et al., 2009). The significance of this study is that chitin synthases are up-regulated in response to cell wall stress and thus are important for fungal survival. In this study, the promoters of *CHS2* and *CHS8* were mutated by site-directed mutagenesis or nested deletions. The selected sites were chosen due to previous studies or by *in silico* analysis. The mutated *Candida* promoter sequences were ligated upstream of the *Streptococcus thremophilus lacZ* gene. If the mutated site in the promoter element were important for a specific cell wall stress, *lacZ* would not be induced; and colonies would be white instead of blue on X-gal-containing medium. (X-gal is 5-bromo-4-chloro-indolyl-β-D-galactopyranoside.) The chosen potential sites reflected induction of several pathways and included known binding motifs. Mutations of individual promoter elements selected by *in silico* analysis had no effect on expression. Consequently, they performed nested deletions using exonuclease III digestion of restriction-digested plasmids containing the CHS2 and CHS8 genes. These mutated promoters were introduced into *Candida* by transformation, and induction of the *lacZ* reporter was assayed after stresses. In this manner they were able to identify regions, but not specific regulatory elements, in the CHS2 and CHS8 promoters that responded to cell wall stressors (Lenardon et al., 2009). In order to determine which signaling pathways acted upon these promoters, the mutated CHS2 and CHS8 constructs were introduced into C. *albicans* signaling pathway deletion mutants. These studies demonstrated that the cell wall integrity, calcineurin, and HOG (osmotic sensing) pathways mediated expression through the CHS2 promoter; only the cell wall integrity pathway affected CHS8-mediated expression (Lenardon et al., 2009). A pitfall of the deletion method is that it does not identify exact residues. It is possible to lose structural consistency, and it may delete other regulatory elements. In order to identify appropriate binding sites, random mutagenesis of CHS2 and CHS8 promoters by XL1-Red (a mutator strain of *Escherichia coli* useful for mutating cloned fragments) could have been performed (Palmer and Sturtevant, 2004). The ensuing transformants could be quickly screened on cell wall stressor media.

In another study, the authors wanted to identify the promoter binding sites in the pH responsive gene, *PHR1*, in C. *albicans* (Ramon and Fonzi, 2003). The pH response pathway had been well researched in A. *nidulans*. However, the promoter binding elements in the *Aspergillus* pH responsive gene (*pacC*) could not be translated to *PHR1*. Therefore, regions of DNA binding were identified *in vitro* by ChIP (Chromatin Immunoprecipitation Assay) and then confirmed by site-directed mutagenesis (Ramon and Fonzi, 2003).

Site-directed mutagenesis and ChIP have also been used to identify the genes that a specific transcription factor binds. A good illustration is the gain-of-function allele of the transcription factor *CAP1* that was constructed by site-directed mutagenesis and then analyzed in ChIP assays (Znaidi et al., 2009).

2.4 Gene function-essential genes

Surprisingly, site-directed mutagenesis has not been used extensively to determine the function of a gene. Gene disruption is routinely the method of choice to study gene function. However, this is not possible when studying the function of essential genes. The use of conditional promoters is often used. Results can sometimes be misleading, since phenotypic testing is performed under suboptimal growth conditions due to promoter-dependent nutritional constraints. Even so, expression of an essential gene under a conditional reporter does not allow complete analysis of multifunctional genes. Very few studies have taken advantage of directed mutagenesis of a specific gene.

The first report of the use of site-directed mutagenesis of an essential and multifunctional gene was the signaling regulatory gene, *BMH1* (14-3-3 gene). There is only one 14-3-3 protein (Bmh1p) in C. *albicans*, and it is essential (Cognetti et al., 2002). Multiple approaches were attempted to express the gene under a regulatable promoter, but they were unsuccessful (Palmer et al., 2004). This may be because *BMH1* regulates multiple cellular processes involved in growth, and the phenotypic studies were performed under suboptimal growth conditions due to promoter-dependent nutritional constraints. Therefore, *BMH1* appeared to be an excellent candidate to test the feasibility of both site-directed and random mutagenesis. Amino acid residues in the 14-3-3 allele required for ligand binding, dimerization, and growth were reported for other eukaryotic species. Due to the high degree of conservation between 14-3-3 proteins, the same residues were selected for substitution in the *C. albicans BMH1* allele. Six sites were chosen. Transformants were screened for filamentation and growth defects (Palmer et al., 2004). Two approaches tested the applicability of random mutagenesis of the *BMH1* allele (Palmer and Sturtevant, 2004). A plasmid containing the *BMH1* allele was propagated in the *E. coli* XL-1 Red strain (Stratagene), which is deficient in multiple primary DNA repair pathways and thus introduces random mutations in the plasmid. Mutagenized plasmids were isolated after 11 to 44 divisions and introduced into the remaining *BMH1* locus in a *BMH1* heterozygote strain. The second random mutagenesis approach was PCR-mediated. (DNA polymerases used for PCR can be mutagenic under certain conditions.) The *BMH1* allele was subject to PCR amplification with an unbalanced nucleoside pool. The PCR products were ligated into a *Candida* transformation vector, and pools were introduced into the *BMH1* heterozygote by transformation, as above. The *E. coli*-mediated mutagenesis resulted in a higher efficiency of correct integration of the mutated allele than did the PCR-mediated method. Around 1400 (1000 – *E. coli*-mediated; 368 – PCR-mediated) *C. albicans* transformants containing randomly mutagenized *BMH1* alleles were screened under a variety of phenotypic stresses. These tests were rapid and easily visible; thus, they translated easily into a screen. Mutant alleles were isolated from transformants that demonstrated altered phenotypes and were sequenced. In the end, from 1000 *E. coli*- and 368 PCR-mutated colonies, 2 and 4 alleles, respectively (0.4%), were identified with altered coding sequences. That these mutations were responsible for the altered phenotypes was validated by constructing *C. albicans* strains isogenic for the site-directed mutations, as described previously. While the efficiency of the random mutagenesis methods was lower than reported for bacteria, non-lethal mutants were identified. Thus, this is a valid approach to study gene function in fungi (Palmer and Sturtevant, 2004). The outcome of the site-directed and random mutagenesis approaches was a set of isogenic strains in which *BMH1* or a mutant *BMH1* allele was expressed under its own promoter at an exogenous locus. These strains were analyzed under a variety of environmental conditions reflecting stresses in the host. It was possible to discriminate between separate pathways involved in filamentation, growth, and survival in the host (Kelly et al., 2009; Palmer et al., 2004; Palmer and Sturtevant, 2004). Additional mutants may have been identified if transformants were screened in additional tests or in *in vivo* models. On the other hand, since *BMH1* is an essential gene, there may be a limited number of amino acids that can be mutated and still result in a non-lethal allele.

Site-directed mutagenesis was also used to decipher the role of the hemoglobin response gene (*HBR1*) in vegetative growth. It was known that *HBR1* induced mating type genes, but mating is not an essential process in *C. albicans* (Peterson et al., 2011). Sites required for optimal growth and the oxidative stress response in the homologous gene in *Saccharomyces* were targeted in the *C. albicans* gene. The mutant alleles were introduced into a *HBR1* heterozygote and were regulated by the *MET3* promoter. This study identified amino acid residues important for mating locus regulation, but not for vegetative growth. Thus, amino acids identified to be important in model fungal species do not always translate to related pathogenic fungi (Peterson et al., 2011).

Essential genes are often prospective drug targets. One such gene is *MET6*. In *C. albicans*, Prasannan et al. (2009) constructed GST fusions of mutated *C. albicans MET6* and expressed the fusion protein in a *MET6 Saccharomyces* mutant. A 3-D model that was modified from the known crystallized structure of the *Arabidopsis* enzyme was used to select sites for mutation in *MET6*. Eight residues were chosen based on conservation across species and probability of being catalytic sites. Site-directed mutagenesis was introduced by the Quikchange kit from Stratagene. The mutant GST-Met6p fusion proteins demonstrated varied enzymatic activity validating the use of this approach in the design of new antifungal drugs (Prasannan et al., 2009).

2.5 Gene function – genes with multiple functions

The transcriptional regulator, *EFG1*, regulates multiple cellular processes in *C. albicans*. *EFG1* is a member of the APSES protein family. Although the domain that defines this family is known, the actual structure-function relationships were not understood. Thus, defined regions within and flanking the APSES domain were deleted. This was mediated by PCR. Instead of amino acid substitution, 15 – 103 nucleotides were deleted from within the *EFG1* gene, similar to what is done for promoter bashing (mutating promoters) (Noffz et al., 2008). A disadvantage to this approach is that it is not possible to discriminate if an altered phenotype is due to the compromise of protein structure or to the absence of protein expression. However, immunoblotting confirmed that the mutated protein was expressed in all mutants. Two mutants did express lower levels of Efg1p that could account for altered phenotypes (Noffz et al., 2008). Thus, it is important to confirm protein expression of mutants. The authors were able to associate specific regions of Efg1p with distinct cellular processes that it regulates. The deletion alleles were also used in over-expression and one-hybrid (gene-fusion technology to identify a DNA-binding domain) experiments. Thus, this approach was successful in determining structure – function relationships of an APSES protein (Noffz et al., 2008).

2.6 Other applications

Site-directed mutagenesis has been used to determine how GPI-tagged proteins discriminate between localization to the plasma membrane and cell wall (Mao et al., 2008). N and C termini of cell wall or plasma proteins were fused to GFP. The termini were subjected to truncation and mutagenesis. Localization of mutant alleles was examined by microscopy. One potential pitfall, however, is that the GFP tag itself can cause protein mislocalization. These experiments identified the omega cleavage site. Further domain exchange and mutagenesis studies identified which residues dictated cell wall or plasma membrane localization (Mao et al., 2008).

2.7 Methodology

Site-directed mutagenesis (*i.e.*, targeted substitution of one or more nucleotides) in a gene was normally performed via overlap PCR and/or the QuikChange Site-Directed Mutagenesis Kit (Stratagene/Agilent Technologies). The principle of these methods is the same. Complementary primers are designed with the nucleotide substitution at the desired site of the mutation. The primers are complementary to the region of the template with the wild-type residue. The template is a double-stranded DNA vector (usually a plasmid) containing a DNA clone of the region of interest. PCR with a high fidelity polymerase results in a plasmid with the mutation of the primer. The product is digested with *Dpn*I, which cleaves only the parental plasmid (template) because *Dpn*I requires fully or hemi-methylated DNA. (The parental plasmid is methylated by the *E. coli* host; DNA amplified by PCR is unmethylated.) The resulting DNA is then introduced into competent cells by transformation. Resulting plasmids are sequenced to confirm the mutation. It is also important to confirm that the mutation does not affect gene expression. Single-site mutagenesis has also been used to introduce silent mutations that result in construction of a restriction enzyme site in order to facilitate genetic manipulation (Cognetti et al., 2002; Schmalhorst et al., 2008).

3. Insertional mutagenesis

Insertional mutagenesis methods are commonly used in model fungi species. Although genomes are similar between model and medically important fungal species, there are still significant differences. Forward screens (screens for new genes that are involved in a phenotype, often using homologs) in model fungi will not identify genes important for pathogenesis, since these species are usually attenuated in virulence or are avirulent. Signaling pathways are shared among fungi, but downstream targets and regulation vary. It is estimated that only 61% of the essential genes in *S. cerevisiae* are also essential in *C. albicans*. There may be even more differences in filamentous fungi (Carr et al., 2010). The advent of improved genetic techniques and whole-genome sequencing has dramatically improved the ability to perform forward screens in the medically important fungi. One major drawback has been diploidism. Ways to circumvent the problem of diploidy have included parasexual genetics (non-meiotic conversion of a diploid to a haploid) (Carr et al., 2010; Firon et al., 2003) and haploid insufficiency (a phenotype resulting from the loss of one allele in a diploid) (Uhl et al., 2003). Additional requirements that are species–specific include a 'mutagen' and an appropriate screen/phenotype. Insertional mutagenesis is normally now facilitated by transposons, but it is still necessary to identify transposons that work efficiently in the fungal species of choice. Much of the initial work demonstrated a bias for insertions, including a bias of non-coding regions. A recent analysis of three transposons has identified Tn7 as having the least insertion bias in *Candida glabrata* (Green et al., 2012). This would probably translate to other fungal species whose genomes are also rich in A/T sequences. Certainly, in the post-genomics era, utilization of forward genetics approaches have increased due to the improved ability to identify the site of insertion.

3.1 Selection of insertion mutants by complementation of auxotrophy

Initial studies used complementation of auxotrophy as a 'mutagen.' Auxotrophic strains were transformed with plasmids carrying an auxotrophic marker (*e.g., URA3/5*). For example, in *C. neoformans*, capsule formation is associated with virulence. Laccase is required for capsule formation. To identify the laccase gene, a *ura*-deficient mutant was transformed multiple times (to obtain independent mutants) with a *URA* construct that has an *E. coli*-specific replicon. When expressed in *C. neoformans,* the construct integrates randomly into the genome and complements the uracil auxotrophy. Transformants were selected for growth on medium lacking uracil. They were then screened on differential media that would identify strains with laccase deficiency due to a pigment change. Out of 1000 transformants, nine strains with an altered phenotype were identified. Plasmid rescue was performed to identify the insertion point. (Plasmid rescue results from cleaving genomic DNA with the appropriate restriction enzyme. The inserted fragment, along with a piece of the interrupted gene, is released. The released DNA can circularize in the presence of ligase and form a plasmid that replicates in *E. coli*. Sequencing of the piece of interrupted gene is easily done and identifies the gene, which can then be cloned intact.). In this manner, a novel virulence attribute was identified, the vacuolar (H+) – ATPase subunit (*VPH1*) (Erickson et al., 2001). In general, the drawbacks to the auxotrophic approach were inefficient integration, integration via homologous rather than non-homologous recombination, and difficulty in identification of the insertion site.

3.2 Signature-tagged mutagenesis (STM)

Signature-tagged mutagenesis is a method originally designed to identify genes required for pathogenesis (Hensel et al., 1995). A large number of mutants were created by insertional mutagenesis. The inserted DNA includes a unique oligonucleotide tag that resembles a 'barcode.' In principle, up to 96 mutants can be inoculated into one host; strains not recovered are thought to harbor a mutation specific for *in vivo* growth (Hensel et al., 1995). This method was first used in *Salmonella* and was modified for *C. glabrata*, *A. fumigatus*, and *C. neoformans* (Brown et al., 2000; Cormack et al., 1999; Nelson et al., 2001). There were certain considerations in translating this approach to fungi, including larger genomes, non-coding DNA, inefficient methods for insertion, selection of the appropriate host environment (Brown et al., 2000), and inoculation parameters (Nelson et al., 2001). These issues were addressed in the studies below (Brown et al., 2000; Nelson et al., 2001).

The first studies were performed prior to the identification of useful transposons. In order to identify virulence factors in *A. fumigatus*, two approaches were used to address random insertion of signature tags (Brown et al., 2000). The first used restriction-mediated integration (REMI). Protoplasts of the recipient strain were transformed with clones with tags in the presence of the restriction enzyme *Kpn*I (96 transformations). The rationale was that these clones would integrate into *Kpn*I sites randomly situated in the genome. The construction of the second library relied on ectopic integration, and *Aspergillus* was transformed with linearized clones (84 transformations). The tags for the transformation constructs for both approaches were generated by PCR using templates developed for *Salmonella typhimurium* and cloned into a fungal transformation vector that carries a gene for hygromycin resistance (Brown et al., 2000). A similar approach was used for *Cryptococcus neoformans*, and the selection of insertions was based on ectopic integration of a linear plasmid conveying hygromycin resistance (Nelson et al., 2001). Further analysis demonstrated that integration was mostly random, except for one hotspot that was the actin/*RPN10* promoter. In both cases, integration efficiency was lower than reported for bacteria.

Many of the medically important fungi can cause different types of infections and/or colonize and infect multiple organs. Unlike bacteria, they do not have true 'virulence factors'; but they do have virulence "attributes." Since *in vivo* murine models are involved, it is important to limit the number of mice used; and thus it is necessary to predetermine the appropriate model, the time points and the organs to harvest. For *Aspergillus fumigatus*, the STM libraries were tested in an immunosuppressed murine inhalation model (Brown et al., 2000). For *C. neoformans*, Nelson et al. (2001) carefully determined the course of infection in a murine model and chose a time point that reflected attenuated or increased virulence based on cfu (colony forming units) counts in the brain (Nelson et al., 2001). They also asked an interesting question: Would a virulent strain allow survival of an attenuated strain? For instance, if the virulent strain damaged endothelium, normally avirulent strains might theoretically have increased abilities to disseminate. They tested this by co-infecting with acapsular (avirulent) and capsular (virulent) strains. The avirulent acapsular strains were not recovered, and they concluded that virulent strains would not help avirulent strains (Nelson et al., 2001). However, this may not be true for all attenuated strains; and a strain's ability to piggyback upon another will depend on its defect. This is a general drawback of STM and confirms that virulence tests with single strains have to be performed.

Another important parameter is the number of strains that can be injected into a mouse and have an equal opportunity to survive. Nelson et al. (2001) did a prescreen with hygromycin and G418 resistant strains (100:1) and ascertained that it was possible to inoculate 100 strains. However, studies with hybridization signals showed that they could not reliably detect more than 80 strains. Experiments were performed with pools of 48 strains. Six hundred seventy-two mutants were screened, and 39 gave different output signals. Twenty-four of the mutants were tested singly in the mouse, and 6 of these had significant changes in virulence (Nelson et al., 2001). Brown et al. (2000) determined that subsequent hybridization efficiency was 80%, so, although they used pools of 96, they always inoculated 2 mice per pool (Brown et al., 2000). In total 4648 tagged strains were screened, and 35 strains (0.8%) gave weak signals in the output pool after two rounds of STM. These strains were tested in a competitive inhibition infection, in which the attenuated strain was present as 50% of the inoculum. Nine strains showed a competitive disadvantage, and two of these demonstrated significantly reduced virulence. The site of the mutation of one strain was not identifiable; the second mutation was upstream of the PABA synthetase gene. Further analysis confirmed that *pabaA* is required for virulence.

Cormack et al. (1999) exploited STM to construct a mutant library in *C. glabrata* (Cormack et al., 1999). Each strain could be easily identified by a distinct tag. Ninety-six unique strains were generated by integrating 96 different tags, flanked by identical primer sites, into the already disrupted *URA3* locus. Since *C. glabrata* has an efficient system of non-homologous recombination, the *Saccharomyces URA3* gene was used for random mutagenesis. Transformants were selected on media minus uracil. Pools of the 96 tagged strains were screened for adherence to human cultured epithelial cells. Out of 4800 mutants (50 pools of 96), 31 mutants demonstrated aberrant adherence. Sixteen of these were non-adherent. Interestingly, 14/16 of these integrated into non-coding sequence upstream of the same gene, *EPA1*. This led to the identification of subtelomeric transcriptional silencing (Cormack et al., 1999). However, this method would not have identified *EPA1* by traditional STM, since *EPA1* null mutants are virulent *in vivo*.

3.3 Transposon-mediated insertional mutagenesis

Transposon technology has been used in pathogenic fungi to construct libraries, add epitope tags, and understand cellular processes. The technology has been adapted for diploid organisms using the parasexual cycle, haploid insufficiency, and homologous recombination (Carr et al., 2010; Davis et al., 2002; Firon et al., 2003; Juarez-Reyes et al., 2011; Spreghini et al., 2003; Uhl et al., 2003). The use of transposons has superseded auxotrophic and STM approaches.

Essential genes are often considered good drug targets. Firon et al. (2003) exploited the parasexual cycle to develop a transposon-mediated insertional mutagenesis protocol to identify essential genes in *A. fumigatus* (Firon et al., 2003). A diploid strain, homozygous auxotrophic for pyrimidines and heterozygous for a spore color marker, was randomly mutagenized with an *imp160::pyrG* transposon. The candidate mutant strains were induced to become haploid by the mitochondrial destabilizer, benomyl. The genotype of the parent strain allowed haploid progeny to be identified by pigmentation. Diploid strains were grey-green, but haploid progenies were white or reddish. Replica plating identified the haploid progeny that harbored transposons. If haploid strains carried a transposon-inactivated allele, they expressed pyrG and grew on both selective (without uridine/uracil) and non-selective media. Conversely, strains without a transposon grew only on non-selective media. If the transposon inactivated an essential gene, the haploid strain did not grow on either medium. With this approach, 3% of the haploid progeny of 2,386 diploid strains were found to be unable to grow on either medium and, therefore, possibly had mutations in essential genes. These strains were propagated further on selective media and haploid progeny could not be obtained from 1.2% of the resultant diploid revertants. The sites of insertion were determined by 2-step PCR using semi-random primers and 5′-end transposon-specific primers (see Section 3.5.1). Ninety percent of insertion sites were identified (Firon et al., 2003). Since the insertion rate of the transposon into essential loci was lower than expected, additional transposon insertion sites were analyzed. Although an insertion site did not depend on genome sequence or chromosomal location, there did appear to be a bias toward noncoding regions (34%) (Firon et al., 2003). Carr et al. (2010), who observed that transposon mobilization could be induced at 10 °C, improved upon this approach. Therefore, using the same screen, 96 additional essential loci were identified. They found no obvious bias of insertion in noncoding regions. Interestingly, only half of the genes had essential homologs in *Saccharomyces*, confirming the necessity for species-specific screening.

Uhl et al. (2003) developed a transposon mutant library in *C. albicans*. Restriction enzyme-digested *C. albicans* gDNA (genomic DNA) was mixed with a linearized donor transposon Tn7–containing plasmid. This plasmid harbored elements for replication in *E. coli*, for selection in both *E. coli* and *C. albicans* and a fungal *lacZ* reporter system. The fragments were ligated and introduced into *E. coli* by transformation. Plasmids were isolated from over 200,000 transformants and batch isolated. Transposon–gDNA junctions were sequenced in plasmids to confirm random integration. *C. albicans* was transformed with the Tn7-gDNA plasmids to give an 18,000-strain transposon mutant library. It was assumed that each strain had an independent insertion. That would mean there was a transposon approximately every 2.5 kb. However, only one allele of a gene was disrupted in these strains. (*C. albicans* is diploid, so one allele remains non-disrupted.) Uhl et al. (2003) exploited haploid insufficiency to screen for filamentation mutants, since heterozygote strains in genes involved in morphogenesis exhibit reduced filamentation. This screen was rapid and successful for identifying processes that required genes sensitive to dosage effects. However, this certainly will not be the case for all genes involved in pathogenesis.

Davis et al. (2002) constructed a transposon mutant library in *C. albicans*, but these strains harbored insertions in both alleles. This approach was based on a homologous recombination model that allowed the disruption of both alleles of *C. albicans* in one transformation step (Enloe et al., 2000). A cassette (UAU), which contains the *URA3* gene disrupted with a functional *ARG4* gene, was inserted into transposon Tn7. the Tn7–UAU transposon was inserted randomly into a *C. albicans* library. Digestion with the appropriate restriction enzyme released DNA fragments that contained *C. albicans* DNA interrupted with the Tn7-UAU transposon. These fragments were used to transform *C. albicans*. Homology from the interrupted DNA allowed replacement of the chromosomal wild-type version by homologous recombination. The chromosomal version was then mutated because it carried the Tn7-UAU transposon. Recombinants could be selected because transformation into the recipient ura- arg- *Candida* strain will confer arginine prototrophy. Occasionally the other intact copy of the gene acquired the transposon. Thus, both copies of the gene were mutated. Using arginine selection, homozygous mutants could not be distinguished from the heterozygotes. However, in a small percentage of ARG+ transformants, the *ARG4* gene is spontaneously looped out. If there were two copies of the transposon and if looping out occurred in one, it gave an ARG+, URA+ strain. Thus, both alleles were disrupted. This allowed for the construction of a large set of mutants, though it was still not as efficient as it would be for a haploid strain. This library is widely used by the *Candida* community (Davis et al., 2002; Norice et al., 2007; Park et al., 2009).

Spreghini et al. (2003) exploited transposon mutagenesis to add an epitope to the putative cell wall protein, Dfg5p. Since conventional epitope tagging of amino and carboxyl termini was not an option, they wanted to identify an internal site which, when disrupted with a tag, did not compromise function. The Tn7 transposon was used to mutagenize the *DFG5* insert in a plasmid and insertions within *DFG5* coding region were confirmed by sequencing. Then the mutagenized plasmid was redigested to get rid of the majority of Tn7, leaving only a 15-bp (base pair) insertion, which resulted in an insertion of 5 amino acids that did not disrupt function and could be recognized by an available antibody. The internally tagged *DFG5* insert was then ligated into *C. albicans* vectors for further study (Spreghini et al., 2003).

In the transposon examples above, mutagenesis was performed *in vitro*, and then mutagenized DNA was introduced into recipient strains. Magrini and Goldman (2001) took a different approach by directly mutagenizing *Histoplasma capsulatum in vivo*. The transformation cassette was a linear telomere vector (because the presence of a telomeric sequence is required for efficient homologous recombination in *Histoplasma*) containing the selection marker *URA5*, the *MOS1* transposase gene regulated by a strong promoter, and the hygromycin resistance gene flanked by *MOS1* terminal repeats to create a synthetic transposon. *Histoplasma* transformants were selected in presence of 5-FOA (5-Fluoroorotic acid, which selects against URA5) to select for loss of the donor plasmid and on hygromycin for the presence of the synthetic transposon, which encodes hygromycin resistance. It is not known if this library has been utilized because T-DNA appears to be more commonly used in *Histoplasma* (see below).

A novel use of random insertion was the analysis of subtelomeric silencing of *C. glabrata* adhesin genes. Learning where silencing occurred was accomplished by randomly placing a *URA3* reporter at different distances from a telomere and examining where *URA3* was silenced. The transposon Tn7–*URA3* was introduced into a subtelomeric sequence of *C. glabrata* cloned on an *E. coli* plasmid. Resulting constructs were integrated into subtelomeric regions of *C. glabrata* by homologous recombination. It was possible to select for 'silenced' *URA3* on 5-FOA media (Juarez-Reyes et al., 2011).

3.4 *Agrobacterium* T-DNA

Agrobacterium tumefaciens carries an approximately 200-kbp (kilobase pair) tumor-inducing (Ti) plasmid. A portion of this plasmid is called T-DNA. In plants, the T-DNA randomly inserts in the genome; and the outcome is a tumorous growth. This plasmid has been modified for genetic manipulation purposes to retain the insertional DNA (T-DNA). The plasmid vector can also replicate in *E. coli* and has cloning sites for additional DNA. T-DNA has been used to construct mutants with increased, reduced, or no expression of genes, depending on the plasmid used (Krysan et al., 1999). In the last decade, insertional mutagenesis via T-DNA has been successfully adapted for medically important fungi. In general, a fungal selectable marker is ligated into the *Agrobacterium* Ti plasmid within the T-DNA region and introduced into *A. tumefaciens* by electroporation. Equal concentrations of *A. tumefaciens* carrying the delivery plasmid and target fungal strain are incubated together for varying lengths of time under conditions that mimic plant wound conditions, which are accomplished by low pH and the addition of acetosyringone. The T-DNA is transferred to the target organism by a conjugation-like mechanism. A mutant that contains an insertion of T-DNA is selected with the appropriate fungal selective marker.

Prior to T-DNA mutagenesis, insertional mutagenesis was attempted by electroporation or biolistic transformation of naked DNA. Researchers have developed protocols that have improved the efficiency of transformation using T-DNA in *C. neoformans* (Idnurm et al., 2004), *Histoplasma* and *Blastomyces* (Brandhorst et al., 2002; Edwards et al., 2011; Gauthier et al., 2010; Laskowski and Smulian, 2010; Marion et al., 2006; Smulian et al., 2007; Sullivan et al., 2002). In addition, T-DNA mutagenesis protocols have been developed for *Coccidioides* (Abuodeh et al., 2000), *Trichoderma* spp. (Cardoza et al., 2006; Dobrowolska and Staczek, 2009; Yamada et al., 2009), and *Penicillium marneffei* (Kummasook et al., 2010; Zhang et al., 2008). In *C. neoformans*, the use of T-DNA improved both the efficiency and the stability of transformation events. The resulting transformants also demonstrated less complicated integrations and less additional gene rearrangements. There did seem, however, to be a bias for promoter sequences. In one study, some of the integration events were not linked to *NAT*, the gene for the Nourseothricin resistance marker on the inserted DNA (Idnurm et al., 2004).

Blastomyces, in particular, is a challenge to transform, since it is multinucleate. Transforming DNA often integrates at multiple sites (Brandhorst et al., 2002). This is usually bypassed by transforming conidia or performing multiple rounds of selection to enrich for homokaryons (all the nuclei are genetically identical). Sullivan et al. (2002) developed a protocol for both *Histoplasma* and *Blastomyces*. Many conditions were tested, including bacteria:yeast ratios, life stage of the recipient strain, and the choice of selectable marker. Interestingly, the efficiency of transformation was 5–10 times higher with uracil selection than with hygromycin selection. Southern analysis confirmed that integration was random, but there were often direct repeat concatemers in *Blastomyces*. There were clear improvements over electroporation, including increased efficiency, ability to use spores as the recipient, and single-site integrations (Sullivan et al., 2002). Additional studies in *Blastomyces* using T-DNA have identified genes involved in phase transition (Gauthier et al., 2010).

T-DNA was used to identify genes in *Histoplasma* involved in pathogenesis in a novel high-throughput macrophage-killing screen (Edwards et al., 2011). Transgenic (a novel gene was introduced) macrophage lines were constructed that constitutively expressed bacterial *lacZ*.

The activity of β- galactosidase, the product of *lacZ*, directly correlated to the number of macrophages. Thus, this line was used as a readout for macrophage killing. Over 2000 *Histoplasma* transformants made from *A. tumefaciens*-treated *Histoplasma* cells were incubated with macrophages and screened for killing activity after 7 days. Three strains were less efficient in killing, and one was significantly inefficient in killing both transgenic and primary macrophages. Flanking sequences were identified by PCR and sequencing. The authors identified a new virulence gene in *Histoplasma*, a homolog of Hsp82 (Edwards et al., 2011).

Marion et al. (2006) performed a more comprehensive analysis of insertional mutagenesis in *Histoplasma capsulatum* using *Agrobacterium*-mediated transformation. Optimal co-incubation times, bacteria:yeast ratios and temperature were determined. Southern hybridization analysis showed that approximately 90% of the insertions were random and at a single site. Inverse PCR and plasmid rescue were used to identify the flanking sequences. Their results indicated that mutagenesis by T-DNA resulted in the absence of chromosomal rearrangements and deletions. The biological relevance of the T-DNA mutants was approached by screening for genes involved in the biosynthesis of α-(1, 3)–glucan, which is posited to be a virulence attribute. The absence of α-(1, 3)–glucan was easily visualized since colonies have a smooth, rather than rough, morphology. Approximately 50,000 insertional mutants were screened, and 25 had smooth morphology. Eighty-eight percent had single insertions and reduced α-(1, 3)–glucan. Five of twenty-two had distinct insertions in the α-(1, 3)–glucan synthase gene (AGS1), which validated their screen. RNAi technology (synthetic inhibitory RNA) was used to confirm the insertion mutant phenotype with the wild-type allele. The phenotypes of the two other mutants were confirmed. One mutation was in *UGP1* (previously reported to play a role in glucan synthesis). The other mutation was in the amylase gene, which was previously unreported to play a role (Marion et al., 2006).

The use of T-DNA in *Histoplasma* has provided additional information. As with all genetic manipulations, it is important to confirm that the mutation is responsible for the ensuing phenotype. Smulian et al. (2007) wanted to make GFP-expressing strains and used hygromycin resistance as a marker and T-DNA as the tool for integration. It turned out that all the transformants were hypervirulent. Site-directed mutagenesis of the hygromycin resistant gene, *hph*, confirmed that the increased virulence was due to the acquisition of hygromycin resistance. One mutant actually gained the ability to form cleistothecia, a mating structure that was not present in the parent strain. This phenotypic trait was not due to the *hph* gene; and, thus, the strain may be used as a tool to study mating in *Histoplasma* (Laskowski and Smulian, 2010).

3.5 Methodologies to identify the site of insertion

3.5.1 Two-step PCR

In the first step of two-step PCR (Chun et al., 1997), sequence on one side of the insertion site is amplified with a degenerate primer and a primer homologous to the sequence in one of the ends of the inserted DNA. (There are two end-specific primers. A primer specific for only one end is used. Note that a tranposon can insert in either orientation.) The degenerate primer contains 20 nucleotides of defined sequence at the 5'-end, 10 nucleotides of degenerate sequence (*i.e.*, all 4 nucleotides are used at each position for synthesis) + GATAT at the 3′-end. The sequence GATAT is predicted to occur every 600 bp in the yeast genome. The second step amplifies the first PCR product with two non-degenerate primers. The forward primer contains the 20 nt (nucleotides) of defined sequence in the degenerate primer. The reverse primer is immediately 3′ (antisense strand) to the insertion-specific primer used in the first PCR reaction to guarantee that the desired DNA is amplified (Chun et al., 1997). This method was originally defined in *Saccharomyces* and was successfully used to identify transposon insertions in *A. fumigatus* (Carr et al., 2010; Firon et al., 2003).

3.5.2 Thermal asymmetric interlaced PCR (TAIL PCR)

TAIL PCR (Liu and Whittier, 1995) is another method to identify sequences flanking insertions. It is a modified version of hemispecific (one-sided) PCR. The purpose is to favor amplification of the desired product. It uses specific primers homologous to DNA in the integrating cassette or plasmid and a degenerate primer that can anneal to the gDNA flanking the insertion. The strategy is that the specific primers are long, nested, and have a high Tm; the degenerate primer is short and has a low Tm. The first five cycles are high stringency cycles to favor annealing to and linear amplification from the specific primer. Then there is one low stringency cycle to allow the degenerate primer to anneal. Because there are now several copies of the gDNA adjacent to the insertion, the chance of the degenerate primer annealing to the desired product is increased. However, other products might form from the primers finding additional annealing sites in the genome. Using a second and a third primer completely homologous to the inserted DNA will favor the desired product that is made from both the specific and degenerate primers instead of either one alone. This is accomplished by interlacing reduced stringency and high stringency cycles.

4. Closing remarks

Site-directed and insertional mutagenesis are techniques that can be used to advance our understanding of the pathogenesis of medically important fungi. The exploitation of these tools has resulted in a better understanding of drug-resistant mechanisms, transcription factors, signaling pathways and vital cellular processes. Site-directed mutagenesis could be better utilized to decipher the functions of essential and multi-functional genes. While all approaches cannot be used in the always-diploid strains, transposon-mediated insertional mutagenesis can be used to construct libraries. Additionally, T-DNA can be used to improve transformation efficiency in dimorphic fungi and in *C. neoformans*.

5. References

Abuodeh, R. O., Orbach, M. J., Mandel, M. A., Das, A., Galgiani, J. N., 2000. Genetic Transformation of *Coccidioides immitis* Facilitated by *Agrobacterium tumefaciens*. Journal of Infectious Diseases. 181,6: 2106-2110.

Alvarez-Rueda, N., Fleury, A., Morio, F., Pagniez, F., Gastinel, L., Le Pape, P., 2011. Amino Acid Substitutions at the Major Insertion Loop of *Candida albicans* Sterol 14alpha-Demethylase Are Involved in Fluconazole Resistance. PLoS ONE. 6,6: e21239.

Bassilana, M., Arkowitz, R. A., 2006. Rac1 and Cdc42 Have Different Roles in *Candida albicans* Development. Eukaryotic Cell. 5,2: 321-329.

Bockmühl, D. P., Ernst, J. F., 2001. A Potential Phosphorylation Site for an A-Type Kinase in the Efg1 Regulator Protein Contributes to Hyphal Morphogenesis of *Candida albicans*. Genetics. 157,4: 1523-1530.

Brakhage, A. A., Langfelder, K., 2002. MENACING MOLD: The Molecular Biology of *Aspergillus fumigatus*. Annual Review of Microbiology. 56,1: 433-455.

Brandhorst, T. T., Rooney, P. J., Sullivan, T. D., Klein, B. S., 2002. Using new genetic tools to study the pathogenesis of *Blastomyces dermatitidis*. Trends in Microbiology. 10,1: 25-30.

Brown, J. S., Aufauvre-Brown, A., Brown, J., Jennings, J. M., Arst, H., Jr., Holden, D. W., 2000. Signature-tagged and directed mutagenesis identify PABA synthetase as essential for *Aspergillus fumigatus* pathogenicity. Mol Microbiol. 36,6: 1371-1380.

Cardoza, R. E., Vizcaino, J. A., Hermosa, M. R., Monte, E., Gutierrez, S., 2006. A comparison of the phenotypic and genetic stability of recombinant *Trichoderma* spp. generated by protoplast- and *Agrobacterium*-mediated transformation. J Microbiol. 44,4: 383-395.

Carr, P. D., Tuckwell, D., Hey, P. M., Simon, L., d'Enfert, C., Birch, M., Oliver, J. D., Bromley, M. J., 2010. The Transposon impala Is Activated by Low Temperatures: Use of a Controlled Transposition System To Identify Genes Critical for Viability of *Aspergillus fumigatus*. Eukaryotic Cell. 9,3: 438-448.

Chun, K. T., Edenberg, H. J., Kelley, M. R., Goebl, M. G., 1997. Rapid Amplification of Uncharacterized Transposon-tagged DNA Sequences from Genomic DNA. Yeast. 13,3: 233-240.

Cognetti, D., Davis, D., Sturtevant, J., 2002. The *Candida albicans* 14-3-3 gene, BMH1, is essential for growth. Yeast. 19,1: 55-67.

Cormack, B. P., Ghori, N., Falkow, S., 1999. An adhesin of the yeast pathogen *Candida glabrata* mediating adherence to human epithelial cells. Science. 285,5427: 578-582.

Davis, D. A., Bruno, V. M., Loza, L., Filler, S. G., Mitchell, A. P., 2002. *Candida albicans* Mds3p, a Conserved Regulator of pH Responses and Virulence Identified Through Insertional Mutagenesis. Genetics. 162,4: 1573-1581.

Dobrowolska, A., Staczek, P., 2009. Development of transformation system for *Trichophyton rubrum* by electroporation of germinated conidia. Curr Genet. 55,5: 537-542.

Edwards, J. A., Zemska, O., Rappleye, C. A., 2011. Discovery of a Role for Hsp82 in *Histoplasma* Virulence through a Quantitative Screen for Macrophage Lethality. Infect. Immun. 79,8: 3348-3357.

Enloe, B., Diamond, A., Mitchell, A. P., 2000. A single-transformation gene function test in diploid *Candida albicans*. J Bacteriol. 182,20: 5730-5736.

Erickson, T., Liu, L., Gueyikian, A., Zhu, X., Gibbons, J., Williamson, P. R., 2001. Multiple virulence factors of *Cryptococcus neoformans* are dependent on VPH1. Mol Microbiol. 42,4: 1121-1131.

Feng, Q., Summers, E., Guo, B., Fink, G., 1999. Ras Signaling Is Required for Serum-Induced Hyphal Differentiation in *Candida albicans*. Journal of Bacteriology. 181,20: 6339-6346.

Firon, A., Villalba, F., Beffa, R., d'Enfert, C., 2003. Identification of Essential Genes in the Human Fungal Pathogen *Aspergillus fumigatus* by Transposon Mutagenesis. Eukaryotic Cell. 2,2: 247-255.

Fox, D. S., Heitman, J., 2005. Calcineurin-Binding Protein Cbp1 Directs the Specificity of Calcineurin-Dependent Hyphal Elongation during Mating in *Cryptococcus neoformans*. Eukaryotic Cell. 4,9: 1526-1538.

Gauthier, G. M., Sullivan, T. D., Gallardo, S. S., Brandhorst, T. T., Wymelenberg, A. J. V., Cuomo, C. A., Suen, G., Currie, C. R., Klein, B. S., 2010. SREB, a GATA transcription factor that directs disparate fates in *Blastomyces dermatitidis* including morphogenesis and siderophore biosynthesis. PLoS Pathogens. 6,4: 1-16.

Green, B., Bouchier, C., Fairhead, C., Craig, N., Cormack, B., 2012. Insertion site preference of Mu, Tn5, and Tn7 transposons. Mobile DNA. 3,1: 3.

Hensel, M., Shea, J. E., Gleeson, C., Jones, M. D., Dalton, E., Holden, D. W., 1995. Simultaneous identification of bacterial virulence genes by negative selection. Science. 269,5222: 400-403.

Hicks, J. K., Bahn, Y.-S., Heitman, J., 2005. Pde1 Phosphodiesterase Modulates Cyclic AMP Levels through a Protein Kinase A-Mediated Negative Feedback Loop in *Cryptococcus neoformans*. Eukaryotic Cell. 4,12: 1971-1981.

Idnurm, A., Reedy, J. L., Nussbaum, J. C., Heitman, J., 2004. *Cryptococcus neoformans* Virulence Gene Discovery through Insertional Mutagenesis. Eukaryotic Cell. 3,2: 420-429.

Juarez-Reyes, A., De Las Penas, A., Castano, I., 2011. Analysis of subtelomeric silencing in *Candida glabrata*. Methods Mol Biol. 734: 279-301.

Kakeya, H., Miyazaki, Y., Miyazaki, H., Nyswaner, K., Grimberg, B., Bennett, J. E., 2000. Genetic analysis of azole resistance in the Darlington strain of *Candida albicans*. Antimicrob Agents Chemother. 44,11: 2985-2990.

Kelly, M. N., Johnston, D. A., Peel, B. A., Morgan, T. W., Palmer, G. E., Sturtevant, J. E., 2009. Bmh1p (14-3-3) mediates pathways associated with virulence in *Candida albicans*. Microbiology. 155,Pt 5: 1536-1546.

Krysan, P., Young, J., Sussman, M., 1999. T-DNA as an insertionalmutagen in *Arabidopsis*. Plant Cell. 11,12: 2283 - 2290.

Kummasook, A., Cooper, C. R., Vanittanakom, N., 2010. An improved *Agrobacterium*-mediated transformation system for the functional genetic analysis of *Penicillium marneffei*. Medical Mycology. 48,8: 1066-1074.

Lamb, D. C., Kelly, D. E., Schunck, W. H., Shyadehi, A. Z., Akhtar, M., Lowe, D. J., Baldwin, B. C., Kelly, S. L., 1997. The mutation T315A in *Candida albicans* sterol 14alpha-demethylase causes reduced enzyme activity and fluconazole resistance through reduced affinity. J Biol Chem. 272,9: 5682-5688.

Lamb, D. C., Kelly, D. E., Baldwin, B. C., Kelly, S. L., 2000. Differential inhibition of human CYP3A4 and *Candida albicans* CYP51 with azole antifungal agents. Chem Biol Interact. 125,3: 165-175.

Laskowski, M. C., Smulian, A. G., 2010. Insertional mutagenesis enables cleistothecial formation in a non-mating strain of *Histoplasma capsulatum*. BMC Microbiology. 10.

Lenardon, M., Lesiak, I., Munro, C., Gow, N., 2009. Dissection of the *Candida albicans* class I chitin synthase promoters. Molecular Genetics and Genomics. 281,4: 459-471.

Li, C. R., Wang, Y. M., Wang, Y., 2008. The IQGAP Iqg1 is a regulatory target of CDK for cytokinesis in *Candida albicans*. EMBO J. 27,22: 2998-3010.

Liu, Y. G., Whittier, R. F., 1995. Thermal asymmetric interlaced PCR: automatable amplification and sequencing of insert end fragments from P1 and YAC clones for chromosome walking. Genomics. 25,3: 674-681.

Magrini, V., Goldman, W. E., 2001. Molecular mycology: a genetic toolbox for *Histoplasma capsulatum*. Trends in Microbiology. 9,11: 541-546.

Mah, J.-H., Yu, J.-H., 2006. Upstream and Downstream Regulation of Asexual Development in *Aspergillus fumigatus*. Eukaryotic Cell. 5,10: 1585-1595.

Mao, Y., Zhang, Z., Gast, C., Wong, B., 2008. C-Terminal Signals Regulate Targeting of Glycosylphosphatidylinositol-Anchored Proteins to the Cell Wall or Plasma Membrane in *Candida albicans*. Eukaryotic Cell. 7,11: 1906-1915.

Marion, C. L., Rappleye, C. A., Engle, J. T., Goldman, W. E., 2006. An alpha-(1,4)-amylase is essential for alpha-(1,3)-glucan production and virulence in *Histoplasma capsulatum*. Mol Microbiol. 62,4: 970-983.

Mellado, E., Garcia-Effron, G., Alcazar-Fuoli, L., Melchers, W. J., Verweij, P. E., Cuenca-Estrella, M., Rodriguez-Tudela, J. L., 2007. A new *Aspergillus fumigatus* resistance mechanism conferring *in vitro* cross-resistance to azole antifungals involves a combination of cyp51A alterations. Antimicrob Agents Chemother. 51,6: 1897-1904.

Menon, V., Li, D., Chauhan, N., Rajnarayanan, R., Dubrovska, A., West, A. H., Calderone, R., 2006. Functional studies of the Ssk1p response regulator protein of *Candida albicans* as determined by phenotypic analysis of receiver domain point mutants. Molecular Microbiology. 62,4: 997-1013.

Morio, F., Loge, C., Besse, B., Hennequin, C., Le Pape, P., 2010. Screening for amino acid substitutions in the *Candida albicans* Erg11 protein of azole-susceptible and azole-resistant clinical isolates: new substitutions and a review of the literature. Diagn Microbiol Infect Dis. 66,4: 373-384.

Nelson, R. T., Hua, J., Pryor, B., Lodge, J. K., 2001. Identification of virulence mutants of the fungal pathogen *Cryptococcus neoformans* using signature-tagged mutagenesis. Genetics. 157,3: 935-947.

Nicholls, S., MacCallum, D. M., Kaffarnik, F. A., Selway, L., Peck, S. C., Brown, A. J., 2011. Activation of the heat shock transcription factor Hsf1 is essential for the full virulence of the fungal pathogen *Candida albicans*. Fungal Genet Biol. 48,3: 297-305.

Noffz, C. S., Liedschulte, V., Lengeler, K., Ernst, J. F., 2008. Functional Mapping of the *Candida albicans* Efg1 Regulator. Eukaryotic Cell. 7,5: 881-893.

Norice, C. T., Smith, F. J., Jr., Solis, N., Filler, S. G., Mitchell, A. P., 2007. Requirement for *Candida albicans* Sun41 in Biofilm Formation and Virulence. Eukaryotic Cell. 6,11: 2046-2055.

Palmer, G. E., Johnson, K. J., Ghosh, S., Sturtevant, J., 2004. Mutant alleles of the essential 14-3-3 gene in *Candida albicans* distinguish between growth and filamentation. Microbiology. 150,Pt 6: 1911-1924.

Palmer, G. E., Sturtevant, J. E., 2004. Random mutagenesis of an essential *Candida albicans* gene. Curr Genet. 46,6: 343-356.

Park, H., Liu, Y., Solis, N., Spotkov, J., Hamaker, J., Blankenship, J. R., Yeaman, M. R., Mitchell, A. P., Liu, H., Filler, S. G., 2009. Transcriptional Responses of *Candida albicans* to Epithelial and Endothelial Cells. Eukaryotic Cell. 8,10: 1498-1510.

Peterson, A. W., Pendrak, M. L., Roberts, D. D., 2011. ATP Binding to Hemoglobin Response Gene 1 Protein Is Necessary for Regulation of the Mating Type Locus in *Candida albicans*. Journal of Biological Chemistry. 286,16: 13914-13924.

Prasannan, P., Suliman, H. S., Robertus, J. D., 2009. Kinetic analysis of site-directed mutants of methionine synthase from *Candida albicans*. Biochemical and Biophysical Research Communications. 382,4: 730-734.

Ramon, A. M., Fonzi, W. A., 2003. Diverged Binding Specificity of Rim101p, the *Candida albicans* Ortholog of PacC. Eukaryotic Cell. 2,4: 718-728.

Sanchez-Martinez, C., Perez-Martin, J., 2002. Gpa2, a G-Protein alpha Subunit Required for Hyphal Development in *Candida albicans*. Eukaryotic Cell. 1,6: 865-874.

Sanglard, D., Ischer, F., Koymans, L., Bille, J., 1998. Amino acid substitutions in the cytochrome P-450 lanosterol 14alpha-demethylase (CYP51A1) from azole-resistant *Candida albicans* clinical isolates contribute to resistance to azole antifungal agents. Antimicrob Agents Chemother. 42,2: 241-253.

Schmalhorst, P. S., Krappmann, S., Vervecken, W., Rohde, M., Muller, M., Braus, G. H., Contreras, R., Braun, A., Bakker, H., Routier, F. H., 2008. Contribution of Galactofuranose to the Virulence of the Opportunistic Pathogen *Aspergillus fumigatus*. Eukaryotic Cell. 7,8: 1268-1277.

Sheng, C., Chen, S., Ji, H., Dong, G., Che, X., Wang, W., Miao, Z., Yao, J., Lü, J., Guo, W., Zhang, W., 2010. Evolutionary trace analysis of CYP51 family: implication for site-directed mutagenesis and novel antifungal drug design. Journal of Molecular Modeling. 16,2: 279-284.

Smulian, A. G., Gibbons, R. S., Demland, J. A., Spaulding, D. T., Deepe, G. S., Jr., 2007. Expression of Hygromycin Phosphotransferase Alters Virulence of *Histoplasma capsulatum*. Eukaryotic Cell. 6,11: 2066-2071.

Snelders, E., Karawajczyk, A., Verhoeven, R. J. A., Venselaar, H., Schaftenaar, G., Verweij, P. E., Melchers, W. J. G., 2011. The structure-function relationship of the *Aspergillus fumigatus* cyp51A L98H conversion by site-directed mutagenesis: The mechanism of L98H azole resistance. Fungal Genetics and Biology. 48,11.

Spreghini, E., Davis, D. A., Subaran, R., Kim, M., Mitchell, A. P., 2003. Roles of *Candida albicans* Dfg5p and Dcw1p Cell Surface Proteins in Growth and Hypha Formation. Eukaryotic Cell. 2,4: 746-755.

Sullivan, T., Rooney, P., Klein, B., 2002. *Agrobacterium tumefaciens* integrates transfer DNA into single chromosomal sites of dimorphic fungi and yields homokaryotic progeny from multinucleate yeast. Eukaryot Cell. 1,6: 895 - 905.

Uhl, M. A., Biery, M., Craig, N., Johnson, A. D., 2003. Haploinsufficiency-based large-scale forward genetic analysis of filamentous growth in the diploid human fungal pathogen *C. albicans*. EMBO J. 22,11: 2668-2678.

VandenBerg, A. L., Ibrahim, A. S., Edwards, J. E., Jr., Toenjes, K. A., Johnson, D. I., 2004. Cdc42p GTPase Regulates the Budded-to-Hyphal-Form Transition and Expression of Hypha-Specific Transcripts in *Candida albicans*. Eukaryotic Cell. 3,3: 724-734.

Yamada, T., Makimura, K., Satoh, K., Umeda, Y., Ishihara, Y., Abe, S., 2009. *Agrobacterium tumefaciens*-mediated transformation of the dermatophyte, *Trichophyton mentagrophytes*: an efficient tool for gene transfer. Med Mycol. 47,5: 485-494.

Zhang, P., Xu, B., Wang, Y., Li, Y., Qian, Z., Tang, S., Huan, S., Ren, S., 2008. *Agrobacterium tumefaciens*-mediated transformation as a tool for insertional mutagenesis in the fungus *Penicillium marneffei*. Mycological Research. 112,8: 943-949.

Znaidi, S., Barker, K. S., Weber, S., Alarco, A.-M., Liu, T. T., Boucher, G., Rogers, P. D., Raymond, M., 2009. Identification of the *Candida albicans* Cap1p Regulon. Eukaryotic Cell. 8,6: 806-820.

10

Site-Directed Mutagenesis Using Oligonucleotide-Based Recombineering

Roman G. Gerlach, Kathrin Blank and Thorsten Wille
Robert Koch-Institute Wernigerode Branch
Germany

1. Introduction

Methods enabling mutational analysis of distinct chromosomal locations, like site-directed mutagenesis, insertion of foreign sequences or in-frame deletions, have become of fast growing interest since complete bacterial genome sequences became available. Various approaches have been described to modify any nucleotide(s) in almost any manner. Some genetic engineering technologies do not rely on the *in vitro* reactions carried out by restriction enzymes and DNA ligases (Sawitzke et al., 2001). Complicated genetic constructs that seem to be impossible to generate *in vitro* can be created within one week using *in vivo* technologies (Sawitzke et al., 2001).

Over several decades, researchers developed and refined various strategies for genetic engineering that make use of the homologous recombination system. Its natural main functions are restoring collapsed replication forks, repairing damage-induced double-strand breaks and maintaining the integrity of the chromosome (Poteete, 2001).

We want to focus on a technique for recombination-mediated genetic engineering ("recombineering", Copeland et al., 2001). Recombineering requires only minimal *in vitro* effort. It has been applied to *Escherichia coli*, *Salmonella*, and a range of other Gram-negative bacteria, as well as to bacteriophages, cosmids and bacterial artificial chromosomes (BACs). It was demonstrated that single-stranded DNA (ssDNA) oligonucleotides can be used as substrates for recombineering in *E. coli* (Ellis et al., 2001, Heermann et al., 2008) and BACs (Swaminathan et al., 2001). However, most commonly linear, double-stranded DNA (dsDNA) has been used as the targeting construct (Maresca et al., 2010), e.g., for chromosomal gene replacement (Murphy, 1998), whole gene disruption (Datsenko et al., 2000) or the development of novel cloning strategies, including subcloning of BAC DNA (Lee et al., 2001).

In the early 1990s, the DNA double-strand break and repair recombination pathway proved to be very efficient for recombining incoming linear DNA with homologous DNA in the yeast *Saccharomyces cerevisiae* (Baudin et al., 1993). For generating *null* alleles of a distinct gene, a PCR-amplified *HIS3* selectable marker flanked by homologous sequences to the ORF (ranging from 35 to 51 nucleotides in length) was used to transform a recipient strain lacking the complete *HIS3* gene or a strain containing the His3Δ200 allele. Due to the auxotrophic selection marker, transformants bearing the expected mutation were among the His+ clones, with up to 80% efficiency (Baudin et al., 1993).

In contrast to *S. cerevisiae*, *E. coli* fails to be readily transformable with linear DNA fragments because of rapid DNA degradation by the intracellular RecBCD exonuclease activity (Lorenz et al., 1994). Mutants defective in the RecBCD nuclease exhibit no degradation of linear DNA in *E. coli*, but unfortunately these strains are also deficient for any recombination events. This recombination defect can be partially rescued in strains with *recA+* background (Jasin & Schimmel, 1984). Other mutants defective in *recBC* (or either *recB* or *recC*) carrying an additional suppressor mutation, *sbc* (suppressor of *recB* and *recC*), possess activation of the RecET recombination pathway (*sbcA*) or enhanced recombination by the RecF pathway (*sbcB*). *recE* and *recT* are found on the defective lambdoid *E. coli* prophage Rac and encode an exonuclease and a ssDNA-binding/strand-exchange protein, respectively (Fouts et al., 1983, Poteete, 2001). Expression of *recET* is induced by few *cis*-acting point mutations, e. g. *sbcA6* (Clark et al., 1994).

One highly applicable RecET-mediated recombination reaction, termed 'ET-cloning', combines a homologous recombination reaction between linear DNA fragments and circular target molecules, like BAC episomes (Zhang et al., 1998). After co-transformation of linear and circular DNA molecules, only *recBC sbcA* recipient strains resulted in the intended recombination products (insertion or deletion). Recombination was more efficient with increasing length of the homology arms, and some constructs showed higher efficiency with increasing distance between the two homology sites (Zhang et al., 1998). For ET-cloning in *recBC+* strains, which are commonly used as hosts for P1, BAC or PAC episomes, a plasmid encoding the recombination functions was constructed. In pBAD-ETγ, *recE* is under the control of an inducible promoter; and *recT* is expressed from a strong constitutive promoter. To inhibit RecBC-mediated degradation of linear DNA, the λ *gam*, encoding the Redγ protein, was incorporated on the plasmid (Zhang et al., 1998). Later, *E. coli* hosts with chromosomally-encoded, inducible recombinases have been developed to allow easy manipulation of BAC DNA (Lee et al., 2001).

1.1 The bacteriophage λ Red recombination system

Besides the mutagenesis pathway described above, Red recombination is one of the most commonly exploited techniques to foster recombination between the bacterial chromosome and linear dsDNA introduced into the cell (Murphy, 1998). The Red recombination system of the bacteriophage λ leads to a precise and rapid approach with greatly enhanced rates of recombination, compared to those found in *recBC*, *sbcB* or *recD* mutants. Its ability to catalyze the incorporation of PCR-generated DNA species led to an immense spread of the system. Numerous groups have developed various methodologies tailored to a variety of scientific questions. Besides the high recombination rate, the biggest advantage of the λ Red system, compared to previously used recombination systems, is that it accepts very short regions of homologous DNA (stretches of less than 100 nucleotides) for recombination. Because fragments of such size can be readily synthesized, there is a high degree of freedom in designing targeting constructs for recombination.

Which components make up the λ Red system? The genes of the Red system, *exo*, *bet*, and *gam*, cluster together in the P_L operon, which is expressed in the early transcriptional program of bacteriophage λ (Poteete, 2001). The three resulting λ Red gene products, Redα, Redβ and Redγ, are necessary to carry out homologous recombination of dsDNA. Redα, whose monomer has a Mr of 24 kDa, is responsible for a dsDNA-dependent exonuclease activity. It degrades one strand at the ends of a linear dsDNA molecule in the 5' to 3' direction. This generates 3' ssDNA overhangs, which are substrates for recombination (Little, 1967). The ring-shaped trimer of Redα passes dsDNA through its center at one end, but only ssDNA emerges from the other end (Kovall et al., 1997). The Redβ protein, whose monomer has a Mr of 28 kDa, acts as a ssDNA-binding protein that anneals complementary single strands and mediates strand exchange, thus finishing the recombination process (Li et al., 1998). When bound to ssDNA, Redβ forms large rings and makes a complex with Redα to promote recombination (Passy et al., 1999). Therefore, the Redα/Redβ pair has a function analogous to that of the RecE/RecT pair. Muyrers et al. (2000b) could highlight that there are specific interactions between the partners of each recombination system. The exonuclease RecE does not form complexes with the strand exchange protein Redβ or vice versa; Redα does not interact with RecT. Finally, the Redγ polypeptide, whose Mr is 16 kDa, protects linear dsDNA from degradation by binding the RecBCD protein and inhibiting its nuclease activities (Murphy, 1991).

1.2 Use of λ Red recombination for manipulation of bacterial genomes

The basic strategy of the λ Red system is the replacement of a chromosomal sequence with a (e.g., PCR-amplified) selectable antibiotic resistance gene flanked by homology extensions of distinct lengths. For genetic engineering in the E. coli chromosome, two efficient λ Red-mediated methods were developed by two independent groups. The first method utilized E. coli strains containing the P_L operon of a defective λ prophage (e.g., deletion of cro to bioA genes, which includes the lytic genes) under control of the temperature-sensitive λ cI-repressor (allele cI857, Yu et al., 2000). To express high levels of the P_L operon, cultures were shifted from repressing conditions at 32°C to inducing conditions at 42°C for 7.5 to 17.5 minutes. This was optimal for achieving maximal recombination levels. By shifting the cells back to 32°C, further expression of the P_L operon was repressed; and cell death was prevented. Furthermore, the absence of the Cro-repressor enabled P_L operon expression to be fully derepressed when the cI-repressor was inactivated under heat induction (Yu et al., 2000). Electroporation of this transiently induced λ lysogen with a short linear, PCR-generated dsDNA segment resulted in efficient recombination events (Poteete, 2001).

The second very efficient λ Red-mediated recombination approach, involved a low-copy plasmid with λ gam, bet, and exo under control of an arabinose-inducible promoter (pKD46, Datsenko et al., 2000). The plasmid pKD46 yielded greatly enhanced recombination events and was preferable to the similar pKD20. However, both harbor the Red system under a well-regulated promoter to avoid undesired reactions under non-inducing conditions and a temperature-sensitive replicon to allow for easy curing of the respective plasmid after recombination (Datsenko et al., 2000). For generating gene disruptions within the E. coli chromosome, the R6Kγ ori-based suicide plasmids pKD3 and pKD4 were used as templates to amplify the chloramphenicol (cat) or kanamycin (aph) resistance gene cassettes, respectively, in a PCR. Using primer pairs with site-specific homology extensions to amplify the resistance cassette, the resulting PCR product was used to transform a freshly competent E. coli expressing λ Red from pKD46. After successful recombination, the resistance gene cassette can be excised by Flp recombinase supplied in trans (1.2.3, Fig. 1).

One example of a possible refinement of the λ Red procedure promotes high-frequency recombination using ssDNA substrates. It has been discovered that only λ Redβ is absolutely required for ssDNA recombination (Ellis et al., 2001). Neither *exo* nor *gam* causes a dramatic effect on the recombination efficiency. Only a minor 5-fold reduction was observed in the *gam* deletion strain, most likely attributed to the single-stranded nuclease activity of the RecBCD protein complex, which is not inhibited in a *gam*-deficient strain (Ellis et al., 2001, Wang et al., 2000). λ Redβ bound to ssDNA is able to protect the DNA segment from nuclease attack, which might explain the recombination events that occured despite the *gam* deletion (Karakousis et al., 1998).

These methods offer a technology for studying bacterial gene functions or even for introducing mutations or markers in the chromosomes of eukaryotic cells, e.g., to provide special "tags" in the DNA of living cells (Ellis et al., 2001).

1.2.1 Gene deletion

λ Red recombination has been successfully used for convenient generation of gene deletions in *E. coli* (Datsenko et al., 2000), *Salmonella enterica* (Hansen-Wester et al., 2002), *Pseudomonas aeruginosa* (Liang et al., 2010, Quénée et al., 2005), *Streptomyces coelicolor* (Gust et al., 2004), *Shigella* spp. (Ranallo et al., 2006), *Yersinia pseudotuberculosis* (Derbise et al., 2003) and *Y. pestis* (Sun et al., 2008).

The first step in generating gene deletions is creating a linear targeting construct which consists usually of a resistance gene ("*res*" in Fig. 1) flanked by terminal extensions homologous to the sequences surrounding the deletion site. In Fig. 1, the homologous regions used for deletion of *orf2* are depicted in blue and red. In the simplest approach, substrates for Red recombination can be generated by adding short homology extensions to PCR primers and using them to amplify a resistance cassette (Datsenko et al., 2000, Yu et al., 2000, Zhang et al., 2000). The length of the homology arms required for efficient recombination depends on the organism. For example, 36-50 bp (base pairs) were shown to work for *E. coli* or *Salmonella enterica* (Datsenko et al., 2000, Hansen-Wester et al., 2002) and 50-100 bp for *Pseudomonas aeruginosa* (Liang et al., 2010). Much longer homology arms were required for efficient Red recombination in *Y. pseudotuberculosis* and *Y. pestis* (~500 bp, Derbise et al., 2003, Sun et al., 2008) and *Vibrio cholerae* (100-1000 bp, Yamamoto et al., 2009). For the latter examples, multiple PCRs have to be done to generate functional targeting constructs.

In the next step, the PCR product is used to transform bacteria expressing λ Red proteins. Homologous recombination results in insertion of the cassette at the precise position determined by the homology extensions (Fig. 1). Transformants can be selected using their acquired antibiotic resistance. Target regions for site-specific recombinases (Fig. 1, yellow triangles) provide the option for subsequent removal of the resistance cassette (see also 1.2.3).

1.2.2 DNA insertion

In addition to removing DNA from bacterial genomes (1.2.1), λ Red recombination can also be applied to precisely insert any DNA within a genome. This approach has been widely used for analyzing bacterial gene expression via the generation of reporter gene fusions (Gerlach et al., 2007a, Lee et al., 2009, Yamamoto et al., 2009) or epitope tagging (Cho et al., 2006, Lee et al., 2009, Uzzau et al., 2001). In a similar approach, promoter sequences can be inserted or exchanged within the genome (Alper et al., 2005, Wang et al., 2009).

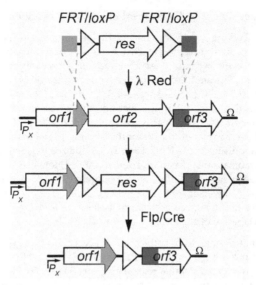

Fig. 1. Deletion of *orf2* using λ Red recombination and subsequent removal of resistance marker (*res*) via Flp or Cre recombinase. Red and blue regions denote regions of homology. The yellow region denotes the target for the site-specific recombinase.

In these cases, the targeting construct includes besides a selectable marker the DNA to be inserted. Using primers with homology extensions, these targeting constructs can be amplified by PCR from sets of template vectors available for different reporter genes (e.g., ß-galactosidase, luciferase, green fluorescent protein (*gfp*) and epitope tags (e.g., Flag®, haemagglutinin (HA), 8xmyc, 6xHis). An interesting alternative that obviates the need for multiple template vectors was developed by Gust et al. (2004). The resistance cassette and the reporter gene were amplified in separate PCRs, using in each reaction only one primer with the homology extension for Red recombination. The other primers included regions of homology to each other to allow them to anneal. The joint molecule was then used in a second round of PCR to generate a fusion cassette.

Depending on the scientific question to be answered, different integration strategies for reporter genes are available. For transcriptional fusions, a promoterless reporter gene is inserted downstream of a promoter of interest. The reporter gene may have optimized translational signals, including an optimized ribosome binding site (RBS) at the optimal distance from the start codon. If such a construct is inserted within an operon, hybrid operons are generated (Gerlach et al., 2007a). We have introduced so-called "start codon fusions," in which the reporter gene is inserted behind the native RBS and start codon of the gene under study, so that expression is assessed in the native genomic context (Gerlach et al., 2007a, Wille et al., 2012). This gene fusion strategy is closely related to translational fusions. The classical Red recombination protocol enables the easy generation of C-terminal fusion proteins, in which the reporter gene or epitope tag is inserted in-frame at any position in an open-reading-frame (ORF). For the generation of N-terminal fusions, a "scarless" recombination protocol (see 2.) has to be applied.

1.2.3 Site-specific recombination for removal of antibiotic resistance genes

Several methods, involving various site-specific recombination systems, have been developed for the removal of unwanted marker sequences from the chromosome. The most frequently used site-specific recombinases for subsequent excision of antibiotic resistance genes are Flp and Cre. Flp and Cre recombinases recognize 34-bp long sequences with palindromic elements (*FRT* and *loxP*, respectively). Both sites, *FRT* and *loxP*, consist of two 13-bp inverted repeats flanking an 8-bp non-palindromic core sequence. The asymmetry of the core sequence determines the polarity of the recombination site and has extensive consequences for the outcome of the recombination. Recombination between directly orientated recombination sites leads to excision of the intervening sequence (Fig. 1, yellow triangles), while recombination between inverted recombination sites results in an inversion of the intervening sequence. Recombination continuously takes place as long as the recombinase is present (Schweizer, 2003, Zhang et al., 2002). However, removal of resistance genes by either one of the site-specific recombinases leaves a *loxP* or *FRT* "scar" sequence within the genome (Datsenko et al., 2000, Lambert et al., 2007).

Although there is limited homology between the scar sequences themselves, they might serve as hotspots for recombination in successive recombination steps, representing a risk for unwanted deletions or chromosomal rearrangements (Datsenko et al., 2000). In addition, these scars might have influence on gene functions when operon structures or intragenic regions were modified (Blank et al., 2011). Mutations within the inverted repeats of *loxP*, resulting in the *lox66* and *lox71* alleles, can minimize genetic instability. Recombination using either generates *lox72*, which has a strongly reduced binding affinity for Cre (Lambert et al., 2007).

2. Site-directed mutagenesis using oligonucleotides

Precise insertion of chromosomal mutations has been established as the "gold standard" for analysis of bacterial gene function. Generation of point mutations, seamless deletions and in-frame gene fusions without leaving selectable markers or a recombination target site (e.g., *loxP* or *FRT*) requires reliable counterselectable markers. The published protocols are usually based on two successive rounds of recombination: (I) integration of a positively selectable marker (e.g., an antibiotic resistance gene) in the first step and (II) selection for marker loss (counterselection) in the final recombination step (Reyrat et al., 1998). Some of the counterselection methods have been shown to function together with the recombination of short synthetic DNA fragments. This circumvents the tedious and time-consuming requirement for PCR-based mutagenesis and cloning to generate mutant alleles for the second recombination step.

2.1 Counterselection with SacB

The *Bacillus subtilis sacB* gene is widely used as a genetic tool for counterselection puposes. SacB confers sucrose sensitivity to a wide range of Gram-negative bacteria (Blomfield et al., 1991, Kaniga et al., 1991). The levansucrase SacB catalyzes transfructorylation from sucrose to various acceptors, hydrolysis of sucrose and synthesis of levans (Gay et al., 1985). The molecular basis of its toxic effect on many Gram-negative bacteria is still not completely understood. It is thought that accumulation of levans in the periplasm or transfer of fructose residues to inappropriate acceptors, subsequently generating toxic compounds, might play an important role (Pelicic et al., 1996).

Linear targeting constructs harboring *sacB* combined with a kanamycin resistance gene (*neo*, Zhang et al., 1998), an ampicillin resistance gene (*bla*, Liang et al., 2010) or a chloramphenicol resistance gene (*cat*, Sun et al., 2008) were used for the first recombination rounds. An interesting marker combining the ability for positive selection and counterselection in one fusion gene is *sacB-neo*. The resulting protein SBn confers kanamycin resistance, as well as sucrose sensitivity, to Gram-negative bacteria.

The latter two methods were used to generate gene deletions within the chromosomes of *P. aeruginosa* and *Y. pestis*, respectively. In these organisms, the λ Red system was used to recombine the targeting constructs with homology extensions of 50-100 bp (Liang et al., 2010) or ~500 bp (Sun et al., 2008) flanking the cassettes.

In the homologous recombination step, clones were selected for the respective antibiotic resistance. Recombinants were selected on medium plates containing 5-7% sucrose to select for loss of the cassette. Exact timing of counterselection is a critical issue when working with SacB or SBn, since *sacB*-inactivating mutations were shown to accumulate. For example, only 10-15% (not 100%) of *sacB*-clones lost their kanamycin resistance after counterselection of SBn on sucrose plates (Muyrers et al., 2000a). Nevertheless, Muyrers and coworkers successfully used this selection scheme to introduce a point mutation within a BAC. In these experiments, the altered base was included in one of the homology arms of both targeting constructs. Interestingly, the investigators observed that the homologous arms for the second targeting construct should be at least 500 bp in length. This excludes the use of a completely synthetic oligonucleotide at that step (Muyrers et al., 2000a).

2.2 Dual selection of recombinants with GalK or ThyA

Besides the fusion protein SBn, *E. coli* galactokinase (GalK) and thymidylate synthase A (ThyA) can each be used in the dual role of both a positive selective marker (selection for the gene) and a negative selective marker (selection for the absence of the gene). Depending on the substrates used for growing recombinants, gain or loss of the markers can be selected very efficiently. Both systems were established for the manipulation of BACs in appropriate *E. coli* host strains that are deficient for *galK* or *thyA* and have the λ Red recombination system present within their genomes (Wong et al., 2005, Warming et al., 2005). These methods cannot be used for site-directed mutagenesis of genomes of bacteria with functional *galK* or *thyA*.

The *galETKM* operon enables *E. coli* to metabolize D-galactose (Semsey et al., 2007). In the initial step, GalK catalyzes the phosphorylation of D-galactose to galactose-1-phosphate. Recombinants were selected after the first recombination step with D-galactose as the sole carbon source. A functional GalK is absolutely required for growth under these conditions. Besides D-galactose, GalK can efficiently phosphorylate a galactose analog, 2-deoxy-galactose (DOG). Because the product 2-deoxy-galactose-1-phosphate cannot be further metabolized, it is enriched to toxic concentrations (Alper et al., 1975). In the second recombination step, an oligonucleotide harboring the desired mutation was used as substrate for Red recombination. Loss of GalK was selected on plates containing glycerol as sole carbon source and DOG (Warming et al., 2005).

A *thyA* mutation results in thymine auxotrophy of *E. coli*, since the enzyme is required for dTTP and, therefore, DNA *de novo* synthesis. Integration of the *thyA*-containing targeting construct in the first recombination step was selected on minimal medium in the absence of thymine. For its function, ThyA requires tetrahydrofolate (THF) as a cofactor. During the process THF is oxidized to dihydrofolate (DHF). The pool of THF is replenished from DHF by dihydrofolate reductase (DHFR). The action of DHFR can be inhibited by the antibiotic trimethoprim. Using a growth medium supplemented with trimethoprim and thymine, loss of *thyA* was selected in the second round of recombination. At this step, a PCR product containing a mutated allele was applied as targeting construct (Wong et al., 2005).

2.3 Counterselection using streptomycin resistance

Several mutants of the ribosomal protein S12 (RpsL) were shown to confer streptomycin resistance (Sm[R], Springer et al., 2001). Strains harboring such an *rpsL* allele (e.g., *rpsL150*) can be used as hosts in another counterselection method. The main principle is based on the fact that streptomycin resistance-conferring mutations are recessive in merodiploid strains (Lederberg, 1951). In the first step, the wild type (wt) allele of *rpsL* was inserted within the gene of interest using λ Red or RecE/T recombination (1.1). A cassette containing an *rpsL-neo* fusion gene was used. Clones exhibited Km[R] and Sm[S] (Heermann et al., 2008). The desired point mutation for the gene of interest had already been introduced into one of the two 50-bp homology arms of the *rpsL-neo* cassette. The *rpsL-neo* cassette was then deleted in a second round of recombination. For this step, a single-stranded oligonucleotide containing no *rpsL* allele but containing arms of homology to the gene of interest were used. To maintain the mutation, one of the homology arms again harbored the desired mutation. Recombinants that deleted the wt *rpsL* could be selected by their resistance to streptomycin. After successful recombination, the wt copy of *rpsL* was removed, and the gene of interest was re-established, now containing the desired mutation (Heermann et al., 2008).

2.4 Selection with the fusaric acid sensitivity system

A counterselection technique developed by Bochner et al. (1980) enables direct selection of tetracycline sensitive (Tc[S]) clones from a predominantly tetracycline resistant (Tc[R]) bacterial population. The method is based on the hypersensitivity of lipophilic Tc[R] cells to chelating agents, like fusaric acid or quinaldic acids. The precise mechanism of tetracycline exclusion is so far unknown and the subject of much speculation. The hypersensitivity seems to be caused by alterations of the host cell membrane, which are evoked from the expression of the tetracycline resistance gene. These alterations interfere, on one hand, with tetracycline permeation to confer tetracycline resistance, but, on the other hand, also to increase susceptibility to other toxic compounds (Bochner et al., 1980). This effect was exploited by using a medium that was effective for the selection of Tc[S] revertants. The counterselection was successful in *Salmonella*; but it was much less effective with most, especially fast-growing, *E. coli* strains (Bochner et al., 1980). Decreasing the nutrient concentration of the selection plates significantly minimized the background of Tc[R] colonies of fast-growing bacteria (Maloy et al., 1981).

The counterselection of Tc[R] clones on Bochner-Maloy plates was sometimes used as the final step in recombineering protocols. Point mutations were inserted in BACs using a combination of λ *gam* (1.1) with RecE/T (1.) to integrate the gene for tetracycline resistance.

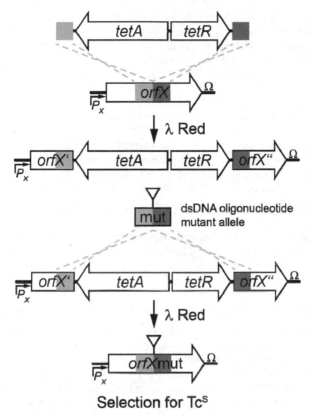

Selection for Tc^S

Fig. 2. Use of a *tet* cassette in conjunction with Bochner-Maloy Tc^S-selective medium for counterselection. Blue and red regions denote regions of homology. "mut" indicates a mutation. The mutation is first in the targetting construct, then recombination transfers it to "*orfX*" ("*orfX*"mut). "P$_X$" designates the promoter for *orfX*; "Ω" indicates a transcriptional terminator.

A PCR product carrying the desired mutation was used in the second recombination step to exchange *tet*. Recombinants (Tc^S) were selected on Bochner-Maloy plates (Nefedov et al., 2000). Similar two-step recombination approaches were also used to manipulate the genome of *Salmonella enterica* serovar Typhimurium ("*S.* Typhimurium") (Gerlach et al., 2009, Karlinsey, 2007). For this technique, a *tetAR* cassette, which encodes Tc^R, was inserted in the first step within a target gene ("*orfX*") with the help of homology extensions (blue and red, Fig. 2) resulting in tetracycline resistant clones. In the second recombination step, either PCR-derived mutant alleles (Karlinsey, 2007) or synthetic oligonucleotides (Gerlach et al., 2009) were used as targeting constructs to remove the *tetAR* cassette. For the latter rationale, no further cloning steps to generate the mutant allele (Fig. 2 "mut") were required. The applicability of this approach was proven with a *Salmonella* virulence-associated gene as an example. *siiF* encodes the putative ATPase of a type I secretion system (T1SS) located within the *Salmonella* Pathogenicity Island 4 (SPI-4, Gerlach et al., 2007b). Previous work on a homologue demonstrated, that a single amino acid exchange within the Walker Box A of the ABC (ATP-binding cassette) motif disrupted the function of the transport ATPase (Koronakis, 1995). For our mutagenesis of *siiF*, we designed oligonucleotides to introduce a silent mutation resulting in a novel *NlaIV* site (2.6.1), as well as a change of Gly at position 500 to Glu (G500E) or a change of Lys at position 506 to Leu (K506L). Both amino acid positions are within the ABC motif. As a control, we introduced a silent mutation to generate a new *SacI* restriction site (2.6.1) within *siiF*. After growth selection on Bochner-Maloy medium plates that favor the growth of Tc^S bacteria, clones were screened for the newly inserted restriction sites by the relevant restriction enzymes. Positive recombinants were subjected to functional analyses. The experiments showed the expected results: (I) no influence of the silent mutations and (II) loss of substrate secretion from the amino acid exchanges within Walker Box A (Gerlach et al., 2009).

The selection efficiency of Bochner-Maloy plates was reported not to exceed 50% (Podolsky et al., 1996). Therefore, the selection procedure was not very stringent. Exact timing of all incubation steps was necessary; but still high background might be observed, making purification of positive clones difficult. Highly increased selection efficiencies were obtained with plates containing 5-7 mM NiCl$_2$, which led to 80-100% positive Tc^S revertants (Podolsky et al., 1996).

2.5 Double-strand breaks introduced by I-SceI can be used to select recombinants

The endonuclease I-*Sce*I of the yeast *Saccharomyces cerevisiae* is a novel tool for counter-selection. I-*Sce*I is an endonuclease with a long recognition sequence of 18 bp, thus ensuring the statistical absence of natural I-*Sce*I recognition sites within bacterial genomes (Monteilhet et al., 1990). Counterselection with I-*Sce*I is based on the induction of lethal double-strand breaks (DSB) within the genome or BAC, thus inhibiting DNA propagation. The Red system promotes enough recombination that recombinants can be obtained by screening survivors.

Several methods for site-directed mutagenesis of BACs and/or bacterial genomes utilizing I-*Sce*I expression have been published. Usually the methodologies are based on the insertion of an I-*Sce*I recognition sequence together with a positively selectable marker near the sequence to be modified. Furthermore, a system allowing for transient expression of the I-*Sce*I restriction enzyme in a coordinated fashion after expression of the λ Red recombination system is required. For the manipulation of BACs, a special *E. coli* host strain was developed to facilitate the independent expression of λ Red and I-*Sce*I. *E. coli* GS1783 harbors within its genome λ Red under control of a temperature-sensitive repressor and I-*Sce*I under control of an arabinose-inducible promoter (Tischer et al., 2010). However, for modification of bacterial genomes the components for recombination and I-*Sce*I have to be provided on one or two plasmid(s). The single plasmid solutions allow for independent inducible expression of both functions using arabinose and tetracycline (pWRG99 and pGETrec3.1; Blank et al., 2011, Jamsai et al., 2003) or arabinose and rhamnose (pREDI, Yu et al., 2008).

For mutagenesis of the genomes of *Salmonella enterica* servoar Enteritidis (*S.* Enteritidis, Cox et al., 2007) and *E. coli* (Kang et al., 2004), setups with the Red components and I-*Sce*I encoded on two separate plasmids were used. Targeting constructs consisting of an I-*Sce*I recognition site, a Km^R cassette and long flanking homology regions (>200 bp) were constructed in a two-step PCR approach. After chromosomal integration of the linear DNA via λ Red-mediated recombination, clones were selected with kanamycin (Cox et al., 2007). A two-step PCR approach was used to combine long extensions homologous to *lamB* with sequences encoding antigen epitopes in the second targeting construct. For production of antibodies against foreign proteins, epitopes were inserted into the outer membrane protein LamB to facilitate surface presentation. This linear PCR product was used in a co-transformation with pBC-I-SceI at a molar ratio of 40:1 (PCR product:plasmid) into λ Red-expressing S. Enteritidis. I-*SceI* was constitutively expressed from plasmid pBC-I-SceI (Kang et al., 2004). Screening for the desired recombinants was based on the inability to grow on kanamycin-containing plates (Cox et al., 2007). Because there is no convenient possibility for plasmid curing, the pBluescript (Stratagene)-based pBC-I-SceI remains in the host strains (Cox et al., 2007, Kang et al., 2004).

GET recombination is a method developed for the manipulation of BACs. It employs λ Gam and RecE/T for recombination (1.1) and I-*SceI* for counterselection (Jamsai et al., 2003). In a study by Jamsai et al. (2003), the I-*SceI* endonuclease gene downstream of a repressed promoter, together with a constitutive gene for Km^R and an I-*SceI* recognition site, was inserted within the gene of interest in the first recombination step. As targeting construct for the second recombination step, a 1708-bp PCR product carrying the desired mutation was inserted in exchange for the I-*SceI*-kanamycin resistance cassette. I-*SceI* expression was induced for 30 minutes with addition of heat-treated chlorotetracycline. Expression was induced from both the inserted I-*SceI*-kanamycin resistance cassette and plasmid pGETrec3.1 (Jamsai et al., 2003). The kanamycin resistance cassette, with its I-*SceI* recognition site, was successfully removed from 23.6% of the colonies surviving expression of I-*SceI* (Jamsai et al., 2003).

Because no specific mechanisms were implemented in pGETrec3.1 and pBC-I-SceI to promote convenient plasmid curing, it might be difficult to get plasmid-free host strains after site-directed mutagenesis. We solved that problem by integrating a tetracycline-inducible I-*SceI* expression cassette from pST98-AS (Pósfai et al., 1999) in the temperature-sensitive λ Red expression plasmid pKD46 (Datsenko et al., 2000) to generate pWRG99 (Blank et al., 2011). In a first recombination step, a Cm^R cassette (*cat*) together with an I-*SceI* recognition site (dark green, Fig. 3A) was integrated within *phoQ* in the genome of S. Typhimurium. For that, extensions of 40 nt length homologous to the regions surrounding the intended mutation site within *phoQ* were added to the primers (Fig. 3A, blue and red). PhoQ is the histidine sensor kinase of the *Salmonella* virulence-associated two-component signaling system PhoPQ. Successful recombinants were selected using chloramphenicol, and correct integration of the resistance cassette was checked by colony PCR and sequencing. For the second round of recombination, 80mer dsDNAs, derived from oligonucleotides, were introduced into pWRG99-harboring cells expressing λ Red recombination genes. These 80mer dsDNAs were designed (I) to delete the *phoQ* gene (not shown) or (II) to introduce a threonine to isoleucine exchange at position 48 (T48I) of PhoQ, together with a new *SacII* restriction site (Fig. 3A). Expression of I-*SceI* was induced with addition of anhydrotetracycline (AHT) leading to lethal DSBs in the clones still harboring the I-*SceI* recognition site. Surviving clones were screened by PCR, restriction analysis using *SacII* as well as phenotypically. The single amino acid exchange T48I within the periplasmic domain of PhoQ results in constitutive activation of the response regulator PhoP (Miller, 1990). Besides virulence attenuation, constitutive PhoP activation leads to overexpression of the nonspecific acid phosphatase PhoN. Successful recombinants could therefore be screened phenotypically by forming blue colonies on 5-bromo-4-chloro-3-indolyl-phosphate toluidine salt (BCIP) plates due to increased PhoN activity. Correct recombination events could be further confirmed by a macrophage infection model, which showed the predicted virulence-attenuated phenotype (Blank et al., 2011).

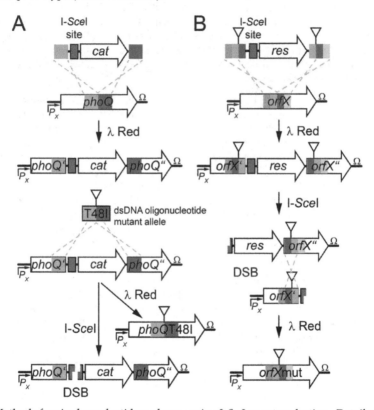

Fig. 3. Methods for single nucleotide exchange using I-*Sce*I counterselection. Details for A and B are in the text. The symbols and colors are the same as those described in Fig. 2. However in A *orfX* is *phoQ*. The antibiotic resistance is Cm[R]; the resistance gene is *cat*. The mutation (inverted triangle) is a point mutation that changes threonine at position 48 in PhoQ to isoleucine (T48I). The green rectangle is the recognition site for the I-*Sce*I endonuclease; "DSB" is the "double-strand break" induced by I-*Sce*I. In B, the symbols are the same as those in A and Fig. 2. Unique to B are "*res*" (antibiotic resistance) and the modular extensions for the primers (orange, blue, red, and light green). All are homologous to *orfX*. The two internal primers (relative to the intact gene) have the blue and red extensions, which will become the duplicated region. The "left" primer has unique homology (orange); the "right" primer has unique homology (light green).

The Red recombination system can anneal single-stranded DNA derived from dsDNA substrates into replicating homologous target sequences (1.1). Usually homologous sequences for recombination are supplied with the homology extensions flanking the targeting construct. In contrast, homologous regions flanking a DSB generated by I-SceI can also act as substrates for Red recombination. This strategy has been used for scarless site-directed mutagenesis of BACs (Tischer et al., 2010, Tischer et al., 2006) and the E. coli genome (Yu et al., 2008). These approaches require the integration of a duplicated sequence stretch in the first recombination round to serve as a substrate for recombination in the second round. For seamless deletions or insertion of point mutations, the duplications can be readily incorporated within the primers used to amplify the positive selection marker. In one study, modular sequence extensions, each ~20 nt in length (Fig. 3B; orange, blue, red and light green), were added to the primers used to amplify a resistance cassette (res). The two primers were unique but shared about 40 nt of sequence. The desired mutation (Fig. 3B, inverted triangles) was included in the duplicated sequence. This resulted in a ~40-bp sequence duplication (red and blue) after integration of the targeting construct into the gene of interest (orfX) via Red recombination (Fig. 3B, Tischer et al., 2010). The duplicated sequences were separated by the I-SceI recognition site and after induction of DSB, the Red-mediated recombination between these sequences led to the reconstitution of orfX in its mutated form orfXmut (Fig. 3B). Figure 3B shows a generalization of this strategy. If longer DNA sequences need to be integrated (1.2.2), a preceding cloning step is required to insert the selectable marker together with the I-SceI recognition site and a sequence duplication into the DNA to be inserted (Tischer et al., 2010). Combining I-SceI-induced DSB with SacB-mediated sucrose sensitivity (2.1) was shown to improve selection for loss of resistance marker within the E. coli genome (Yu et al., 2008).

One major problem of the I-SceI counterselection approach is the accumulation of point mutations within or deletion of the I-SceI recognition site during the selection process. This effect demands tight regulation of I-SceI expression to minimize selection pressure before the final (markerless) recombination takes place. It was important to optimize the procedures to maximize counterselection after the 2nd round of recombination.

2.6 Screening methods for recombinants

An underestimated problem is the screening effort needed to identify correct recombinants when using seamless recombination techniques. Although PCR fragment length polymorphism can be used in case of deletions and insert-specific PCRs in case of DNA insertion, successful single nucleotide exchanges are hard to detect. Direct phenotypical screening or the parallel introduction of novel restriction sites together with the nucleotide exchange are solutions of the problem.

2.6.1 Introduction of silent mutations to generate novel restriction sites

A screening problem arises if mutations introduced via recombineering have no direct or indirect impact on the phenotype or if the phenotypic test required is very time-consuming. Introduction of a novel restriction site adjacent to the mutation was proven to be very useful for colony screening. Designing the oligonucleotides for λ Red recombination offers the prospect of introducing silent mutations in the target region. Identification of silent mutations generating novel restriction sites can be done *in silico* using WatCut [an online tool for restriction analysis, silent mutation scanning and SNP-RFLP analysis (http://watcut.uwaterloo.ca/)] or other DNA analysis software (e.g., Clone Manager). As mentioned before, this screening method has been used successfully [e.g., for screening *siiF* recombinants by introducing a new *Nla*IV and/or *Sac*II restriction site (2.4), as well as for the *pho*QT48I mutation by generating a novel *Sac*II restriction site (2.5)].

2.6.2 Phenotypical screening

If available, phenotypical screening is the fastest way for selecting recombinants with the desired mutation. The screening is based on phenotypic differences between the mutant and the wt. In the simplest case, activity of an integrated reporter gene like *gfp* or *lacZ'* might be detected (Gerlach et al., 2007a). Another possibility for screening is the ability (gain of function) or inability (loss of function, auxotrophy) of the mutant to grow under specific conditions. In the latter case it is necessary to define permissive conditions in which the mutant can grow and restrictive conditions in which it cannot. Moreover, mutations might lead to differences in cell morphology that can be identified by microscopic examination. Last but not least, there may be a difference in the ability to utilize a chromogenic substrate, like BCIP, *p*-nitrophenyl-phosphate (pNPP) or 5-bromo-4-chloro-3-indolyl-β-D-galactopyranoside (X-Gal). These could be used for screening of positive recombinants. All of the phenotypes can originate directly from the activity of the mutated gene product or indirectly by influencing the activity and/or expression level of other proteins. One example is the PhoQ T48I mutant, which causes overexpression of PhoN. The overexpression can be monitored using the chromogenic substrate BCIP (2.5, Blank et al., 2011).

3. Conclusion

Two successive recombination steps catalyzed by the phage λ Red or phage Rac RecE/T recombination systems in combination with a negative selection procedure provide a venue for scarless mutagenesis within bacterial genomes and BACs. The outstanding ability of these enzymes to use homologous sequences as short as 35 bp as substrates for recombination allows the use of linear DNA derived from synthetic oligonucleotides as targeting constructs. The limiting step of this rationale is the availability of a reliable counterselection method. Here we gave an overview about recombination and the counterselection techniques successfully applied to the manipulation of bacterial genomes, as well as BACs.

4. References

Alper, H., Fischer, C., Nevoigt, E. & Stephanopoulos, G. (2005). Tuning genetic control through promoter engineering. *Proc Natl Acad Sci U S A*, Vol.102, No.36, pp. 12678-12683, ISSN 0027-8424

Alper, M.D. & Ames, B.N. (1975). Positive selection of mutants with deletions of the *gal-chl* region of the *Salmonella* chromosome as a screening procedure for mutagens that cause deletions. *J Bacteriol*, Vol.121, No.1, pp. 259-266, ISSN 0021-9193 (Print)

Baudin, A., Ozier-Kalogeropoulos, O., Denouel, A., Lacroute, F. & Cullin, C. (1993). A simple and efficient method for direct gene deletion in *Saccharomyces cerevisiae*. *Nucleic Acids Res*, Vol.21, No.14, pp. 3329-3330, ISSN 0305-1048

Blank, K., Hensel, M. & Gerlach, R.G. (2011). Rapid and Highly Efficient Method for Scarless Mutagenesis within the *Salmonella enterica* Chromosome. *PLoS One*, Vol.6, No.1, pp. e15763, ISSN 1932-6203

Blomfield, I.C., Vaughn, V., Rest, R.F. & Eisenstein, B.I. (1991). Allelic exchange in *Escherichia coli* using the *Bacillus subtilis* sacB gene and a temperature-sensitive pSC101 replicon. *Mol Microbiol*, Vol.5, No.6, pp. 1447-1457, ISSN 0950-382X

Bochner, B.R., Huang, H.C., Schieven, G.L. & Ames, B.N. (1980). Positive selection for loss of tetracycline resistance. *J Bacteriol*, Vol.143, No.2, pp. 926-933, ISSN 0021-9193

Cho, B.K., Knight, E.M. & Palsson, B.O. (2006). PCR-based tandem epitope tagging system for *Escherichia coli* genome engineering. *Biotechniques*, Vol.40, No.1, pp. 67-72, ISSN 0736-6205

Clark, A.J., Satin, L. & Chu, C.C. (1994). Transcription of the *Escherichia coli* recE gene from a promoter in Tn5 and IS50. *J Bacteriol*, Vol.176, No.22, pp. 7024-7031, ISSN 0021-9193

Copeland, N.G., Jenkins, N.A. & Court, D.L. (2001). Recombineering: a powerful new tool for mouse functional genomics. *Nat Rev Genet*, Vol.2, No.10, pp. 769-779, ISSN 1471-0056

Cox, M.M., Layton, S.L., Jiang, T., Cole, K., Hargis, B.M., Berghman, L.R., Bottje, W.G. & Kwon, Y.M. (2007). Scarless and site-directed mutagenesis in *Salmonella enteritidis* chromosome. *BMC Biotechnol*, Vol.7, pp. 59, ISSN 1472-6750

Datsenko, K.A. & Wanner, B.L. (2000). One-step inactivation of chromosomal genes in *Escherichia coli* K-12 using PCR products. *Proc Natl Acad Sci U S A*, Vol.97, No.12, pp. 6640-6645, ISSN 0027-8424

Derbise, A., Lesic, B., Dacheux, D., Ghigo, J.M. & Carniel, E. (2003). A rapid and simple method for inactivating chromosomal genes in *Yersinia*. *FEMS Immunol Med Microbiol*, Vol.38, No.2, pp. 113-116, ISSN 0928-8244

Ellis, H.M., Yu, D., DiTizio, T. & Court, D.L. (2001). High efficiency mutagenesis, repair, and engineering of chromosomal DNA using single-stranded oligonucleotides. *Proc Natl Acad Sci U S A*, Vol.98, No.12, pp. 6742-6746, ISSN 0027-8424

Fouts, K.E., Wasie-Gilbert, T., Willis, D.K., Clark, A.J. & Barbour, S.D. (1983). Genetic analysis of transposon-induced mutations of the Rac prophage in *Escherichia coli* K-12 which affect expression and function of recE. *J Bacteriol*, Vol.156, No.2, pp. 718-726, ISSN 0021-9193

Gay, P., Le Coq, D., Steinmetz, M., Berkelman, T. & Kado, C.I. (1985). Positive selection procedure for entrapment of insertion sequence elements in gram-negative bacteria. *J Bacteriol*, Vol.164, No.2, pp. 918-921, ISSN 0021-9193

Gerlach, R.G., Hölzer, S.U., Jäckel, D. & Hensel, M. (2007a). Rapid engineering of bacterial reporter gene fusions by using red recombination. *Appl Environ Microbiol*, Vol.73, No.13, pp. 4234-4242, ISSN 0099-2240

Gerlach, R.G., Jäckel, D., Hölzer, S.U. & Hensel, M. (2009). Rapid oligonucleotide-based recombineering of the chromosome of *Salmonella enterica*. *Appl Environ Microbiol*, Vol.75, No.6, pp. 1575-1580, ISSN 1098-5336

Gerlach, R.G., Jäckel, D., Stecher, B., Wagner, C., Lupas, A., Hardt, W.D. & Hensel, M. (2007b). *Salmonella* Pathogenicity Island 4 encodes a giant non-fimbrial adhesin and the cognate type 1 secretion system. *Cell Microbiol*, Vol.9, No.7, pp. 1834-1850, ISSN 1462-5814

Gust, B., Chandra, G., Jakimowicz, D., Yuqing, T., Bruton, C.J. & Chater, K.F. (2004). λ red-mediated genetic manipulation of antibiotic-producing *Streptomyces*. *Advances in applied microbiology*, Vol.54, pp. 107-128, ISSN 0065-2164

Hansen-Wester, I. & Hensel, M. (2002). Genome-based identification of chromosomal regions specific for *Salmonella* spp. *Infect Immun*, Vol.70, No.5, pp. 2351-2360, ISSN 0019-9567

Heermann, R., Zeppenfeld, T. & Jung, K. (2008). Simple generation of site-directed point mutations in the *Escherichia coli* chromosome using Red®/ET® Recombination. *Microbial cell factories*, Vol.7, pp. 14, ISSN 1475-2859

Jamsai, D., Orford, M., Nefedov, M., Fucharoen, S., Williamson, R. & Ioannou, P.A. (2003). Targeted modification of a human β-globin locus BAC clone using GET Recombination and an I-*Sce*I counterselection cassette. *Genomics*, Vol.82, No.1, pp. 68-77, ISSN 0888-7543

Jasin, M. & Schimmel, P. (1984). Deletion of an essential gene in *Escherichia coli* by site-specific recombination with linear DNA fragments. *J Bacteriol*, Vol.159, No.2, pp. 783-786, ISSN 0021-9193

Kang, Y., Durfee, T., Glasner, J.D., Qiu, Y., Frisch, D., Winterberg, K.M. & Blattner, F.R. (2004). Systematic mutagenesis of the *Escherichia coli* genome. *J Bacteriol*, Vol.186, No.15, pp. 4921-4930, ISSN 0021-9193

Kaniga, K., Delor, I. & Cornelis, G.R. (1991). A wide-host-range suicide vector for improving reverse genetics in gram-negative bacteria: inactivation of the *blaA* gene of *Yersinia enterocolitica*. *Gene*, Vol.109, No.1, pp. 137-141, ISSN 0378-1119

Karakousis, G., Ye, N., Li, Z., Chiu, S.K., Reddy, G. & Radding, C.M. (1998). The beta protein of phage λ binds preferentially to an intermediate in DNA renaturation. *Journal of molecular biology*, Vol.276, No.4, pp. 721-731, ISSN 0022-2836

Karlinsey, J.E. (2007) λ-Red genetic engineering in *Salmonella enterica* serovar Typhimurium. In: *Advanced Bacterial Genetics: Use of Transposons and Phage for Genomic Engineering*, K.T. Hughes, S.R. Maloy (eds.), pp. 199-209, Academic Press, ISBN 0076-6879, New York

Kovall, R. & Matthews, B.W. (1997). Toroidal structure of λ-exonuclease. *Science*, Vol.277, No.5333, pp. 1824-1827, ISSN 0036-8075

Lambert, J.M., Bongers, R.S. & Kleerebezem, M. (2007). Cre-*lox*-based system for multiple gene deletions and selectable-marker removal in *Lactobacillus plantarum*. *Appl Environ Microbiol*, Vol.73, No.4, pp. 1126-1135, ISSN 0099-2240

Lederberg, J. (1951). Streptomycin resistance: a genetically recessive mutation. *J Bacteriol*, Vol.61, No.5, pp. 549-550, ISSN 0021-9193)

Lee, D.J., Bingle, L.E., Heurlier, K., Pallen, M.J., Penn, C.W., Busby, S.J. & Hobman, J.L. (2009). Gene doctoring: a method for recombineering in laboratory and pathogenic *Escherichia coli* strains. *BMC Microbiol*, Vol.9, pp. 252, ISSN 1471-2180

Lee, E.C., Yu, D., Martinez de Velasco, J., Tessarollo, L., Swing, D.A., Court, D.L., Jenkins, N.A. & Copeland, N.G. (2001). A highly efficient *Escherichia coli*-based chromosome engineering system adapted for recombinogenic targeting and subcloning of BAC DNA. *Genomics*, Vol.73, No.1, pp. 56-65, ISSN 0888-7543

Li, Z., Karakousis, G., Chiu, S.K., Reddy, G. & Radding, C.M. (1998). The beta protein of phage λ promotes strand exchange. *Journal of molecular biology*, Vol.276, No.4, pp. 733-744, ISSN 0022-2836

Liang, R. & Liu, J. (2010). Scarless and sequential gene modification in *Pseudomonas* using PCR product flanked by short homology regions. *BMC Microbiol*, Vol.10, No.1, pp. 209, ISSN 1471-2180

Little, J.W. (1967). An exonuclease induced by bacteriophage λ. II. Nature of the enzymatic reaction. *J Biol Chem*, Vol.242, No.4, pp. 679-686, ISSN 0021-9258

Lorenz, M.G. & Wackernagel, W. (1994). Bacterial gene transfer by natural genetic transformation in the environment. *Microbiol Rev*, Vol.58, No.3, pp. 563-602, ISSN 0146-0749

Maloy, S.R. & Nunn, W.D. (1981). Selection for loss of tetracycline resistance by *Escherichia coli*. *J Bacteriol*, Vol.145, No.2, pp. 1110-1111, ISSN 0021-9193

Maresca, M., Erler, A., Fu, J., Friedrich, A., Zhang, Y. & Stewart, A.F. (2010). Single-stranded heteroduplex intermediates in λ Red homologous recombination. *BMC Mol Biol*, Vol.11, No.1, pp. 54, ISSN 1471-2199

Monteilhet, C., Perrin, A., Thierry, A., Colleaux, L. & Dujon, B. (1990). Purification and characterization of the *in vitro* activity of I-*Sce* I, a novel and highly specific endonuclease encoded by a group I intron. *Nucleic Acids Res*, Vol.18, No.6, pp. 1407-1413, ISSN 0305-1048

Murphy, K.C. (1991). λ Gam protein inhibits the helicase and □-stimulated recombination activities of *Escherichia coli* RecBCD enzyme. *J Bacteriol*, Vol.173, No.18, pp. 5808-5821, ISSN 0021-9193

Murphy, K.C. (1998). Use of bacteriophage λ recombination functions to promote gene replacement in *Escherichia coli*. *J Bacteriol*, Vol.180, No.8, pp. 2063-2071, ISSN 0021-9193

Muyrers, J.P., Zhang, Y., Benes, V., Testa, G., Ansorge, W. & Stewart, A.F. (2000a). Point mutation of bacterial artificial chromosomes by ET recombination. *EMBO Rep*, Vol.1, No.3, pp. 239-243, ISSN 1469-221X

Muyrers, J.P., Zhang, Y., Buchholz, F. & Stewart, A.F. (2000b). RecE/RecT and Redα/Redβ initiate double-stranded break repair by specifically interacting with their respective partners. *Genes Dev*, Vol.14, No.15, pp. 1971-1982, ISSN 0890-9369

Nefedov, M., Williamson, R. & Ioannou, P.A. (2000). Insertion of disease-causing mutations in BACs by homologous recombination in *Escherichia coli*. *Nucleic Acids Res*, Vol.28, No.17, pp. E79, ISSN 1362-4962

Passy, S.I., Yu, X., Li, Z., Radding, C.M. & Egelman, E.H. (1999). Rings and filaments of β protein from bacteriophage λ suggest a superfamily of recombination proteins. *Proc Natl Acad Sci U S A*, Vol.96, No.8, pp. 4279-4284, ISSN 0027-8424

Pelicic, V., Reyrat, J.M. & Gicquel, B. (1996). Expression of the *Bacillus subtilis sacB* gene confers sucrose sensitivity on mycobacteria. *J Bacteriol*, Vol.178, No.4, pp. 1197-1199, ISSN 0021-9193

Podolsky, T., Fong, S.T. & Lee, B.T. (1996). Direct selection of tetracycline-sensitive *Escherichia coli* cells using nickel salts. *Plasmid*, Vol.36, No.2, pp. 112-115, ISSN 0147-619X

Pósfai, G., Kolisnychenko, V., Bereczki, Z. & Blattner, F.R. (1999). Markerless gene replacement in *Escherichia coli* stimulated by a double-strand break in the chromosome. *Nucleic Acids Res*, Vol.27, No.22, pp. 4409-4415, ISSN 1362-4962

Poteete, A.R. (2001). What makes the bacteriophage λ Red system useful for genetic engineering: molecular mechanism and biological function. *FEMS microbiology letters*, Vol.201, No.1, pp. 9-14, ISSN 0378-1097

Quénée, L., Lamotte, D. & Polack, B. (2005). Combined *sacB*-based negative selection and *cre-lox* antibiotic marker recycling for efficient gene deletion in *Pseudomonas aeruginosa*. *Biotechniques*, Vol.38, No.1, pp. 63-67, ISSN 0736-6205

Ranallo, R.T., Barnoy, S., Thakkar, S., Urick, T. & Venkatesan, M.M. (2006). Developing live *Shigella* vaccines using λ Red recombineering. *FEMS Immunol Med Microbiol*, Vol.47, No.3, pp. 462-469, ISSN 0928-8244

Reyrat, J.M., Pelicic, V., Gicquel, B. & Rappuoli, R. (1998). Counterselectable markers: untapped tools for bacterial genetics and pathogenesis. *Infect Immun*, Vol.66, No.9, pp. 4011-4017, ISSN 0019-9567

Sawitzke, J. & Austin, S. (2001). An analysis of the factory model for chromosome replication and segregation in bacteria. *Mol Microbiol*, Vol.40, No.4, pp. 786-794, ISSN 0950-382X

Schweizer, H.P. (2003). Applications of the *Saccharomyces cerevisiae* Flp-*FRT* system in bacterial genetics. *Journal of molecular microbiology and biotechnology*, Vol.5, No.2, pp. 67-77, ISSN 1464-1801

Semsey, S., Krishna, S., Sneppen, K. & Adhya, S. (2007). Signal integration in the galactose network of *Escherichia coli*. *Mol Microbiol*, Vol.65, No.2, pp. 465-476, ISSN 0950-382X

Springer, B., Kidan, Y.G., Prammananan, T., Ellrott, K., Böttger, E.C. & Sander, P. (2001). Mechanisms of streptomycin resistance: selection of mutations in the 16S rRNA gene conferring resistance. *Antimicrobial agents and chemotherapy*, Vol.45, No.10, pp. 2877-2884, ISSN 0066-4804

Sun, W., Wang, S. & Curtiss, R., 3rd (2008). Highly efficient method for introducing successive multiple scarless gene deletions and markerless gene insertions into the *Yersinia pestis* chromosome. *Appl Environ Microbiol*, Vol.74, No.13, pp. 4241-4245, ISSN 1098-5336

Swaminathan, S., Ellis, H.M., Waters, L.S., Yu, D., Lee, E.C., Court, D.L. & Sharan, S.K. (2001). Rapid engineering of bacterial artificial chromosomes using oligonucleotides. *Genesis*, Vol.29, No.1, pp. 14-21, ISSN 1526-954X

Tischer, B.K., Smith, G.A. & Osterrieder, N. (2010) *En passant* mutagenesis: a two step markerless red recombination system. In: *In Vitro Mutagenesis Protocols: Third Edition*, J. Braman (ed.), pp. 421-430, Humana Press, ISBN 1940-6029, New York

Tischer, B.K., von Einem, J., Kaufer, B. & Osterrieder, N. (2006). Two-step red-mediated recombination for versatile high-efficiency markerless DNA manipulation in *Escherichia coli. Biotechniques*, Vol.40, No.2, pp. 191-197, ISSN 0736-6205

Uzzau, S., Figueroa-Bossi, N., Rubino, S. & Bossi, L. (2001). Epitope tagging of chromosomal genes in *Salmonella. Proc Natl Acad Sci U S A*, Vol.98, No.26, pp. 15264-15269, ISSN 0027-8424

Wang, H.H., Isaacs, F.J., Carr, P.A., Sun, Z.Z., Xu, G., Forest, C.R. & Church, G.M. (2009). Programming cells by multiplex genome engineering and accelerated evolution. *Nature*, Vol.460, No.7257, pp. 894-898, ISSN 1476-4687

Wang, J., Chen, R. & Julin, D.A. (2000). A single nuclease active site of the *Escherichia coli* RecBCD enzyme catalyzes single-stranded DNA degradation in both directions. *J Biol Chem*, Vol.275, No.1, pp. 507-513, ISSN 0021-9258

Warming, S., Costantino, N., Court, D.L., Jenkins, N.A. & Copeland, N.G. (2005). Simple and highly efficient BAC recombineering using *galK* selection. *Nucleic Acids Res*, Vol.33, No.4, pp. e36, ISSN 1362-4962

Wille, T., Blank, K., Schmidt, C., Vogt, V. & Gerlach, R.G. (2012). *Gaussia princeps* Luciferase as a Reporter for Transcriptional Activity, Protein Secretion, and Protein-Protein Interactions in *Salmonella enterica* Serovar Typhimurium. *Appl Environ Microbiol*, Vol.78, No.1, pp. 250-257, ISSN 1098-5336

Wong, Q.N., Ng, V.C., Lin, M.C., Kung, H.F., Chan, D. & Huang, J.D. (2005). Efficient and seamless DNA recombineering using a thymidylate synthase A selection system in *Escherichia coli. Nucleic Acids Res*, Vol.33, No.6, pp. e59, ISSN 1362-4962

Yamamoto, S., Izumiya, H., Morita, M., Arakawa, E. & Watanabe, H. (2009). Application of λ Red recombination system to *Vibrio cholerae* genetics: Simple methods for inactivation and modification of chromosomal genes. *Gene*, Vol.438, No.1-2, pp. 57-64, ISSN 1879-0038

Yu, B.J., Kang, K.H., Lee, J.H., Sung, B.H., Kim, M.S. & Kim, S.C. (2008). Rapid and efficient construction of markerless deletions in the *Escherichia coli* genome. *Nucleic Acids Res*, Vol.36, No.14, pp. e84, ISSN 1362-4962

Yu, D., Ellis, H.M., Lee, E.C., Jenkins, N.A., Copeland, N.G. & Court, D.L. (2000). An efficient recombination system for chromosome engineering in *Escherichia coli. Proc Natl Acad Sci U S A*, Vol.97, No.11, pp. 5978-5983, ISSN 0027-8424

Zhang, Y., Buchholz, F., Muyrers, J.P. & Stewart, A.F. (1998). A new logic for DNA engineering using recombination in *Escherichia coli. Nat Genet*, Vol.20, No.2, pp. 123-128, ISSN 1061-4036

Zhang, Y., Muyrers, J.P., Testa, G. & Stewart, A.F. (2000). DNA cloning by homologous recombination in *Escherichia coli. Nat Biotechnol*, Vol.18, No.12, pp. 1314-1317, ISSN 1087-0156

Zhang, Z. & Lutz, B. (2002). Cre recombinase-mediated inversion using lox66 and lox71: method to introduce conditional point mutations into the CREB-binding protein. *Nucleic Acids Res*, Vol.30, No.17, pp. e90, ISSN 1362-4962

11

Using Cys-Scanning Analysis Data in the Study of Membrane Transport Proteins

Stathis Frillingos
University of Ioannina Medical School
Greece

1. Introduction

Membrane transport proteins represent a core group of gene products in all known genomes and play crucial roles in either human physiology and disease or diverse environmental adaptations of microorganisms. Despite their importance, analyses of such proteins have long been hindered by a lack of high-resolution models, which reflects the inherent problems of studying membrane transport proteins outside the membrane. Recently, however, progress with crystallography and structural modeling of membrane transport proteins and expanding information on new sequence entries from genome analyses have set the stage for more systematic mutagenesis approaches to link high-resolution structural data with functional evidence.

The current structural insight on secondary transporters and their classification raises new fundamental questions on the relationships of structure with function. A crucial aspect is that many of the functionally divergent homologs, or even separate families of secondary transporters, are evolutionarily and structurally related. Thus, the majority of known structures fall in only two common folds: lactose permease (LacY) and the neurotransmitter-sodium symporter prototype (LeuT). Other transporters with different folds still display the core feature of organization in structural repeats that coordinate to form the dynamic binding site (Boudker & Verdon, 2010). The binding site operates through the commonly accepted alternating access mechanism, which is now explained as involving both rocking movements of domains pivoting at the binding site and local motions of outside and inside gates flanking the binding site, leading to alterations between outward-facing, inward-facing and intermediate substrate-occluded conformation states (Forrest et al., 2011). Apart from the structural knowledge, systematic insight on the secondary transport mechanisms requires the concerted use of structural and functional approaches, as was achieved in the seminal case of lactose permease (Kaback et al., 2011). This is important to note, since the X-ray structures represent static snapshots of highly dynamic proteins outside their native membrane environment; and, with few exceptions (Weyand et al., 2011), interpretations have been based on compilations of different snapshots from different structural homologs.

In the context of the recent structural evidence, it is striking that transport protein families tend to display high evolutionary conservation in sequence and overall structure; but they also display high functional variations between homologs, implying that relatively few side-chain changes may account for key local effects on active-site conformation and function.

However, mutations responsible for such evolutionary plasticity are rarely discernible in the background of many optimizing, permissive or near-neutral mutations that have accumulated in the present-day sequences over evolutionary time (Harms & Thornton, 2010). Traditional structure-function analysis often fails to annotate such important residues because the mutations may be uninformative or lead to complete loss-of-function/structure phenotypes in the native background. Therefore, a rationally designed mutagenesis study of one particular homolog may yield important information on the overall active-site architecture and mechanism; but it is not enough to reveal the spectrum of molecular determinants dictating the different specificity trends within the family.

In practice, it is common that an evolutionarily-broad transporter family consists of several hundreds or even thousands of related members. They share the same structural fold and binding-site architecture based on X-ray crystallography of one, usually distal, homolog. Few of the members might have been characterized extensively with respect to function, and only one or two might have been studied rigorously with site-directed mutagenesis approaches. How then can we explore the whole spectrum of specificities in such families and derive more essential information on the molecular basis of the different functional profiles?

One approach to this research problem is based on data derived from Cys-scanning analysis. Cys-scanning mutagenesis has already proven to be an essential tool (often referred to as the "gold standard") for the study of structure-function relationships in membrane proteins (Frillingos et al., 1998). In several separate examples of membrane transporters, it has provided valuable insight on active site conformation and function, even before an X-ray structure has become available (Chen & Rudnick, 2000; Kaback et al., 2001; Sorgen et al., 2002; Zomot et al., 2002). Revisiting Cys-scanning evidence after elucidation of a corresponding high-resolution structure has yielded important novel implications on the mechanism (Crisman et al., 2009; Forrest et al., 2008; Guan & Kaback, 2006; Kaback et al., 2007, 2011). The site-specific knowledge derived from Cys-scanning analysis of a specific transporter can also be used, in principle, to design rapid and cost-effective mutagenesis strategies for the functional analysis of different, structurally-related homologs.

The rationale of this approach is to use data from a systematic Cys-scanning analysis of one homolog (the study prototype) in combination, when applicable, with homology-modeling to select new homologs for targeted mutagenesis studies. This approach is based on differences and the extent of conservation, not in the overall sequence, but in the subset of residues delineated as putatively important from the Cys-scanning data. These putatively important residues correspond to positions of the study prototype where a native amino acid is irreplaceable or replaceable with only few side chains or is sensitive to site-specific alkylation of a substituted Cys leading to inactivation. In several experimental paradigms, such positions have been shown to be (a) relatively few (less than 15% of the total number of residues, in general); (b) much more highly conserved among homologs than in the rest of the protein and often mapping within conserved motif sequences (for example, see Georgopoulou et al., 2010); and (c) linked either directly or indirectly with the substrate binding site conformation and function (Kaback et al., 2007). This latter property allows the use of this set of positions as targets of rationalized mutagenesis in new homologs, in order not only to provide a measure of the functional conservation between different homologs, but also to delineate determinants responsible for particular switches in substrate preference or specificity.

In this chapter, I first analytically explain the rationale, discrete steps and aims of the Cys-scanning-based approach. I then describe applications in two families of ion gradient-driven membrane transporters. The first is the oligosaccharide-proton symporter family (OHS), which includes the lactose permease LacY and other closely related sugar transporters. The second is the nucleobase-ascorbate transporter family (NAT/NCS2), which is evolutionarily ubiquitous and encompasses more distantly related homologs and specificities. It is important to emphasize that such an experimental approach, although conceptually sound, has seen limited applications in the field of membrane transport proteins to date. A more systematic and generalized application of this approach is expected to have a major impact on the field, since it should allow rapid and effective mutagenesis designs. The impact will occur even in the absence of a high-resolution model, provided only that Cys-scanning analysis has been performed for one of the transporter homologs.

2. Rationale, discrete steps, and aims of the approach

Cys-scanning mutagenesis is a well-established strategy for structure-function analysis of proteins. It has proven particularly useful and provided valuable insight for the analysis of polytopic membrane proteins and, in particular, membrane transporters. Cys-scanning protocols rely on the engineering and availability of functional protein variants that are devoid of all or part of the native Cys residues (Cys-less or Cys-depleted versions, respectively) and the use of these Cys-less or Cys-depleted versions as a background for site-specific mutagenesis to introduce new single-Cys replacements at selected positions. The term *scanning* derives from the common application of this strategy to individually replace each amino acid residue in a contiguous sequence portion or even in the whole sequence of a protein with Cys and create an extensive library of single-Cys replacement mutants for this protein. In addition, a battery of different site-directed techniques can be applied to probe specific features of each Cys-substituted position (accessibility to solvent, relevance to substrate binding, sensitivity to the conformational changes of turnover, proximity to other sites in the protein) with appropriate sulfhydryl-specific reagents. Thus, Cys-scanning analysis often yields a wealth of data that are used to build comprehensive structure-mechanistic models for the protein under study, even in the absence of high-resolution crystallographic evidence (Frillingos et al., 1998; Sorgen et al., 2002; Tamura et al., 2003).

Nevertheless, the research potential of Cys-scanning analysis is not limited to the systematic study of structure-function relationships of individual proteins. The evidence derived from Cys-scanning analysis of a study prototype can serve as a basis to materialize rapid and effective mutagenesis designs in new, previously unknown or unstudied proteins that are related to the study prototype by sequence or structure homology. In this way, evolutionarily broad families of related proteins can be studied with respect to their active site consensus architecture and function, the spectrum of different specificity trends and mechanistic deviations, and the molecular determinants responsible for these differences. New homologs and mutagenesis targets must be selected appropriately, on the basis of the prototypic Cys-scanning evidence. Such a Cys-scanning-based approach may prove extremely contributory in the field of membrane transporters. It can increase the low representation of experimentally characterized homologs in most transporter families and reduce the paucity of crystallographic structural models, as described in the *Introduction* section.

Five discrete steps of the approach are analyzed below, with emphasis on the rationale referring to membrane transport proteins. The probable outcomes from application of this strategy (as delineated in 2.5) are substantiated further by two research paradigms presented in section 3. The two paradigms are rather seen as pilot experimental studies which demonstrate the applicability and importance of such an approach for dissecting substrate recognition and selectivity determinants in new, *ab initio* studied transporters.

2.1 Capitalizing on the Cys-scanning mutagenesis legacy

Cys-scanning mutagenesis and site-directed cysteine modification has been widely used to elucidate structure-function relationships in membrane transport proteins. The reasons for this broad application are both practical and conceptual.

Practical reasons include:

a.   the feasibility of engineering of many bacterial transporters devoid of native Cys residues (Cys-less versions) that are functionally equivalent to wild type (for example, Culham et al., 2003; Jung, H. et al., 1998; Sahin-Tóth et al., 2000; Slotboom et al., 2001; van Iwaarden et al., 1991; Weissborn et al., 1997) and represent an ideal substrate for Cys-scanning analyses, as well as the robust evidence that the vast majority of single-Cys replacement mutants do not affect dramatically the transporter expression, structural integrity or function (see Frillingos et al., 1998; Tamura et al., 2003);

b.   the availability of a diverse compendium of thiol-specific reagents from established companies such as Molecular Probes, Toronto Research Chemicals, and others;

c.   the range of thiol-specific reagents and strategies developed for membrane proteins, such as substituted-cysteine accessibility method (SCAM) using hydrophilic methanethiosulfonate (MTS) derivatives (Akabas et al., 1992) or other reagents (Yan & Maloney, 1993), cysteine-cysteine cross linking protocols (Wu & Kaback, 1996), site-directed fluorescence spectroscopy (Jung, K. et al., 1993; Wu et al., 1995), site-directed spin labeling (SDSL) (Voss et al., 1995), and site-directed alkylation in situ with radioactive (Frillingos & Kaback, 1996; Guan & Kaback, 2007) or fluorescent probes (Georgopoulou et al., 2010; Jiang et al., 2011);

d.   the obvious advantage of Cys-scanning technologies over strategies, like Ala-scanning mutagenesis, which do not allow further site-specific derivatization of the substituted amino acid;

e.   the fact that high-resolution crystallographic models did not appear for membrane transport proteins until the last two decades, *e.g.*, the ion gradient-driven transporters, for which the first X-ray structure appeared less than a decade ago (see Abramson et al., 2003), due to inherent difficulties with these hydrophobic, integral in the membrane and conformationally dynamic proteins. This delay allowed sufficient time for Cys-scanning applications to expand and provide alternative low-resolution approaches to the study of structure and mechanism (see Kaback & Wu, 1997).

Conceptual reasons include:

a.   the success of systematic Cys-scanning analyses in revealing important residues of membrane transport proteins, including irreplaceable residues, binding-site residues or residues that are important conformationally for the mechanism of energy coupling. This is based on two parameters: (i) the utility of using single-Cys mutants in indicating positions of low significance for the mechanism (active and alkylation-insensitive Cys mutants) and, at the same time, delineating the relatively few residues of potentially major significance (inactive or alkylation-sensitive Cys mutants) for more extensive study with site-directed mutagenesis; (ii) the diverse array of specific Cys modification reagents and protocols that have been developed and used to probe accessibility to solvent, relevance to substrate binding, sensitivity to the conformational changes of turnover, proximity to other sites or other functional properties for each Cys-substituted position (Frillingos et al., 1998);

b. the fact that low-resolution models derived for membrane transport proteins with Cys-scanning approaches continue to provide insight for this class of proteins, even in the post-crystallization era of research (see Kaback et al., 2007, 2011). Most characteristically, the information on the conformational dynamics of an active transport protein deduced from appropriate site-directed Cys modification assays is a valuable complement to the static crystal-structure images. Such information is always needed for an integrated insight on the transport mechanism (Kaback et al., 2011).

The wealth of data derived from the library of single-Cys, paired-Cys and other site-directed mutants produced for a particular transporter in the course of a Cys-scanning mutagenesis study can be used to design new approaches for the analysis of other homologs that might be poorly studied or even not characterized previously with respect to function. In many cases, the availability of at least one high-resolution structural prototype (from a solved X-ray structure for one homolog representing an evolutionarily broad family or group of families with structurally related transporters) might provide valuable additional information and allow the formulation of preliminary structural models. However, even in the absence of such models, the information from Cys-scanning analysis of a prototypic homolog *per se* is sufficient to guide selection of new homologs for study and of amino acid targets for effective site-directed mutagenesis designs in these homologs. The selection of new homologs depends, of course, on the research question asked. However, a common theme is to interrogate what is the structure-functional basis of particular differences in substrate selectivity or in the specificity profile for the recognition of ligands. These questions are highly relevant to the current state of the art in the field of membrane transporters, since many different transport proteins appear to be evolutionarily and structurally related; and rather small sequence changes at key residues are expected to dictate major functional differences (Boudker & Verdon, 2010; Forrest et al., 2011; Lu et al., 2011; Weyand et al., 2011; Yousef & Guan, 2009).

2.2 Selecting new homologs and mutagenesis targets to study

The first and "rate-determining" step in the Cys-scanning-based approach for the analysis of new transporter homologs is the selection of new homologs and mutagenesis targets to study. This selection is based essentially on the set of residues delineated as important from the Cys-scanning analysis of the prototypic homolog and depends, as a consequence, on the extent and results of analysis of the study prototype (Figure 1).

In principle, the Cys-scanning data refer to the functional properties of single-Cys, paired-Cys and other site-directed mutants engineered in the course of the relevant studies. A scanning experiment has two parts at minimum:

1. Analysis of Cys-replacement mutants: the scope is to delineate positions where a mutant is inactive, of very low activity or sensitive to inactivation by specific alkylating agents such as the relatively small and membrane permeable N-ethylmaleimide (NEM), which is commonly used to scan for alkylation-sensitive cysteines. Selected mutants are analyzed for site-directed alkylation in the presence or absence of substrate or in other conditions pertinent to appropriate mechanistic questions;

2. Further site-directed mutagenesis at positions where a single-Cys mutant presents with very low or negligible activity or high sensitivity to inactivation upon alkylation with NEM: the scope is to delineate positions where several replacements yield very low activity or different kinetics or specificity than wild type, and define a pattern of permissive and non-permissive replacements (taking into account the bulk, hydrophobicity, polarity, geometry or other properties of the side chain changes).

Overall, the two lines of experiments are expected to delineate a set of residues which are crucial for the transport mechanism of the study prototype in various respects. For example, positions where bulky replacements or alkylation of a substituted cysteine with the maleimidyl adduct lead to inactivation may reflect important conformational constraints and interactions with other parts of the protein that are essential for the permease turnover (Jiang et al., 2011; Tavoulari & Frillingos, 2008). On the other hand, residues which are replaceable with few or no other side chains and, at the same time, accommodate site-specific mutants of impaired affinity or distorted specificity for substrate may be crucial for substrate recognition and binding, while positions of Cys replacements which are protected from alkylation in the presence of substrate may be at the vicinity of the binding site. It is generally true that such important residues fall in one of the following three categories: (a) irreplaceable; (b) replaceable with few other side chains; (c) sensitive to inactivation of the Cys replacement by NEM. These three potential properties can be used to define the set of important residues of the study prototype, as deduced from Cys-scanning analysis data (Karena & Frillingos, 2011; Papakostas & Frillingos, 2012). It is also unequivocally true that this set of residues represents positions with a higher degree of side chain conservation than the rest of the protein and, in cases of transporters that have been studied thoroughly, correspond to a small percentage of the total amino acids in the sequence, usually 10-15% (Frillingos et al., 1998; Georgopoulou et al., 2010; Mermelekas et al., 2010; Tamura et al., 2003). These features make this clearly defined set of residues suitable for use as a basis to select (i) new homologs for *ab intio* study and (ii) amino acid targets for site-specific mutagenesis in these homologs (Figure 1). Homology modeling is also of great value in selecting mutagenesis targets for the new homologs provided that a prototypic crystal structure is available (for examples, see section 3). For reasons explained in the previous section, the Cys-scanning approaches yield additional dynamic information on the role of specific residues that cannot be provided by the structural models *per se*.

More explicitly, the first step of the approach involves a homology search referring not to the whole coding sequence but to the set of the important residues of the study prototype (as defined above). The aim of this search is to select new homologs (from the unknown or poorly-characterized pool of sequence entries) on the basis of specific differences in sequence, implying distinct conserved patterns that might correlate with shifts in specificity. This process resembles the search for characteristic sequence motifs that are conserved as a consensus throughout a transporter family. However, it is more effective in practice, as it is reinforced with experimental data (see Georgopoulou et al., 2010; Karatza et al., 2006; Kasho et al., 2006; Papageorgiou et al., 2008). If the homolog under investigation is characterized functionally and turns out to be different in specificity from the prototype in an assayable manner, amino acid targets for site-directed mutagenesis of the new homolog are selected from the set of important residues of the prototype. The method is as follows:

1. Residues that are invariant between the new homolog and the study prototype are used as targets of conservative replacements or more extensive mutagenesis to provide a measure of the functional conservation for key conserved residues between the two transporters
2. Residues that differ between the new homolog and the study prototype are used as targets of replacements of each relevant position of the new homolog with the corresponding amino acid found in the prototype, in an attempt to modulate specificity or other substrate-recognition properties and draw comprehensive conclusions on the determinants of the different transporter preferences. Such mutagenesis designs may need to include combinatorial replacements and/or construction of cross-homolog chimeras, as described in the next section.

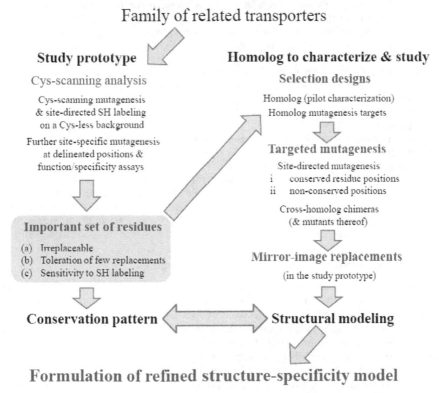

Fig. 1. Flowchart of the Cys-scanning analysis-based strategy for *ab initio* study of new transporter homologs. A detailed account of the individual steps and aims of this strategy is presented in section 2.

2.3 Exploring the use of cross-homolog chimeras and mutants

To dissect the molecular basis of different substrate selectivity trends between closely related transporters, the study of individual site-directed replacements is usually not sufficient; and more combinatorial approaches, involving multiple mutagenesis targets, are needed. The reason is that small contributions from relatively low-effect side chain changes may be crucial for the functional profile outcome, depending on the molecular background used for *in vitro* mutagenesis. In other words, the same replacements may have different effects when combined with different pre-existing mutations on the same transporter background.

Clearly, therefore, combinatorial replacements are important; and both site-specific replacement mutants and cross-homolog chimeras (replacing larger sequence regions and motifs that contain the single-amino acid targets with the homologous ones of the study prototype) can be used in this respect. In particular, the engineering of cross-homolog chimeras between transporters with different substrate profiles often yields variants that are relatively unstable or promiscuous with respect to recognition of substrates. The properties of a chimera may represent an advantageous starting point for exploring a series of mutational events leading to the evolution of new specificities (Tokuriki & Tawfik, 2009). For example, the engineered chimeras might lead to low expression, low activity or promiscuous specificity profiles (Papageorgiou et al., 2008; Papakostas et al., 2008). Chimeras with such properties can be subjected to further site-directed mutagenesis to re-introduce amino acids of the original set of important residues at specified positions. This further mutagenesis might reveal important determinants of uptake and specificity that are not evident in the native transporter backgrounds. The hypothesis is based on two pieces of evidence. (a) Mutations responsible for key functional switches in proteins often yield uninformative phenotypes. This is thought to be due to the effects of other, less important, mutations (a phenomenon known as conformational epistasis) (Harms & Thornton, 2010). (b) More promiscuous and conformationally dynamic proteins exhibit greater evolvability, or potential for divergence of new functions, as supported recently by studies involving molecular evolution, ancient gene resurrection and directed evolution (Tokuriki & Tawfik, 2009). The applicability of such combinatorial and evolution-directed studies is rather limited, at present, and has not been explored extensively in membrane transport proteins. However, this field is rapidly progressing and will probably have a significant impact on the rationale of related site-directed mutagenesis designs in the near future (Morange, 2010).

2.4 Mirror-image replacements in the prototypic transporter

Depending on the results from functional and specificity-profile analysis of the new homolog mutants and/or cross-homolog chimeras, mirror-image replacements can be designed and engineered at the corresponding positions of the study prototype to test whether a particular shift in specificity of the new homolog can be achieved in the inverse direction by replacements of the study prototype at the same residue sites. This line of experimentation is important because it reveals the extent to which a particular side chain influences the substrate recognition profile and may distinguish residues that have a major role on substrate preference from less influential ones. In addition, the comparison between mutants bearing replacements at the same site, but in different native (or chimeric) backgrounds, provides information on the functionally significant interactions of a key specificity mutation with other positions in the different related transporters.

2.5 Formulation of structure-specificity homology models

The sum of data from the comparative analysis of mutants and chimeras between the two related transporters with different specificities (*i.e.*, the new homolog and the initial study prototype) can be used to formulate refined structure-function models highlighting particular aspects of specificity. In addition, the conservation pattern of the residues involved in specificity can be taken into account to draw more generalized structure-specificity conclusions on a group of homologous transporters of the family. This process is facilitated most appropriately when an X-ray structure is available for at least one structural homolog of the transporter family under study, as is the case with the two research paradigms described in the following section.

3. Applications of the Cys-scanning-based approach

The use of a Cys-scanning analysis-based approach (as outlined in Figure 1) has seen limited application to membrane transport proteins to date. In this section, we present two paradigms of use of such an approach to address research questions on the differential substrate preference between closely related transporter homologs. The first paradigm (3.1, MelY) refers to homologs of the well known and thoroughly-studied lactose permease from *Escherichia coli* (LacY), which is a reference protein for all secondary (ion gradient-driven) active transporters. The lessons derived from the detailed site-directed analysis of LacY, which refers to frontline research contributions over almost three decades, are essential for understanding any transporter of this class (Guan & Kaback, 2006; Jiang et al., 2011). The X-ray structures solved for LacY and related homologs are also seminal, as the LacY structural fold typifies the Major Facilitator Superfamily (MFS), encompassing one fourth of all transporters, as well as other more distantly related families (Kaback et al., 2011). The second paradigm (3.2, UacT) refers to the evolutionarily ubiquitous family of nucleobase transporters NAT/NCS2, which has been studied with respect to structure-function relationships only in two members, UapA (Amillis et al., 2011) and XanQ (Karena & Frillingos, 2011). This family of transporters is modeled on a newly described, rather unusual, structural fold (Lu et al., 2011), which has important implications for the binding-site architecture and mechanisms of active transporters. It is also important with respect to potential biomedical applications concerning the development of pathogen-selective cytotoxic nucleobase analogs for targeted antimicrobial therapies (Köse & Schiedel, 2009).

3.1 Substrate selectivity of MelY (in the oligosaccharide-proton symporter family OHS)

The lactose permease from *Escherichia coli* (LacY) is a prototypic example for the study of secondary active transporters. It has been analyzed extensively with cysteine-scanning and site-directed mutagenesis leading to the delineation of residues that are crucial for the mechanism of β-galactoside:H$^+$ symport. X-ray structures of LacY solved in the presence or absence of substrate (Abramson et al., 2003; Chaptal et al., 2011; Guan et al., 2007; Mirza et al., 2006) have confirmed many of the conclusions derived from the biochemical and biophysical studies. The protein is composed of two domains, one N-terminal and one C-terminal, each representing a bundle of six transmembrane alpha-helices and designated N6 and C6, respectively (Abramson et al., 2003). Symmetrical movements between the two domains are associated with alternating opening and closure of the binding site to either side of the membrane during the mechanism of active transport (the alternating access model) (Kaback et al., 2007). High-resolution structural evidence from three homologous transporters crystallized in the inward-facing conformation and one (the fucose permease FucP) captured in an outward-facing conformation provided additional strong support for the mechanism of alternating access in lactose permease and related transport proteins (Dang et al., 2010).

LacY belongs to the Oligosaccharide:H^+ Symporter (OHS) family, a member of the Major Facilitator Superfamily (MFS) (Kasho et al., 2006). The OHS family includes several functionally characterized bacterial proteins, such as the lactose permeases of *Citrobacter freundii* and of *Klebsiella pneumoniae*, the melibiose permease (MelY) of *Enterobacter cloacae*, the raffinose permease (RafB) of *E. coli*, and the sucrose permease (CscB) of *E. coli* (Kasho et al., 2006; Vadyvaloo et al., 2006). Although specificity profiles between members of this family are often closely related, mutagenesis studies to examine the basis of substrate selectivity in members other than LacY are rare. One such example refers to *E. cloacae* MelY, a symporter that had not been analyzed for structure-function relationships prior to application of a Cys-scanning based approach (Tavoulari & Frillingos, 2008).

3.1.1 The research question: subtle selectivity difference between LacY and MelY

MelY (GenBank BAA19154) exhibits 57% identity and 75% similarity in sequence with *E. coli* LacY (UniProtKB P02920). Both proteins transport lactose or melibiose or the monosaccharide galactose (with K_m values ranging from 0.2 mM to 0.6 mM), but MelY is unable to transport the analog methyl-1-thio-β,D-galactopyranoside (TMG), that is a very efficient substrate for LacY (K_m 0.54 mM). However, MelY recognizes TMG as a ligand and conserves Cys148 (of helix TM5) in the sugar binding site as a TMG-binding residue (Tavoulari & Frillingos, 2008). Therefore, there is a subtle difference in specificity between the two galactoside transporters. The difference concerns the inability of MelY to catalyze the active transport of TMG, although it can bind this substrate analog with high affinity, comparable to LacY. (The K_i values for competitive inhibition of the lactose uptake by TMG are in the range of 1-2 mM for both transporters.)

Homology alignment and threading of MelY into the known structure of LacY (Protein Data Bank ID: 1PV7) shows that the organization of residues in the putative MelY sugar-binding site is the same as in LacY, and residues irreplaceable for the symport mechanism are conserved. Moreover, MelY differs from LacY in only 15% of the subset of residues at which either a single-Cys mutant is inactivated by site-directed alkylation or few amino acid replacements are tolerated (Figure 2). These observations provide a basis for a systematic site-directed mutagenesis study of MelY aiming at identifying subtle-selectivity determinants within the set of important LacY residues (Frillingos et al., 1998; Kaback et al., 2001), which differ in MelY. Such an approach was taken recently (Tavoulari & Frillingos, 2008) for the dissection of side chain determinants of the substrate profile in MelY relative to the well known LacY.

The difference between the two transporters concerns the uptake of one particular analog (TMG). Since the difference is clearly not in binding per se but in the subsequent translocation reactions, that inability of MelY is probably a substrate-specific impairment of binding from the conformational changes needed to complete turnover. In other words, it appears that, with TMG as a substrate, the conformational movements of MelY permease are inefficient. Thus, TMG binds but is not transported by wild-type MelY to any significant extent. To elucidate the structure-functional basis of this property in detail, both individual and combinatorial replacements were employed, involving rationally designed mutagenesis and analysis of several cross-ortholog chimeras, as described in the next sections.

3.1.2 Selection of mutagenesis targets

The initial targets for site-specific mutagenesis included residues of the important set of LacY (Figure 2) that are conserved or not conserved in MelY. Mutagenesis of MelY at conserved positions showed that irreplaceable residues of LacY are also irreplaceable in MelY (Asp-131, Arg-149, Glu-274, Arg-307, His-327, Glu-330), while Cys-153 (corresponding to Cys-148, which has been extensively utilized as a binding-site reporter in LacY) displays similar properties to those of Cys-148. Most notable is that Cys-153 of MelY, like Cys-148 of LacY, is highly sensitive to alkylation by N-ethylmaleimide (NEM) leading to inactivation of the single-Cys-153 mutant. Like in LacY, the presence of substrate fully reverses this inactivation. These initial observations established that the mechanism of galactoside transport is very similar between MelY and LacY and, in particular, that key residues of the binding site (like Cys-148/Cys-153) have the same functional role in both transporters. In addition, it was established that, although non-transportable, the analog TMG is specifically recognized as a ligand and is bound by MelY and that this binding involves the same functionally conserved residues as in LacY (Tavoulari & Frillingos, 2008).

In a second round of mutagenesis, non-conserved residues of the important set of LacY were replaced with the corresponding amino acid found in MelY or vice versa, aiming at modulating specificity to the counter homolog direction. These non-conserved positions are Leu-65/Val-70, Gly-96/Ala-101, Ala-122/Ser-127, Val-264/Als-269, Ala-279/Ser-284, Cys-355/Gln-360 and Val-367/Ala-372 (Figure 2). In the progress of studies, which involved functional analysis of mutants with respect to their efficiency for active transport of lactose, melibiose and TMG (Tavoulari & Frillingos, 2008), both combinatorial mutagenesis and chimera engineering were employed in an attempt to fully convert the one selectivity type (TMG-permissive or TMG-abortive) to the other (TMG-abortive or TMG-permissive, respectively) by replacing multiple selectivity-related targets simultaneously or larger sequence regions or domains of the transporters.

3.1.3 Cross-homolog chimeras

An interesting feature of transporters of the MFS superfamily, which highlights the conformational autonomy of the two domains (Figure 3), is that *in vivo* expression of the gene in two segments (after splitting the sequence at the central cytoplasmic loop between N6 domain and C6 domain) leads to functional complementation (Bibi & Kaback, 1990). Such functional complementation has also been observed with LacY splits at other loop sites, as depicted in Figure 2 (Kaback & Wu, 1997; Kaback et al., 2001). It was then not surprising that many of cross-homolog chimeras engineered between LacY and MelY at sites corresponding to active split junctions were also active with respect to galactoside transport (Figure 2; S. Frillingos, unpublished information). These observations emphasize the conformational flexibility allowed between the two domain repeats for establishing the dynamic binding site in LacY-type transporters. These observations are consistent with structural and modeling information as well (Radestock & Forrest, 2011). In the course of the Cys-scanning analysis-based approach for studying the selectivity profile of MelY (Tavoulari & Frillingos, 2008), active MelY/LacY chimeras were used as a background for mutagenesis and proved crucial for implementing selectivity switches from the one profile to the other, as explained in the next section.

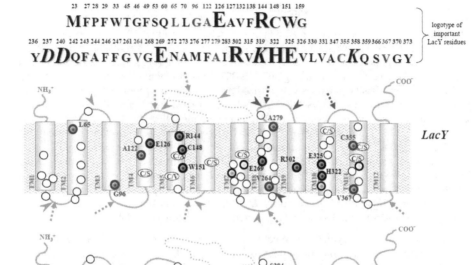

Fig. 2. **Topology models of LacY and MelY and the important set of residues in LacY.** A logotype of the important residues deduced from the Cys-scanning analysis of LacY is shown on top, with larger-size letters indicating functionally irreplaceable residues, medium size indicating residues that are replaceable with few alternative side chains and involved in binding, smaller size indicating residues where a Cys replacement is sensitive to inactivation by N-ethylmaleimide and *italics* denoting two pairs of Asp-Lys which are irreplaceable with respect to the charge-pair balance and/or orientation in the membrane (Abramson et al., 2003; Frillingos et al., 1998). Topology models are derived from the X-ray structure of LacY (PDB 1PV7) in combination with prediction algorithms and experimental data on the accessibility of loops to reagents or sequence insertions/deletions (Kaback et al., 2001) and homology threading of MelY (Tavoulari & Frillingos, 2008). The α-helical segments are indicated in rectangles (*blue* and *orange* in LacY denote helices of the N6 and the C6 domain, respectively) and the large intracellular loop between N6 and C6 is shown with a dashed line. *Arrowheads* and *broken arrows* in the LacY model denote positions of splits (the permease gene is split in two coding sequences which are expressed separately in

the same cell to test for functional complementation) or junctions of LacY/MelY chimeras, respectively; splits or chimeras are shown in *teal* (active constructs) and in *red* (inactive). The positions of the important LacY residues are shown in both models with *circles* and mutagenesis targets are *bolded* and shown in *black* (conserved residues) or in *red* (residues that differ in MelY). The eight native Cys residues of LacY and the amino acid replacing each Cys in the Cys-less permease version are shown in ellipses.

3.1.4 Site-directed mutagenesis of LacY and mirror-image replacements in MelY

The practical aim of mutagenesis studies targeted at positions of the non-conserved set of important LacY residues (Figure 2) was to establish side-chain requirements for converting the one transporter profile to the other, with focus on the criterion of whether a transporter variant can take up TMG. The switch between the two selectivity profiles was accomplished, to a major extent, by using combinations of site-specific replacements with cross-homolog chimeras interchanging the N6 and C6 domains of the two transporters (Tavoulari & Frillingos, 2008).

More analytically, the experimental strategy was applied as follows:

1. Switch from LacY to the transport profile of MelY
    *   Engineering and analysis of site-specific replacements of LacY residues:

        Site-directed mutagenesis was performed to replace each one of the important LacY residues that are not conserved in MelY (Figure 2) with the corresponding amino acid found in MelY and assay the mutants for active transport of lactose, melibiose and TMG. One of these mutants (V367A), as well as a double-replacement combining this mutation with another important-site mutation in TM11 (V367A/C355Q), showed negligible TMG uptake and a transport profile that resembles the one of MelY. The remaining six mutants (L65V, G96A, A122S, V264A, A279S and C355Q) transport lactose, melibiose or TMG at high rates and show no deviation from the LacY profile (Tavoulari & Frillingos, 2008).

    *   Analysis of chimeras that exchange domains N6 and C6 between LacY and MelY:

        The alternating access mechanism in LacY entails dynamic movements of domains N6 and C6 relative to each other to implement the conformational changes of turnover. N6-C6 chimeras, which exchange these two domains between LacY and its closely related homolog MelY, are highly active, implying a considerable flexibility in promoting such conformational changes between the two domains. Detailed functional analysis, however, revealed that both N6-C6 chimeras (with N6 from LacY and C6 from MelY or vice versa) were incapable of transporting TMG although they recognized TMG as a lactose-competitive ligand; on the other hand, the two chimeras were equally efficient to transport lactose or melibiose as LacY and MelY (Tavoulari & Frillingos, 2008). Thus, the interchange of the two domains in these chimeras represents a mutagenesis strategy different from the single-replacement mutations in helix TM11 to convert the LacY selectivity profile to that of MelY.

2. Switch from the transport profile of MelY to the profile of LacY
    *   Mirror-image replacements of residues in MelY:

Focusing on position Ala-372/Val-367 (see 1a), site-directed replacements involving mutation A372V in combination or not with Q360C (TM11) were made in the background of MelY. These mirror-image replacements failed to restore significant TMG uptake and showed no deviation from the transport profile of MelY, implying that a single change at position Ala-372/Val-367 is insufficient to yield the TMG-permissive phenotype.

- Mutagenesis in chimeric N6-C6 backgrounds:

Combination of the mutations V367A (1a) or A372V (2a) with the corresponding N6-C6 background was examined to see whether the TMG-uptake activity can be restored by manipulating flexibility between the two domains. Strikingly, the N6(LacY)-C6(MelY/A372V) mutant showed high affinity and capacity for TMG uptake and a selectivity profile that is indistinguishable from the profile of LacY (Tavoulari & Frillingos, 2008), implying that the TMG transport cycle is restored by the interaction of Val-372(367) with residues of the N6 domain (Figure 3). The reverse mutant, N6(MelY)-C6(LacY/V369A), was indistinguishable in substrate selectivity from MelY and the parental N6-C6 chimeras.

3.1.5 The refined structure-function-selectivity model

An obvious conclusion from the above results would be that efficient transport of TMG requires fine-tuned coordination between the N6 domain and TM11 in C6 domain; this interaction is probably mediated through interactions of an alkylation-sensitive and solvent-accessible face of TM11 (Jiang et al., 2011) with residues of the N6 domain and Val-367 at the periplasmic end of TM11 might have an important contribution in this respect (Tavoulari & Frillingos, 2008).

Homology modeling of MelY in comparison with LacY (Figure 3) shows that Val-367 (TM11) is close to Ala-50 (TM2) in LacY and forms a hydrophobic network that might contribute to the functional inward-facing conformation along with other side chains from TM1, TM2 and TM5 of the N6 domain (Figure 3F). Such a network might be more crucial for TMG than for lactose or melibiose; the non-galactosyl moiety of TMG (which is small, hydrophobic and aglycon) is oriented differently in the binding pocket of LacY than are the non-galactosyl moieties of the disaccharides lactose or melibiose. The non-galactosyl moiety of TMG may promote a slightly different inward-facing conformation, in which hydrophobic interactions might play a pivotal role. On the other hand, orientation of the galactosyl moiety, which determines specificity and is bound by highly conserved and irreplaceable residues (Met-23, Glu-126, Arg-144, Cys-148, Trp-151, Glu-269), is the same for all substrates (Abramson et al., 2003). In MelY all galactosyl-binding residues are conserved; but the putative hydrophobic network at the periplasmic side is disrupted, with less bulky and/or less hydrophobic residues (V367A, A50S, I48V, F30L, A25T, V158G, I157T) (Figure 3E). This difference may account for the fact that MelY binds TMG, but fails to couple this binding to any significant transport. Interactions of the methyl group of TMG in the binding pocket of MelY might promote a tight closure of helices to the periplasmic side incompatible with active transport.

Formation of an efficient hydrophobic network will depend on the side chain contributions from the N6 half. Thus, despite the presence of a Val at position 367, packing of the helices at the periplasmic side of mutant MelY(A367V) or chimera N6(MelY)-C6(LacY), which contribute less bulky and less hydrophobic side chains from the N6 domain, might still be refractory for an efficient coupling with TMG transport. On the other hand, optimal contacts between helices allow proper formation of the inward-facing conformation and progress of turnover for the TMG. Uptake appears to be restored when Ala-367 is replaced with Val in the N6(LacY)-C6(MelY) chimera, which reintroduces the side chains of LacY in TM1, TM2 and TM5 of the N6 domain. This might be due to reestablishment of hydrophobic interaction of Val-367 with residue(s) of the N6 domain at the periplasmic side and reconstitution of an efficient hydrophobic network.

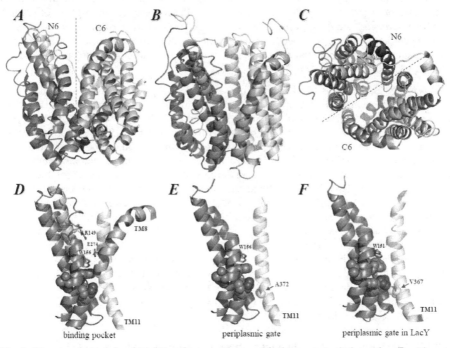

Fig. 3. **Structural models of MelY and comparison with the prototypic homolog (LacY).** The sequence of MelY was threaded on the known X-ray structure of LacY (PDB 1PV7) (Abramson et al., 2003) using the SWISSPROT modeling server. The structural models were displayed with PyMOL v1.4. The overall helix packing model is shown in three different views (*A-C*). Views *A* (from the side of the membrane) and *C* (from the periplasm) highlight the axis of pseudosymmetry (broken line), which defines the two domains (N6 and C6). Domain N6 contains the bundle of transmembrane helices TM1 (*violet*), TM2 (*raspberry*), TM4 (*wheat*), TM5 (*salmon*) and two peripheral ones, TM3 and TM6 (*blue*). Domain C6 contains the bundle of helices TM7 (*split pea green*), TM8 (*smudge green*), TM10 (*pale green*), TM11 (*yellow orange*) and the peripheral TM9 and TM12 (*grey*). View *B* highlights the central position of TM11 at the interface between the N6 and C6 bundles. The arrangement of TM11 with respect to TM8 and to the N6 bundle of TM1, TM2, TM4 and TM5 is shown more clearly in *D-F*. Panel *D* highlights key galactoside-binding residues, which are invariant

between LacY and MelY. Panel *E* (and *F*, showing LacY) highlights the cluster of residues (at TM11, TM5, TM2 and TM1), which gate the periplasmic substrate pathway and differ between MelY and LacY (see text). The residue implicated in the change of specificity (Ala-372/Val-367) is indicated with an *arrow* and *red label*.

3.2 Ab initio analysis of UacT (in the nucleobase-cation symporter-2 family NAT/NCS2)

The Nucleobase-Ascorbate Transporter (NAT) or Nucleobase-Cation Symporter-2 (NCS2) family is evolutionarily ubiquitous and includes more than 2,000 putative members in all major taxa of organisms. Despite their relevance to the recognition and uptake of several frontline purine-related drugs, only 16 members have been characterized experimentally to date. These are specific for the cellular uptake of uracil, xanthine or uric acid (microbial, plant and non-primate mammalian genomes) or vitamin C (mammalian genomes) (Gournas et al., 2008; Yamamoto et al., 2010).

The NAT/NCS2 family is of particular interest in two respects. First, in an evolutionary perspective, it encompasses transporters with largely different substrate profiles that model on a novel, unprecedented structural fold, as revealed recently (Lu et al., 2011). Second, in a biomedical perspective, it offers important possibilities for translation of the structure-function knowledge to the rational design of targeted antimicrobial drugs, based on the fact that the human homologs do not recognize nucleobases or related cytotoxic compounds (Yamamoto et al., 2010). An additional research challenge is that only 15 of the ~2000 predicted members have been identified functionally and only two members have been studied rigorously with respect to analysis of structure-function relationships, namely the xanthine permease XanQ of *E. coli* (Georgopoulou et al., 2010; Karena & Frillingos, 2011) and the uric acid/xanthine permease UapA of *A. nidulans* (Amillis et al., 2011; Papageorgiou et al., 2008). It is notable that mutagenesis data from both lines of study have revealed striking similarities between the two transporters, reinforcing the idea that few residues conserved throughout the family may be invariably critical for function and underlie specificity differences. Most of these residues are also highlighted as active-site relevant in models built on the recently released X-ray structure of the uracil permease homolog UraA (Protein Data Bank ID: 3QE7) (Lu et al., 2011).

3.2.1 The research question: distinction between xanthine and uric acid (8-oxy-xanthine)

Most of the experimentally known members of the NAT/NCS2 family have been characterized as purine nucleobase transporters, which are specific for the proton gradient-driven uptake of xanthine, uric acid (8-oxy-xanthine) or both. This group of related transporters include 11 bacterial, fungal or plant homologs, namely the xanthine transporters XanQ (UniProtKB accession number P67444) and XanP (P0AGM9) from *Escherichia coli* and PbuX (P42086) from *Bacillus subtilis*, the uric-acid transporters UacT (or YgfU) (Q46821) from *E. coli* and PucK (O32140) and PucJ (O32139) from *B. subtilis*, and the dual-selectivity uric-acid/xanthine transporters UapA (Q07307) and UapC (P487777) from the filamentous fungus *Aspergillus nidulans*, AfUapA (XP748919) from its pathogenic relative *A. fumigatus*, Xut1 (AAX2221) from the yeast *Candida albicans*, and Lpe1 (AAB17501) from maize (*Zea mays*) (see Karena & Frillingos, 2011). Based on the spectrum of their known specificities, a major research challenge is to understand the mechanism of differential recognition between xanthine and uric acid (8-oxy-xanthine) and between different binding-site preferences for xanthine analogs with variations at the imidazole moiety (8-methylxanthine, 8-azaxanthine, oxypurinol) (Goudela et al., 2005; Karatza & Frillingos, 2005).

Inspection of conserved sequence motifs and sequence alignment analysis of the different xanthine and/or uric acid-transporting homologs indicated interesting patterns of correlation with changes between xanthine-selective and xanthine/uric acid dual-selectivity NAT transporters, especially at a characteristic sequence region of transmembrane segment TM10 known as the NAT-signature motif (Georgopoulou et al., 2010). However, such sequence differences did not correlate with clear-cut changes in substrate selectivity of corresponding mutants that were made in either XanQ (Georgopoulou et al., 2010; Karatza et al., 2006) or UapA (Papageorgiou et al., 2008; Koukaki et al., 2005). In particular, the most pronounced change in XanQ was accomplished with replacement of Gly-333 to Arg (at the carboxyl-terminal end of the motif sequence) yielding aberrant recognition of 8-methylxanthine (which is not a wild-type ligand), but without affecting the selectivity preference for xanthine (Georgopoulou et al., 2010). Recognition of 8-methylxanthine has also been observed with a number of other single-replacement XanQ mutants and even with UapA/XanQ chimeras (Papakostas et al., 2008), implying that several changes at different sites in this xanthine-specific transporter can confer a degree of promiscuity for the recognition of analogs at the imidazole moiety of xanthine. However, since none of these changes resulted in a clear selectivity change (most notably, none allowed recognition or uptake of uric acid), it is evident that the strict preference of XanQ for xanthine is not easily modifiable and a more systematic approach is needed to address the basis of xanthine/8-oxy-xanthine selectivity differences. Such an approach is offered through the exploitation of evidence from a systematic Cys-scanning analysis of XanQ and the elucidation of the function of a new, uric-acid selective homolog (UacT), as described in the next section.

3.2.2 Selection of the homolog to study and the mutagenesis targets

The xanthine-specific permease XanQ has been subjected to a systematic Cys-scanning and site-directed mutagenesis study to address the role of each amino acid residue (Georgopoulou et al., 2010; Karatza et al., 2006; Karena & Frillingos, 2009, 2011; Mermelekas et al., 2010; Papakostas et al., 2008). Of more than 180 residues analyzed to date, a small set emerges as crucial for the mechanism at positions at which a native residue is functionally irreplaceable, replaceable with a limited number of side chains or sensitive to alkylation of a substituted Cys with N-ethylmaleimide leading to inactivation (Figure 4). Homology modeling showed that these functionally important residues could be implicated in substrate binding (Glu-272, Gln-324, Asp-276, Ala-323) or involved in crucial hydrogen bonding (Asn-325, His-31) or disposed to the cytoplasmic halves of TM10 and TM8, which contain key binding residues (Karena & Frillingos, 2011). Site-directed alkylation analysis of XanQ has suggested that Gln-324 and Asn-325 may participate directly in the XanQ binding site (Georgopoulou et al., 2010), while His-31 and Asn-93 are essential for the proper binding affinity and selectivity, as evidenced from ligand inhibition assays (Karena & Frillingos, 2009). In the light of the homologous UraA structure (Lu et al., 2011), it appears that the effect of His-31 (TM1) might be indirect through its interaction with Asn-325

(TM10), while Asn-93 (TM3) is at the binding pocket and may be involved in more direct interactions with substrate or substrate-binding residues (Karena & Frillingos, 2011).

The information derived from the Cys-scanning analysis of XanQ provides a basis to study structure-function relationships in other related members of the NAT/NCS2 family, which are not yet characterized or are poorly studied. In this respect, of particular interest are new homologs with distinct selectivity profiles relative to XanQ. One such homolog is UacT (more commonly known as YgfU), a low-affinity uric acid transporter from *E. coli* characterized recently (Papakostas & Frillingos, 2012). UacT is a proton-gradient-dependent, low-affinity (K_m 0.5 mM) and high-capacity transporter for uric acid that also transports xanthine, but with disproportionately low capacity. Although UacT shares low sequence homology with XanQ (28% identity of residues), it retains most of the residues of the important set identified in XanQ with Cys-scanning analysis (Figure 4). It thus offers a good substrate to apply the Cys-scanning-based approach for elucidation of changes involved in the switch of substrate preference from xanthine (XanQ) to uric acid (UacT).

To delineate targets of mutagenesis in UacT, we have taken into account residues of the important set of XanQ that are conserved or not conserved in the different-selectivity homolog, as depicted in Figure 4. The initial round of mutagenesis in UacT included (i) conservative replacements of residues that are invariant and functionally irreplaceable in XanQ (Glu-270, Asp-298, Gln-318, Asn-319); (ii) rationally designed replacements of side chains that correspond to affinity- or specificity-related residues in XanQ (His-37, Thr-100); (iii) replacements of non-conserved residues of the important set with the corresponding amino acid found in XanQ (T259V, M274D, L278T, V282S, S317A, V320N, R327G, S426N).

3.2.3 Cross-homolog chimeras and mutants

In their majority, the set of important residues identified for XanQ cluster at contiguous regions of transmembrane segments TM8 and TM10, as well as at specific sites in TM1, TM3 and TM14 (Figure 4). In a previous attempt to replace extended sequence portions containing multiple important residues of XanQ with the corresponding regions of the dual-selectivity UapA transporter from *A. nidulans* and search for deviations in substrate preference, it was striking that most of the engineered chimeric constructs were unstable and failed to express in the membrane (see Figure 4). Only one chimera of this set was expressible, but without displaying any transport activity. It was the one that replaced TM14 with the corresponding segment of UapA in the background of XanQ (Papakostas et al., 2008). Interestingly, this chimera could be rescued for active xanthine uptake with reintroduction of two residues from the important set (Asn-430, Ile-432) in the UapA-derived graft (Papakostas et al., 2008). In addition, further combinatorial replacements in this region progressively lead to restoration of full activity and the wild-type profile for xanthine selectivity and ligand specificity (Georgopoulou, K., Botou, M. & Frillingos, S., in preparation).

The difficulty in obtaining structurally stable and active chimeric constructs between the two different NAT transporters (XanQ, UapA) cannot be accounted for by the heterologous origin of the fungal UapA sequence. The engineered chimeras are transferred, induced for expression and tested in an *E. coli* K-12 host, yet similar difficulty is observed with cross-

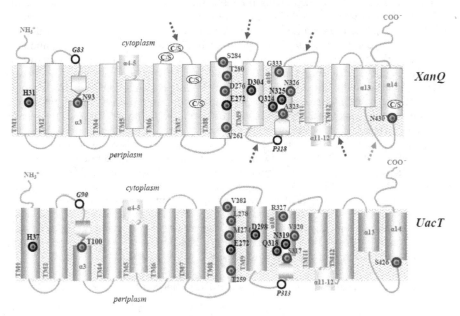

Fig. 4. Topology models of XanQ and UacT and the important set of residues in XanQ. A logotype of the important residues deduced from Cys-scanning analysis of XanQ is shown on top, with larger-size letters indicating irreplaceable residues, medium size indicating residues that are replaceable with few alternative side chains, smaller size indicating residues at which a Cys replacement is sensitive to inactivation by *N*-ethylmaleimide (IC$_{50}$ < 0.1 mM) and *italics* denoting residues that are crucial (smaller size) or irreplaceable (larger size) for expression in the membrane (Karena & Frillingos, 2009, 2011; Georgopoulou et al., 2010; Mermelekas et al., 2010). Topology models are derived from the X-ray structure of UraA (PDB 3QE7) in combination with prediction algorithms and experimental data on the accessibility of loops to hydrophilic reagents (Georgopoulou et al., 2010) and homology threading of XanQ and UacT (Karena & Frillingos, 2011; Lu et al., 2011). The α-helical segments are indicated in rectangles (*blue* and *orange* in XanQ denote helices of the core and the gate domain, respectively). *Broken arrows* in the XanQ model denote junctions of XanQ/UapA chimeras (Papakostas et al., 2008). [UapA is a fungal homolog with dual selectivity, *i.e.*, for both uric acid and xanthine.] The activities of the chimeras are denoted with *teal* (expressed in the membrane and activated upon reintroduction of particular residues; see text) and *red* (not expressed in the membrane). The positions of the important XanQ residues are shown in both models, with *circles* and mutagenesis targets *bolded* and shown in *black* (conserved residues) or in *red* (residues that differ in UacT). The five native Cys residues of XanQ and the amino acid replacing each Cys in the Cys-less permease version are shown in ellipses.

homolog chimeras between XanQ and its *E. coli* paralog UacT, although the phylogenetic distance between XanQ and UacT (28% sequence identity) is equivalent to the one between XanQ and UapA (30% sequence identity) (Georgopoulou, K. & Frillingos, S., in preparation). A more plausible interpretation stems from the intertwined-domain organization of the NAT transporters that was revealed recently from the crystal structure of the UraA homolog (Lu et al., 2011). These transporters are organized in a core and a gate domain, which are composed of two separate contiguous regions each (Figure 4). Although discontinuous in sequence, each domain represents a pair of internal repeats and forms a distinct fold in the structure (Figure 5). The core domain is thought to be pivotal for substrate binding and proton symport, and the gate domain is thought to be crucial for the conformational changes that allow access and release of substrate from the binding site (Lu et al., 2011). However, the relative arrangement of the helices of each domain, which interlace between the repeats to form the dynamic binding site, is highly sensitive to deregulation by changes at key sites. Deregulations leading to instability and loss of the protein expression can be introduced by discontinuities between TM8, TM9, TM10 and TM11 in the chimeric constructs or even by single amino acid changes at the beginning of the crucial TM10 (Pro-318) (Karatza et al., 2006) or TM3 (Gly-83) (Karena & Frillingos, 2011). Thus, sequence rearrangements within the gate domain, which is intimately associated with the binding site architecture, can be grossly deregulating or detrimental for the structural fold. The situation is different with the chimeras involving homologs of LacY (section 3.1) because each domain in the LacY fold is contiguous in sequence and the dynamic binding site is formed at the interface between the two bundles of helices, allowing more flexibility (Kaback et al., 2001).

3.2.4 Site-directed mutagenesis of UacT and mirror-image replacements in XanQ

The most significant conclusions from the analysis of individual site-directed replacements of UacT at the positions of putatively important residues (Figure 4) are that (a) functionally irreplaceable residues of XanQ (such as the substrate binding-relevant Glu-272 and Gln-324) are also irreplaceable in UacT, highlighting the functional conservation of the purine binding site in different-selectivity homologs, and (b) replacements lowering the bulk and polarity of the side chain at one position (Thr-100; TM3) allow conversion of the uric acid-selective UacT to a dual-selectivity variant (mutant with Ala in lieu of Thr-100) that transports both uric acid and xanthine (Papakostas & Frillingos, 2012). Thus, the side chain of Thr-100 at the middle of TM3 is associated directly with defining the purine substrate selectivity with respect to position 8 of the imidazole moiety. This conclusion is strengthened by mirror-image replacements made in XanQ, including an extensive site-directed mutagenesis at the corresponding amino acid found in TM3 (Asn-93) (Karena & Frillingos, 2009, 2011). Mutagenesis at Asn-93 revealed replacements that allow conversion of the xanthine-selective XanQ to dual-selectivity variants (mutants with Ala or Ser in lieu of Asn-93), even though these variants transported the non-wild-type substrate (uric acid) with very low capacity (Karena & Frillingos, 2011). The above considerations are reinforced by the fact that no other single-replacement mutants of either XanQ or UacT has been shown to convert the native transporter to a dual-selectivity one (Karena & Frillingos, 2011). In further support, a similar specificity-related effect is observed with mutants replacing the corresponding TM3 residue (Ser-154) in the fungal homolog UapA, in which introduction of an Ala in lieu of Ser-154 leads to higher affinity for xanthine relative to uric acid, thus shifting the dual-selectivity profile to the xanthine-selective direction (Amillis et al., 2011).

In summary, a major conclusion is that a polar side chain at positions Thr-100/Asn-93/Ser-154 at the middle of TM3 (Figure 4) is associated with the xanthine/uric acid selectivity. However, the selectivity changes observed with the relevant mutants are not sufficiently dramatic to emulate the properties of the other, different-selectivity homologs (Karena & Frillingos, 2011). In this respect, combinatorial replacements are needed to lead to more clear-cut shifts, involving, for example, other sites at which or sequence regions in which mutations have been shown to modify the specificity profile with respect to the imidazole moiety of the substrate (for example, recognition of 8-methylxanthine by XanQ mutants) to a lesser extent (Georgopoulou et al., 2010; Karatza et al., 2006; Karena & Frillingos, 2009).

3.2.5 The refined structure-function-selectivity models

The apparently unique selectivity-related role of Thr-100/Asn-93 in UacT (and XanQ) can be explained by taking into account the distinct conservation pattern of this residue in NAT transporters and homology modeling on the template of the UraA structure (Karena & Frillingos, 2011). In general, Asn-93 is poorly conserved as an amidic side chain even in close XanQ relatives (Karena & Frillingos, 2011); and the same is true of Thr-100 with respect to UacT relatives. However, the polar character of Thr-100/Asn-93 is conserved invariably in the known nucleobase-transporting NAT members (Asn, Thr or Ser), while the ascorbate-transporting SVCTs have an Ala at this position. Furthermore, all the dual-selectivity uric-acid/xanthine transporters (Xut1, UapA, UapC, AfUapA, Lpe1) have a Ser at the corresponding position. To understand the structural relevance of this difference, we have built structural models for the uric acid-selective UacT and the dual-selectivity fungal homologs and compared them with the one of the xanthine-selective XanQ (Figure 5 and data not shown). First of all, the models indicate that this position of TM3 is vicinal to the presumed substrate binding site formed between residues of the middle parts of TM3, TM8 and TM10 in NAT transporters (Figure 5). Strikingly, however, in UacT and all dual-selectivity NATs, the Thr or Ser replacing Asn-93 is distal from the conserved, substrate-relevant glutamate of TM8 (minimal distance between oxygen atoms, 6.0 Å), while Asn-93 in XanQ is significantly closer (distance between oxygen atoms, 4.5 Å). This difference is most prominent in the models of UacT (Figure 5) or UapA (Karena & Frillingos, 2011), which conserve nearly all the other side chains of functionally important residues of TM1, TM3, TM8 or TM10, except Asn-93.

In the dual-selectivity UapA, Ser-154 (corresponding to Asn-93 of TM3) is oriented away from the carboxyl group of Glu-356 (corresponding to Glu-272 of TM8) and leaves more space between TM3 and TM8 in the substrate binding pocket (Amillis et al., 2011; Karena & Frillingos, 2011). Thus, occupation of the Asn-93 position by Ser may relax a constraint for the recognition of analogs modified at position 8 of the imidazole moiety of xanthine and allow binding and transport of uric acid (8-oxy-xanthine), which modifies the NAT selectivity towards a less stringent, dual-substrate profile. Accordingly, the XanQ mutants replacing Asn-93 with Ser (or Ala) yield efficient recognition of 8-methylxanthine and low, but significant, uptake of uric acid, mimicking in part the fungal, dual-selectivity NATs (Karena & Frillingos, 2011).

In the uric acid transporter UacT, Thr-100 (corresponding to Asn-93) is oriented away from the carboxyl group of Glu-270 (corresponding to Glu-272 of TM8) leaving more space in the substrate binding pocket; but, at the same time, the pKa of Glu-270 may be distorted significantly relative to the corresponding carboxylic acid in XanQ due to its proximity to hydrophobic groups from Thr-100 and Met-274 (Figure 5E). These changes on the substrate binding glutamate Glu-272/Glu-270 (Lu et al., 2011) might account for the selectivity difference between UacT (uric acid) and XanQ (xanthine). Interestingly, however, replacement of Asn-93 with Thr in XanQ cannot imitate the UacT profile, but leads to low affinity for all xanthine analogs, possibly due to interference of the methyl group of Thr-93 in the vicinity of the essential Glu-272 that is not counterbalanced by other permissive mutations (Karena & Frillingos, 2009, 2011). Based on this observation, it is evident that further combinatorial replacements are needed to promptly convert XanQ to the UacT selectivity profile or vice versa; and targets for such replacements have to be selected from the residues of the important set (Figure 4), which correspond to binding site-relevant positions (Karena, E., Papakostas, K. & Frillingos, S., in preparation).

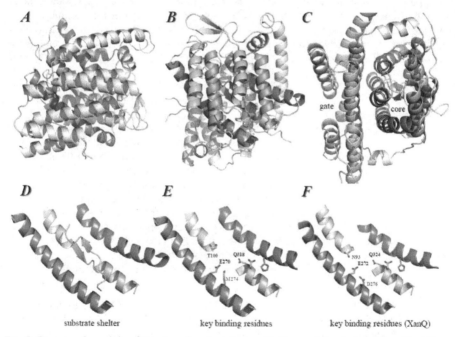

Fig. 5. **Structural models of UacT and comparison with the prototypic homolog (XanQ).** The sequence of UacT (*A-E*) or XanQ (*F*) was threaded on the known X-ray structure of UraA (PDB 3QE7) (Lu et al., 2011) using the SWISSPROT modeling server, and the structural models were displayed with PyMOL v1.4. The overall helix packing model of UacT is shown in three different views (*A-C*). View *C* (from the periplasm) highlights the two domains (core and gate), which are interplexed and not readily discerned in the side views (*A* and *B*). Transmembrane segments of the core domain (associated with substrate binding) are shown in *blue* (TM1), *wheat* (TM3), *teal* (TM10), *salmon* (TM8), *pea green* (TM2, TM4) and *forest green* (TM9, TM11), while the gate domain (associated with the conformational changes allowing access and release of substrate from the binding site) is shown in *grey*. The arrangement of the four substrate-coordinating segments (TM1, TM3,

TM8 and TM10) is shown more clearly in *D-F*. Panel *D* highlights the central position of the two short antiparallel β-strands of TM3 and TM10, which provide a shelter for the nucleobase substrate (Lu et al., 2011). Panel *E* (and *F* showing XanQ) highlights key binding-site residues, with residues differing between UacT and XanQ indicated with a *red label* (Thr-100/Asn-93, implicated in the purine selectivity preference) and an *orange label* (Met-274/Asp-276). For clarity, only the helical segments of TM3 and TM10 are shown in *E* and *F*.

4. Conclusion and perspectives

The two paradigms described above highlight the applicability of approaches that employ Cys-scanning analysis data from a reference molecule (the study prototype) to guide the *ab initio* analysis of structure-function relationships of new transporters in evolutionarily conserved families of structurally related homologs. In particular, they provide a strategy for effective site-directed mutagenesis designs to dissect the molecular determinants underlying substrate selectivity shifts between closely related homologs. In parallel, they explore the mechanism of change to novel selectivity profiles through a combination of site-specific replacements in native and chimeric transporter backgrounds. Apart from the obvious contributions to the research of groups of transporters with high potential for translation to biomedical and other applications, a more systematic and generalized application of this strategy would certainly have a major methodological impact in the field. It should allow rapid and cost-effective mutagenesis designs on newly identified membrane transport proteins, even in cases in which a high-resolution model is unavailable.

5. Acknowledgments

The experimental work presented in this chapter has been supported in part by European Community and National Funds within the frameworks of programs NONEU (Collaborations with Research and Technology Organizations outside Europe; Greece-USA) and PENED (Reinforcement Programme of Human Research Manpower) and by a Fulbright Senior Research Fellowship to the author. I wish to thank H. Ronald Kaback, Tomofusa Tsuchiya and Gérard Leblanc for support on project 3.1 (MelY) and George Diallinas and Kenneth Rudd for helpful discussions on project 3.2 (UacT). I am grateful to Sotiria Tavoulari, Panayiotis Panos, Panayiota Karatza, Ekaterini Georgopoulou, George Mermelekas, Konstantinos Papakostas and Ekaterini Karena for key research contributions during their occupation in my laboratory at Ioannina, Greece.

6. References

Abramson, J., Smirnova, I., Kasho, V., Verner, G., Kaback, H.R. & Iwata, S. (2003). Structure and mechanism of the lactose permease of *Escherichia coli*. *Science*, Vol.301, No.5633, (August 2003), pp. 610-615, ISSN 0036-8075

Akabas, M.H., Stauffer, D.A. & Karlin, A. (1992). Acetylcholine receptor channel structure probed in cysteine-substitution mutants. *Science*, Vol.258, No.5080, (October 1992), pp. 307-310, ISSN 0036-8075

Amillis, S., Kosti, V., Pantazopoulou, A., Mikros, E. & Diallinas, G. (2011). Mutational analysis and modeling reveal functionally critical residues in transmembrane segments 1 and 3 of the UapA transporter. *Journal of Molecular Biology*, Vol.411, No.3, (August 2011), pp. 567-580, ISSN 0022-2836

Bibi, E. & Kaback, H.R. (1990). In vivo expression of the *lacY* gene in two segments leads to functional lac permease. *Proceedings of the National Academy of Sciences of the U.S.A.*, Vol.87, No.11, (June 1990), pp. 4325-4329, ISSN 0027-8424

Boudker, O. & Verdon, G. (2010). Structural perspectives on secondary active transporters. *Trends in Pharmacological Sciences*, Vol.31, No.9, (September 2010), pp. 418-426, ISSN 0165-6147

Chaptal, V., Kwon, S., Sawaya, M.R., Guan, L., Kaback, H.R. & Abramson, J. (2011). Crystal structure of lactose permease in complex with an affinity inactivator yields unique insight into sugar recognition. *Proceedings of the National Academy of Sciences of the U.S.A.*, Vol.108, No.23, (June 2011), pp. 9361-9366, ISSN 0027-8424

Chen, J.G. & Rudnick, G. (2000). Permeation and gating residues in serotonin transporter. *Proceedings of the National Academy of Sciences of the U.S.A.*, Vol.97, No.3, (February 2000), pp. 1044-1049, ISSN 0027-8424

Crisman, T.J., Qu, S., Kanner, B.I. & Forrest, L.R. (2009). Inward-facing conformation of glutamate transporters as revealed by their inverted-topology structural repeats. *Proceedings of the National Academy of Sciences of the U.S.A.*, Vol.106, No.49, (December 2009), pp. 20752-20757, ISSN 0027-8424

Culham, D.E., Hillar, A., Henderson, J., Ly, A., Vernikovska, Y.I., Racher, K.I., Boggs, J.M. & Wood, J.M. (2003). Creation of a fully-functional cysteine-less variant of osmosensor and proton-osmoprotectant symporter ProP from *Escherichia coli* and its application to assess the transporter's membrane orientation. *Biochemistry*, Vol.42, No.40, (October 2003), pp. 11815-11823, ISSN 0006-2960

Dang, S., Sun, L., Huang, Y., Lu, F., Liu, Y., Gong, H., Wang, J. & Yan, N. (2010). Structure of a fucose transporter in an outward-open conformation. *Nature* Vol.467, No.7316, (October 2010), pp. 734-738, ISSN 0028-0836

Forrest, L.R., Zhang, Y.W., Jacobs, M.T., Gesmonde, J., Xie, L., Honig, B.H. & Rudnick, G. (2008). Mechanism for alternating access in neurotransmitter transporters. *Proceedings of the National Academy of Sciences of the U.S.A.*, Vol.105, No.30, (July 2008), pp. 10338-10343, ISSN 0027-8424

Forrest, L.R., Krämer, R. & Ziegler, C. (2011). The structural basis of secondary active transport mechanisms. *Biochimica et Biophysica Acta (BBA) – Bioenergetics*, Vol.1807, No.2, (February 2011), pp. 167-188, ISSN 0005-2728

Frillingos, S. & Kaback, H.R. (1996). Probing the conformation of the lactose permease of *Escherichia coli* by in situ site-directed sulfhydryl modification. *Biochemistry*, Vol.35, No.13, (April 1996), pp. 3950-3956, ISSN 0006-2960

Frillingos, S., Sahin-Tóth, M., Wu, J. & Kaback, H.R. (1998). Cys-scanning mutagenesis : a novel approach to a structure-function relationships in polytopic membrane proteins. *FASEB Journal*, Vol.12, No.13, (October 1998), pp. 1281-1299, ISSN 0892-6638

Georgopoulou, E., Mermelekas, G., Karena, E. & Frillingos, S. (2010). Purine substrate recognition by the nucleobase-ascorbate transporter motif in the YgfO xanthine permease: Asn-325 binds and Ala-323 senses substrate. *Journal of Biological Chemistry*, Vol.285, No.25, (June 2010), pp. 19422-19433, ISSN 0021-9258

Goudela, S., Karatza, P., Koukaki, M., Frillingos, S. & Diallinas, G. (2005). Comparative substrate recognition by bacterial and fungal purine transporters of the

NAT/NCS2 family. *Molecular Membrane Biology*, Vol.22, No.3, (May-June 2005), pp. 263-275, ISSN 0968-7688

Gournas, C., Papageorgiou, I. & Diallinas, G. (2008). The nucleobase-ascorbate transporter (NAT) family: genomics, evolution, structure-function relationships and physiological role. *Molecular Biosystems*, Vol.4, No.5, (May 2008), pp. 404-416, ISSN 1742-206X

Guan, L. & Kaback, H.R. (2006). Lessons from lactose permease. *Annual Reviews of Biophysics and Biomolecular Structure*, Vol.35, (June 2006), pp. 67-91, ISSN 1936-122X

Guan, L. & Kaback, H.R. (2007). Site-directed alkylation of cysteine to test solvent accessibility of membrane proteins. *Nature Protocols*, Vol.2, No.8, (August 2007), pp. 2012-2017, ISSN 1754-2189

Guan, L., Mirza, O., Verner, G., Iwata, S. & Kaback, H.R. (2007). Structural determination of wild-type lactose permease. *Proceedings of the National Academy of Sciences of the U.S.A.*, Vol.104, No.39, (September 2007), pp. 15294-15298, ISSN 0027-8424

Harms, M.J. & Thornton, J.W. (2010). Analyzing protein structure and function using ancestral gene resurrection. *Current Opinion in Structural Biology*, Vol.20, No.3, (June 2010), pp. 360-366, ISSN 0959-440X

Jiang, X., Nie, Y. & Kaback, H.R. (2011). Site directed alkylation studies with LacY provide evidence for the alternating access model of transport. *Biochemistry*, Vol.50, No.10, (March 2011), pp. 1634-1640, ISSN 0006-2960

Jung, H., Rübenhagen, R., Tebbe, S., Leifker, K., Tholema, N., Quick, M. & Schmid R. (1998). Topology of the Na+/proline transporter of *Escherichia coli*. *Journal of Biological Chemistry*, Vol.273, No.41, (October 1998), pp. 26400-26407, ISSN 0021-9258

Jung, K., Jung, H., Wu, J., Privé, G.G. & Kaback, H.R. (1993). Use of site-directed fluorescence labeling to study proximity relationships in the lactose permease of *Escherichia coli*. *Biochemistry*, Vol.32, No.46, (November 1993), pp. 12273-12278, ISSN 0006-2960

Kaback, H.R. & Wu, J. (1997). From membrane to molecule to the third amino acid from the left with a membrane transport protein. *Quarterly Reviews of Biophysics*, Vol.30, No.4, (November 1997), pp. 333-364, ISSN 0033-5835

Kaback, H.R., Sahin-Tóth, M. & Weinglass, A.B. (2001). The kamikaze approach to membrane transport. *Nature Reviews Molecular Cell Biology*, Vol.2, No.8, (August 2001), pp. 610-620, ISSN 1471-0072

Kaback, H.R., Dunten, R., Frillingos, S., Venkatesan, P., Kwaw, I., Zhang, W. & Ermolova, N. (2007). Site-directed alkylation and the alternating access model for LacY. *Proceedings of the National Academy of Sciences of the U.S.A.*, Vol.104, No.2, (January 2007), pp. 491-494, ISSN 0027-8424

Kaback, H.R., Smirnova, I., Kasho, V., Nie, Y. & Zhou, Y. (2011). The alternating access transport mechanism in LacY. *Journal of Membrane Biology*, Vol.239, No.1-2, (January 2011), pp. 85-93, ISSN 0022-2631

Karatza, P. & Frillingos, S. (2005). Cloning and functional characterization of two bacterial members of the NAT/NCS2 family in *Escherichia coli*. *Molecular Membrane Biology*, Vol.22, No.3, (May-June 2005), pp. 251-261, ISSN 0968-7688

Karatza, P., Panos, P., Georgopoulou, E. & Frillingos, S. (2006). Cysteine-scanning analysis of the nucleobase-ascorbate transporter signature motif in YgfO permease of *Escherichia coli*: Gln-324 and Asn-325 are essential and Ile-329–Val-339 form an alpha-helix. *Journal of Biological Chemistry*, Vol.281, No.52, (December 2006), pp. 39881-39890, ISSN 0021-9258

Karena, E. & Frillingos, S. (2009). Role of intramembrane polar residues in the YgfO xanthine permease: His-31 and Asn-93 are crucial for affinity and specificity, and Asp-304 and Glu-272 are irreplaceable. *Journal of Biological Chemistry*, Vol.284, No.36, (September 2009), pp. 24257-24268, ISSN 0021-9258

Karena, E. & Frillingos, S. (2011). The role of transmembrane segment TM3 in the xanthine permease XanQ of *Escherichia coli*. *Journal of Biological Chemistry*, Vol.286, No.45, (November 2011), pp. 39595-39605, ISSN 0021-9258

Kasho, V.N., Smirnova, I.N. & Kaback, H.R. (2006). Sequence alignment and homology threading reveals prokaryotic and eukaryotic proteins homologous to lactose permease. *Journal of Molecular Biology*, Vol.358, No.4, (May 2006), pp. 1060-1070, ISSN 0022-2836

Köse, M. & Schiedel, A.C. (2009). Nucleoside/nucleobase transporters: Drug targets of the future? *Future Medicinal Chemistry*, Vol.1, No.2, (May 2009), pp. 303-326, ISSN 1756-8919.

Koukaki, M., Vlanti, A., Goudela, S., Pantazopoulou, A., Gioule, H., Tournaviti, S. & Diallinas, G. (2005). The nucleobase-ascorbate transporter (NAT) signature motif in UapA defines the function of the purine translocation pathway. *Journal of Molecular Biology*, Vol.350, No.3, (July 2005), pp. 499-513, ISSN 0022-2836

Lu, F., Li, S., Jiang, Y., Jiang, J., Fan, H., Lu, G., Deng, D., Dang, S., Zhang, X., Wang, J. & Yan, N. (2011). Structure and mechanism of the uracil transporter UraA. *Nature* Vol.472, No.7342, (April 2011), pp. 243-246, ISSN 0028-0836

Mirza, O., Guan, L., Verner, G., Iwata, S. & Kaback, H.R. (2006). Structural determination of wild-type lactose permease. *EMBO Journal*, Vol.25, No.6, (March 2006), pp. 1177-1183, ISSN 0261-4189

Morange, M. (2010). How evolutionary biology presently pervades cell and molecular biology. *Journal for General Philosophy of Science*, Vol.41, No.1, (June 2010), pp. 113-120, ISSN 0925-4560

Papageorgiou, I., Gournas, C., Vlanti, A., Amillis, S., Pantazopoulou, A. & Diallinas, G. (2008). Specific interdomain synergy in the UapA transporter determines its unique specificity for uric acid among NAT carriers. *Journal of Molecular Biology*, vol.382, No.5, (October 2008), pp. 1121-1135, ISSN 0022-2836

Papakostas, K., Georgopoulou, E. & Frillingos, S. (2008). Cysteine-scanning analysis of putative helix XII in the YgfO xanthine permease: Ile-432 and Asn-430 are important. *Journal of Biological Chemistry*, Vol.283, No.20, (May 2008), pp. 13666-13678, ISSN 0021-9258

Papakostas, K., & Frillingos, S. (2012). Substrate selectivity of YgfU, a uric acid transporter from *Escherichia coli*. *Journal of Biological Chemistry*, vol.287, No.19, (May 2012), pp. 15684-15695, ISSN 0021-9258

Radestock, S. & Forrest, L.R. (2011). The alternating-access mechanism of MFS transporters arises from inverted-topology repeats. *Journal of Molecular Biology*, vol.407, No.5, (April 2011), pp. 698-715, ISSN 0022-2836

Mermelekas, G., Georgopoulou, E., Kallis, A., Botou, M., Vlantos, V. & Frillingos, S. (2010). Cysteine-scanning analysis of helices TM8, TM9a, TM9b and intervening loops in the YgfO xanthine permease: a carboxyl group is essential at position Asp-276.

Journal of Biological Chemistry, Vol.285, No.45, (November 2010), pp. 35011-35020, ISSN 0021-9258

Sahin-Tóth, M., Frillingos, S., Lawrence, M.C. & Kaback, H.R. (2000). The sucrose permease of Escherichia coli: functional significance of cysteine residues and properties of a cysteine-less transporter. *Biochemistry*, Vol.39, No.20, (May 2000), pp. 6164-6169, ISSN 0006-2960

Slotboom, D.J., Konings, W.N. & Lolkema, J.S. (2001). Cysteine-scanning mutagenesis reveals a highly amphipathic, pore-lining membrane-spanning helix in the glutamate transporter GltT. *Journal of Biological Chemistry*, Vol.276, No.14, (April 2001), pp. 10775-10781, ISSN 0021-9258

Sorgen, P.L., Hu, Y., Guan, L., Kaback, H.R. & Girvin, M.E. (2002). An approach to membrane protein structure without crystals. *Proceedings of the National Academy of Sciences of the U.S.A.*, Vol.99, No.22, (October 2002), pp. 14037-14040, ISSN 0027-8424

Tamura, N., Konishi, S. & Yamaguchi, A. (2003). Mechanisms of drug/H+ antiport: complete cysteine-scanning mutagenesis and the protein engineering approach. *Current Opinion in Chemical Biology*, Vol.7, No.5, (October 2003), pp. 570-579, ISSN 1367-5931

Tavoulari, S. & Frillingos, S. (2008). Substrate selectivity of the melibiose permease (MelY) from *Enterobacter cloacae*. *Journal of Molecular Biology*, vol.376, No.3, (February 2008), pp. 681-693, ISSN 0022-2836

Tokuriki, N. & Tawfik, D.S. (2009). Protein dynamism and evolvability. *Science*, Vol.324, No.5924, (April 2009), pp. 203-207, ISSN 0036-8075

van Iwaarden, P.R., Pastore, J.C., Konings, W.N. & Kaback, H.R. (1991). Construction of a functional lactose permease devoid of cysteine residues. *Biochemistry*, Vol.30, No.40, (October 1991), pp. 9595-9600, ISSN 0006-2960

Voss, J., Hubbell, W.L. & Kaback, H.R. (1995). Distance determination in proteins using designed metal ion binding sites and site-directed spin labeling: application to the lactose permease of *Escherichia coli*. *Proceedings of the National Academy of Sciences of the U.S.A.*, Vol.92, No.26, (December 1995), pp. 12300-12303, ISSN 0027-8424

Weissborn, A.C., Botfield, M.C., Kuroda, M., Tsuchiya, T. & Wilson, T.H. (1997). The construction of a cysteine-less melibiose carrier from *E. coli*. *Biochimica et Biophysica Acta (BBA) – Biomembranes*, Vol.1329, No.2, (October 1997), pp. 237-244, ISSN 0005-2736

Weyand, S., Shimamura, T., Beckstein, O., Sansom, M.S., Iwata, S., Henderson, P.J. & Cameron, A.D. (2011). The alternating access mechanism of transport as observed in the sodium-hydantoin transporter Mhp1. *Journal of Synchrotron Radiation*, Vol.18, No.1, (January 2011), pp. 20-23, ISSN 0909-0495

Wu, J., Frillingos, S. & Kaback, H.R. (1995). Dynamics of lactose permease of Escherichia coli determined by site-directed chemical labeling and fluorescence spectroscopy. *Biochemistry*, Vol.34, No.26, (July 1995), pp. 8257-8263, ISSN 0006-2960

Wu, J. & Kaback, H.R. (1996). A general method for determining helix packing in membrane proteins in situ: helices I and II are close to helix VII in the lactose permease of *Escherichia coli*. *Proceedings of the National Academy of Sciences of the U.S.A.*, Vol.93, No.25, (December 1996), pp. 14498-14502, ISSN 0027-8424

Vadyvaloo, V., Smirnova, I.N., Kasho, V.N. & Kaback, H.R. (2006). Conservation of residues involved in sugar/H+ symport by the sucrose permease of *Escherichia coli* relative to lactose permease. *Journal of Molecular Biology*, Vol.358, No.4, (May 2006), pp. 1051-1059, ISSN 0022-2836

Yamamoto, S., Inoue, K., Murata, T., Kamigaso, S., Yasujima, T., Maeda, J., Yoshida, Y., Ohta, K. & Yuasa, H. (2010). Identification and functional characterization of the first nucleobase transporter in mammals: implication in the species difference in the intestinal absorption mechanism of nucleobases and their analogs between higher primates and other mammals. *Journal of Biological Chemistry*, Vol.285, No.9, (February 2010), pp. 6522-6531, ISSN 0021-9258

Yan, R.T. & Maloney, P.C. (1993). Identification of a residue in the translocation pathway of a membrane carrier. *Cell*, Vol.75, No.1, (October 1993), pp. 37-44, ISSN 0092-8674

Yousef, M.S. & Guan, L. (2009). A 3D structure model of the melibiose permease of *Escherichia coli* represents a distinctive fold for Na+ symporters. *Proceedings of the National Academy of Sciences of the U.S.A.*, Vol.106, No.36, (September 2009), pp. 15291-15296, ISSN 0027-8424

Zomot, E., Zhou, Y. & Kanner, B.I. (2002). Proximity of transmembrane domains 1 and 3 of the gama-aminobutyric acid transporter GAT-1 inferred from paired cysteine mutagenesis. *Journal of Biological Chemistry*, Vol.280, No.27, (July 2005), pp. 25512-25516, ISSN 0021-9258

12

Recombineering and Conjugation as Tools for Targeted Genomic Cloning

James W. Wilson[1], Clayton P. Santiago[1],
Jacquelyn Serfecz[1] and Laura N. Quick[2]
[1]Villanova University,
[2]Children's Hospital of Philadelphia,
USA

1. Introduction

The ability to obtain DNA clones of genes that normally reside in microbial genomes was a huge technical advance in molecular biology. At first, cloning genes utilized approaches involving the complementation of mutants or the screening of genomic libraries to find sequences that hybridized to homologous DNA probes. Typically, this involved using restriction enzymes to clone random genomic fragments followed by subcloning of a smaller piece of the original clone. Then the development of PCR and genomic sequencing allowed specific genomic sequences to be amplified and cloned with more convenience. Now genes are able to be synthesized "from scratch" and ordered from various companies or institutions. However, if many genes contained on a contiguous large genomic segment are required to be cloned, significant technical barriers exist. For the purposes of this discussion, we will establish that a "large" genomic segment constitutes greater than 10 kilobases, since PCR and man-made DNA synthesis become technically challenging and/or costly above this DNA size. Therefore, a convenient, reproducible, and cost-efficient technique to clone large sections of microbial genomes would be highly advantageous.

Frequently bacteria organize genes that work together for a common function as a continuous, physically-linked series across a genome. Large genomic fragments containing many genes that work together for a specific function are very useful for the following reasons: (1) bacteria are able to be engineered for specific purposes in a "quantum leap" using such DNA clones; and (2) basic evolutionary questions are able to be answered using large genomic clones, such as: "Can the cloned gene set be expressed and functional outside of the context of the original genome/species?" These approaches extend the study of genomics by identifying potentially interesting parts of genomes identified via sequencing and studying them in different strain backgrounds. A clear example of this approach is the cloning of protein secretion systems and the subsequent study of these clones (Blondel et al. 2010; Ham et al. 1998; Hansen-Wester, Chakravortty, and Hensel 2004; McDaniel and Kaper 1997; Wilson, Coleman, and Nickerson 2007; Wilson and Nickerson 2006). However, many other gene systems can be studied in this way, with examples including polysaccharide secretion pathways (for capsule and LPS synthesis) and metabolic pathways (anabolism and/or catabolism of key molecules, such as those used in bioremediation). Our ability to extend genomics beyond sequencing to the utilization of newly-identified multi-gene pathways to engineer bacteria will depend upon our ability to clone, manipulate, and transfer large genomic fragments.

A recent strategy that exploits recombineering and conjugation provides a convenient approach to cloning large bacterial genomic fragments (Blondel et al. 2010; Santiago, Quick, and Wilson 2011; Wilson, Figurski, and Nickerson 2004; Wilson and Nickerson 2007). This approach involves insertion of recombinase sites (*e.g.*, FRT, *loxP*) at positions flanking a targeted genomic region, followed by subsequent recombinase-mediated excision of the region as a non-replicating circular molecule (Fig. 1). Then the excised region is "captured" via either site-specific or homologous recombination onto a conjugative plasmid (such as the broad-host-range IncP plasmid R995) that allows the transfer and isolation of the desired construct in a fresh recipient strain (Fig. 1). The advantages of this approach are (1) the highly specific targeting of exact cloning endpoints using recombineering and (2) the use of conjugation to allow the desired construct to be isolated away from the donor strain (in which the recombination events take place). In addition, except for the synthesis of recombineering PCR products, this protocol takes place entirely in bacterial cells, using basic, low-cost microbiological techniques. Though early approaches used subcloned DNA fragments to allow homologous recombination, the use of recombineering for both the introduction of target flanking sites and the capture on R995 alleviates the need for this subcloning.

2. Targeted cloning of large bacterial genomic fragments

2.1 The VEX-Capture technique

The original technique using this approach is termed VEX-Capture (Wilson, Coleman, and Nickerson 2007; Wilson, Figurski, and Nickerson 2004; Wilson and Nickerson 2006, 2006, 2007). The pVEX series of suicide plasmids was used to introduce *loxP* sites into regions flanking targeted genomic regions via homologous recombination (Fig. 2) (Ayres et al. 1993). Cre recombinase (expressed from a plasmid) was used to excise the targeted region and homologous recombination was used to capture the excised circle (Fig. 2). Note that the homologous recombination is driven by the endogenous bacterial RecA-mediated mechanism. A series of *Salmonella typhimurium* genomic islands ranging from 26 to 50 kilobases in size were targeted for cloning using this technique (Wilson, Coleman, and Nickerson 2007; Wilson, Figurski, and Nickerson 2004; Wilson and Nickerson 2006, 2006). Since these islands contain genes that are unique to *S. typhimurium*, one of the initial basic applications of these clones was to study their gene expression patterns in different bacteria (Wilson, Figurski, and Nickerson 2004; Wilson and Nickerson 2006). Though some *S. typhimurium* genes on the tested genomic island were expressed in all bacteria, several genes displayed genus-specific expression patterns (Fig. 3). This indicated that the mechanisms used to express these genes are absent or function differently in certain bacterial genera. These mechanisms could be the focus of study to understand gene expression functions that work only in certain bacterial groups, such as pathogens or environmental bacteria.

Two separate *S. typhimurium* type III secretion systems were cloned using the VEX-Capture approach (Wilson, Coleman, and Nickerson 2007; Wilson and Nickerson 2006). These systems are encoded at the *Salmonella* pathogenicity island 1 and 2 regions (SPI-1 and SPI-2, respectively) of the *S. typhimurium* genome (McClelland et al. 2001). Both clones are

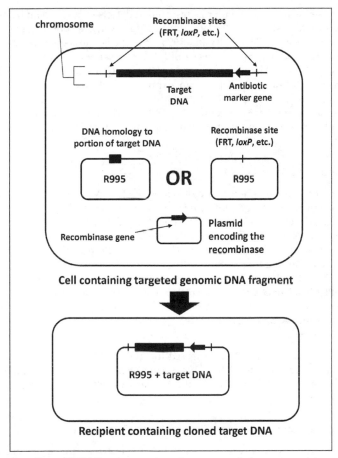

Fig. 1. General outline of VEX-Capture to clone large genomic fragments. A large fragment of a bacterial genome (generally considered as greater than 10 kilobases) is targeted for excision and cloning by inserting recombinase sites at flanking positions. At least one antibiotic marker gene is required to be associated with the target DNA for subsequent selection. The self-transmissible IncP plasmid R995 serves as a cloning vector that will capture the excised genomic fragment using either a small region of DNA homologous to the excised fragment or a corresponding recombinase site. Also co-resident in the same cell is a plasmid expressing the recombinase that recognizes the recombinase sites. Expression of the recombinase results in excision of the target DNA as a non-replicating circular molecule. This circular molecule will be inserted into R995 via homologous recombination or via the recombinase activity. This construct is conveniently isolated away from the target strain via conjugation to a differentially-marked recipient strain and selection for the appropriate markers. In the recipient strain, structural confirmation of the construct and testing for gene expression and function can occur. In addition, transfer to new bacterial recipients can be performed.

functional and serve to complement protein secretion defects in *S. typhimurium* mutants that are deleted for each SPI-1 and SPI-2 island (Fig. 4). However, the authors found remarkably different results between R995 + SPI-1 and R995 + SPI-2 when tested for expression in other Gram-negative bacteria (Fig. 5). The R995 + SPI-2 clone readily displays expression of SPI-2 (indicated using Western blot analysis of the SseB protein) in other Gram-negative genera, while the R995 + SPI-1 clone displays an expression defect outside of *S. typhimurium* (assayed using Western blot analysis of the SipA and SipC proteins). This result suggests that the regulatory mechanisms controlling SPI-1 and SPI-2 expression have evolved differently and in such a way that manifests itself upon transfer to new bacterial backgrounds.

2.2 VEX-Capture modified

A modification of VEX-Capture was used to clone the type VI secretion system encoded at *Salmonella* pathogenicity island 19 (SPI-19) in the *S. gallinarum* genome (Blondel et al. 2010). In this approach, the *loxP* sites and markers (for chloramphenicol and spectinomycin resistance) were PCR-amplified from the pVEX vectors and inserted into flanking positions using phage λ Red recombination (Fig. 6). The SPI-19 region was excised via Cre recombinase and captured onto R995 using homologous recombination (Fig. 6). The resulting R995 + SPI-19 clone was used to complement the colonization defect of the *S. gallinarum* SPI-19 deletion strain in a chicken infection model (Blondel et al. 2010). In addition, the authors transferred the R995 + SPI-19 clone into *S. enteriditis*, a species that contains significant sequence deviation in SPI-19 relative to *S. gallinarum*, to test if the presence of the *S. gallinarum* SPI-19 would increase *S. enteriditis* chicken colonization.

Interestingly, the presence of SPI-19 decreased the ability of *S. enteriditis* to colonize in this infection model (Blondel et al. 2010). This is consistent with the observations described above that demonstrate genomic island phenotypes can differ greatly, depending on the bacterial background.

2.3 New R995 derivatives allow an "all recombinase" approach

Recently an entirely recombinase-based approach for this techninque has been described using modified R995 plasmids (Santiago, Quick, and Wilson 2011). The new series of R995 derivatives encode a range of different marker combinations to increase utility in situations where several markers are used or are already present in the strain background. In addition, these R995 derivatives contain FRT sites that can facilitate the capture of genomic regions that have been excised using the Flp/FRT system (Fig. 7). A major advantage to this approach is that no regions of homology are needed to be cloned into any plasmids. Thus, the only step that takes place outside of cells is the amplification of the PCR products used for λ Red insertion of FRT sites into the flanking positions in the genome. This technique was demonstrated by cloning 20-kilobase regions from the *S. typhimurium* and *Escherichia coli* genomes (Santiago, Quick, and Wilson 2011).

2.4 Catalogue of reagents

Table 1 serves as a summary list of reagents used for the recombinase/conjugation-based cloning of genomic fragments. The PCR template plasmids are suicide plasmids and can

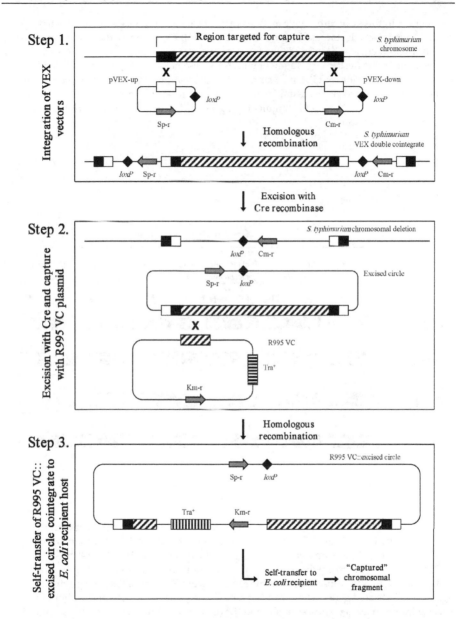

Fig. 2. The VEX-Capture system. Excision and capture of a section of the *S. typhimurium* genome is depicted to illustrate the functioning of the VEX-Capture system. In step one, differentially-marked pVEX vectors containing DNA fragments homologous to the ends of the targeted genomic region are integrated at the desired locations to form a double

cointegrate. In this structure, single *loxP* sites are located on either side of the targeted region. In step two, the targeted region is excised from the genome by the Cre recombinase, and the excised circle is "captured" via homologous recombination with the R995 VC plasmid. Note that the capture fragment on R995 VC is shown as targeted to one end of the excised genomic region, but it can be targeted to any location on the excised region. In step 3, the R995 VC-excised circle plasmid is transferred to an *E. coli* recipient to create a strain containing the captured genomic fragment. Diagram not drawn to scale. Reprinted from (Wilson and Nickerson 2007).

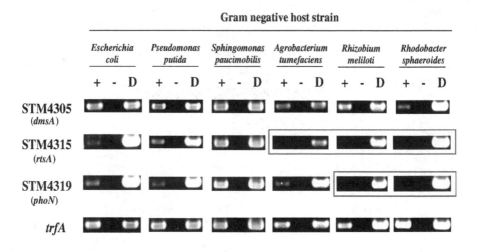

Fig. 3. RT-PCR analysis of *S. typhimurium* island 4305 after transfer to different Gram-negative hosts. The indicated Gram-negative strains containing R995 + *S. typhimurium* island 4305 were analyzed for expression of island genes STM4305, STM4315, STM4319 and the R995 replication gene *trfA* (which serves a positive control). Total RNA from each strain was isolated and reversed transcribed, and the samples were PCR-amplified using primers against the indicated genes. The (+) and (-) lanes indicate samples with and without the reverse transcriptase step, respectively, and the (D) lane indicates where R995 + island 4305 DNA isolated from each was used as template. PCR samples were run on agarose gels and stained with ethidium bromide. The boxed pictures indicate where expression of the gene is not detectable. This figure demonstrates genus-specific expression patterns for those island genes. Reprinted from (Wilson and Nickerson 2006).

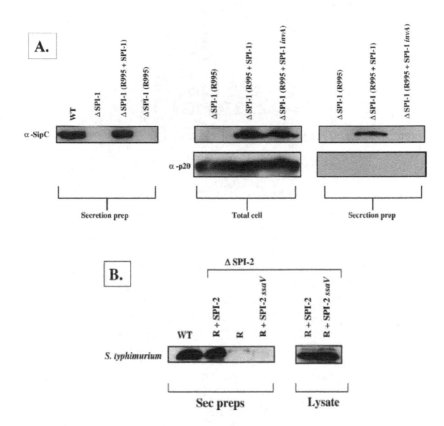

Fig. 4. R995 + SPI-1 and R995 + SPI-2 clones complement corresponding *S. typhimurium* SPI-1 and SPI-2 deletion mutants for substrate protein secretion. Panel A: Western blot analysis of protein secretion preparations and total cell lysates from *S. typhimurium* delta SPI-1 strains containing either R995, R995 + SPI-1, or R995 + SPI-1 *invA* plasmids. The last plasmid contains a mutation in the *invA* gene encoding a SPI-1 type III system protein that is essential for SPI-1-mediated secretion. Antibodies against the SPI-1 secreted substate SipC and the non-secreted bacterial cellular protein p20 are used. Panel B: Western blot analysis as in Panel A but using *S. typhimurium* delta SPI-2 strains containing either R995, R995 + SPI-2, or R995 + SPI-2 *ssaV* (mutation for the *ssaV* gene essential for SPI-2 secretion activity). Antibodies against the SPI-2 protein substrate SseB are used. The results of both panels demonstrate that the cloned SPI-1 and SPI-2 regions on R995 are functional and complement deleted SPI-1 and SPI-2 secretion systems. Reprinted from (Wilson, Coleman, and Nickerson 2007; Wilson and Nickerson 2006).

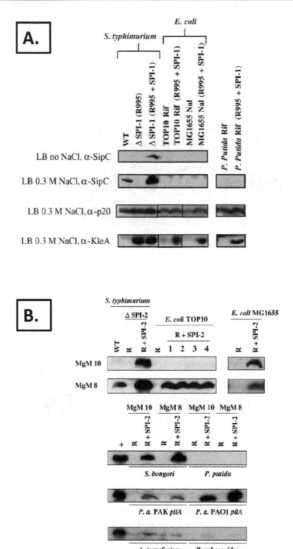

Fig. 5. Different expression patterns for SPI-1 and SPI-2 in different Gram-negative bacterial genera. Panel A: Plasmid R995 + SPI-1 was analyzed for expression of the SPI-1 protein SipC via Western blot analysis in *S. typhimurium*, *E. coli*, and *Pseudomonas putida*. In addition, the samples were also probed for the bacterial housekeeping p20 protein and the R995-encoded protein KleA as controls. The samples shown are total cell lysates of each strain. SipC expression is not detectable in *E. coli*, *P. putida*, attentuated in *P. aeruginosa* and *Agrobacterium tumefaciens* (the last two species not shown). Panel B: Plasmid R995 + SPI-2 expression was analyzed via Western blot assay against the SPI-2 protein SseB in various Gram-negative

bacteria. In contrast to SPI-1, expression of SPI-2 was readily detected in a range of bacterial backgrounds. Two points are of particular note: (1) In *S. typhimurium*, SPI-2 expression is regulated by growth media conditions, such that 10 mM $MgCl_2$ and pH 7.5 repress expression (MgM 10 media) and 8 μM $MgCl_2$ and pH 5.5 activate expression (MgM 8 media). However, expression from R995 + SPI-2 does not follow this regulation, except in the *E. coli* strain TOP10. R995 + SPI-1 expression shows a similar result in *S. typhimurium* in relation to its regulation by sodium chloride; and (2) *P. putida* appears to be recalcitrant to both SPI-1 and SPI-2 expression. Reprinted from (Wilson, Coleman, and Nickerson 2007; Wilson and Nickerson 2006).

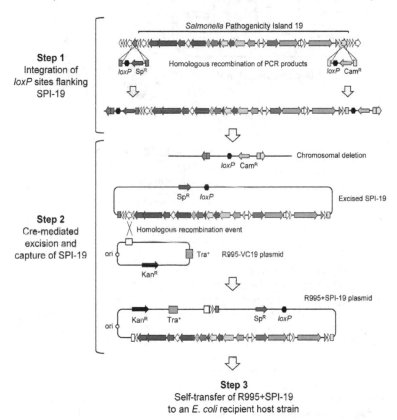

Step 1
Integration of
loxP sites flanking
SPI-19

Step 2
Cre-mediated
excision and
capture of SPI-19

Step 3
Self-transfer of R995+SPI-19
to an *E. coli* recipient host strain

Fig. 6. Schematic representation of the capture of SPI-19 from *S. gallinarum* 287/91 using a modified VEX-Capture method. To clone the type VI secretion system from the *S. gallinarum* genome, Blondel *et. al.* PCR-amplified markers and *loxP* sites from pVEX vectors and inserted them into flanking positions using phage λ Red recombination. The Cre-excised circular molecule was captured by R995 via homologous recombination, and the construct was isolated upon conjugation to an *E. coli* recipient. This construct was used for complementation analysis in a chicken model of infection using *S. gallinarum* and *S. enteriditis* strains and demonstrates the utility of R995 capture plasmids for *in vivo* pathogenesis studies. Reprinted from (Blondel et al. 2010).

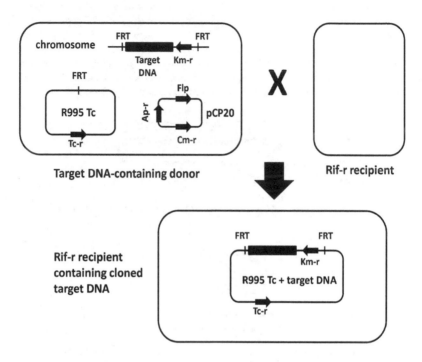

Fig. 7. An "all recombinase" approach to cloning large genomic DNA fragments to R995. This procedure utilizes specially designed R995 derivatives containing FRT sites that can be used as insertion points for a genomic fragment excised using the Flp/FRT system. A targeted DNA region in a bacterial genome is flanked by FRT sites and an antibiotic resistance marker as diagrammed using λ Red recombination. To accomplish this, the "unmarked" FRT site (to the left of the target DNA in the chromosome) is introduced via standard λ Red recombination markers (in Table 1) followed by Flp-mediated deletion of the marker to leave the single, unmarked FRT site. Next, the second flanking FRT site is introduced using a PCR fragment designed with a marker and single FRT site, such that the marker is located between the FRT site and the target DNA. In this example, the marker encodes kanamycin resistance. An R995 derivative containing an FRT site (and encoding tetracycline resistance in this example) is transferred to this strain via conjugation, and then the Flp-expressing plasmid pCP20 is introduced via electroporation. The electroporation outgrowth culture can be used directly as a donor for conjugation with a rifampicin (Rif)-resistant recipient strain. Alternatively, the electroporation can be plated on media containing tetracycline (Tc) and kanamycin (Km) and the colonies can be used as donor. The conjugation is plated on media containing Rif, Tc, and Km to select recipients that have obtained the cloned target DNA on R995. The transconjugants can be used to confirm the clone. A transconjugant can also be used as a donor for transfer of the clone to other bacterial strains for subsequent studies. Reprinted from (Santiago, Quick, and Wilson 2011).

Plasmid or strain	Chacteristics	Reference
Template plasmids		
pKD3	FRT:Cm-r:FRT, R6K ori	Datsenko, 2000
pKD4	FRT:Km-r:FRT, R6K ori	Datsenko, 2000
pJW101	FRT:Tp-r:FRT, R6K ori	Quick, 2010
pJW102	FRT:Sp-r:FRT, R6K ori	Quick, 2010
pJW107	FRT:Tc-r:FRT, R6K ori	This chapter
pVEX1212	loxP:Sp-r, P1 ori	Ayres, 1993
pVEX2212	loxP:Cm-r, R6K ori	Ayres, 1993
Lambda Red plasmids		
pKD46	Ap-r	Datsenko, 2000
pJW103	Km-r	Quick, 2010
pJW104	Cm-r	Quick, 2010
pJW105	Tp-r	Quick, 2010
pJW106	Sp-r	Quick, 2010
Recombinase plasmids		
pCP20	Flp, Ap-r, Cm-r	Datsenko, 2000
pEKA30	Cre, Ap-r, Tc-r	Wilson, 2004
pEKA16	IPTG-inducible Cre, Ap-r, Tc-r	Wilson, 2004
R995 derivatives		
R995	Km-r, Tc-r	Pansegrau, 1994
R995 Km Cm	$\Delta tetRA$::FRT-Cm-FRT	Santiago, 2011
R995 Km Tp	$\Delta tetRA$::FRT-Tp-FRT	Santiago, 2011
R995 Km Sp	$\Delta tetRA$::FRT-Sp-FRT	Santiago, 2011
R995 Tc Cm	$\Delta aphA$::FRT-Cm-FRT	Santiago, 2011
R995 Tc Tp	$\Delta aphA$::FRT-Cm-FRT	Santiago, 2011
R995 Tc Sp	$\Delta aphA$::FRT-Cm-FRT	Santiago, 2011
R995 Km	$\Delta tetRA$::FRT	Santiago, 2011
R995 Tc	$\Delta aphA$::FRT	Santiago, 2011
R995 Cm	$\Delta tetRA$::FRT, $\Delta aphA$::Cm-r	Santiago, 2011
R995 Tp	$\Delta tetRA$::FRT, $\Delta aphA$::Tp-r	Santiago, 2011
R995 Sp	$\Delta tetRA$::FRT, $\Delta aphA$::Sp-r	Santiago, 2011
Strains		
AS11	pir+ (for R6K ori plasmids)	Ayres, 1993
EKA260	repA+ (for P1 ori plasmids)	Ayres, 1993
TOP10 Rif	Rif-r recipient	Wilson, 2004

Table 1. Catalogue of reagents for recombinase/conjugation cloning. Please note that the template plasmids are suicide plasmids that require either AS11 or EKA260 for replication and that the λ Red plasmids and pCP20 are temperature-sensitive for replication (requiring 30 degrees C). The pJW plasmids are derived from either pKD3 (pJW101 and pJW102) or pKD46 (pJW103, pJW104, pJW105, and pJW106) (Quick, Shah, and Wilson 2010).

only replicate in corresponding strains that encode either the R6K Pir protein or P1 RepA protein (Ayres et al. 1993; Datsenko and Wanner 2000). This allows the PCR reaction to be directly electroporated into target cells with no background problems caused by the replication of the templates. It is worthwhile to note the PCR template plasmids with FRT sites contain two such sites flanking a given antibiotic resistance marker. Thus, care must be taken to amplify products containing only one FRT site for the second flanking insertion into the genome to avoid marker loss problems upon Flp expression (please refer to Fig. 7 for more details). It is also worthwhile to note that the self-transmissible IncP plasmid R995 displays a remarkably broad-host-range for both its conjugation and replication system (Adamczyk and Jagura-Burdzy 2003; Pansegrau et al. 1994; Thorsted et al. 1998). This facilitates R995 conjugative transfer to a wide variety of Gram-negative and Gram-positive bacteria and replication in almost all Gram-negative bacteria. Any other conjugative plasmid could be used for this procedure. However, IncP plasmid R995 and related plasmids are excellent options due to their broad-host-range, fully sequenced genomes, and high degree of characterization (especially for the IncPα plasmids R995, RK2, RP4, etc.).

3. Conclusion

Recombineering and conjugation can be exploited to provide a convenient, reproducible, and cost-effective technique for cloning large bacterial genomic fragments. This technique can be performed using easily obtained PCR products, readily available plasmids and strains, and simple, basic microbiology protocols. One question regarding the use of this system is: how large a genomic fragment can be accommodated by R995? So far, the biggest fragment cloned using this technique has been about 50 kilobases, but the upper limits of size have not yet been tested in any systematic way. To make genomic clones more amenable to medical or environmental applications, removal of antibiotic resistance markers and the conjugative transfer system would need to be accomplished. We are currently pursuing the development of alternative selection schemes and removable conjugation systems to address this issue. Overall, the use of the recombinase/conjugation cloning approach is currently underdeveloped as a technique and could expand the field of genomics by providing experiment-based strategies to answer important evolutionary questions.

4. Acknowledgment

We acknowledge the advice, technical help, and overall support of Dr. David Figurski, Dr. Cheryl Nickerson, and the Villanova University Biology Department.

5. References

Adamczyk, M., and G. Jagura-Burdzy. 2003. Spread and survival of promiscuous IncP-1 plasmids. *Acta Biochim Pol* 50 (2):425-53.

Ayres, E. K., V. J. Thomson, G. Merino, D. Balderes, and D. H. Figurski. 1993. Precise deletions in large bacterial genomes by vector-mediated excision (VEX). The *trfA* gene of promiscuous plasmid RK2 is essential for replication in several gram-negative hosts. *J Mol Biol* 230 (1):174-85.

Blondel, C. J., H. J. Yang, B. Castro, S. Chiang, C. S. Toro, M. Zaldivar, I. Contreras, H. L. Andrews-Polymenis, and C. A. Santiviago. 2010. Contribution of the type VI secretion system encoded in SPI-19 to chicken colonization by *Salmonella enterica* serotypes Gallinarum and Enteritidis. *PLoS One* 5 (7):e11724.

Datsenko, K. A., and B. L. Wanner. 2000. One-step inactivation of chromosomal genes in *Escherichia coli* K-12 using PCR products. *Proc Natl Acad Sci U S A* 97 (12): 6640-5.

Ham, J. H., D. W. Bauer, D. E. Fouts, and A. Collmer. 1998. A cloned *Erwinia chrysanthemi* Hrp (type III protein secretion) system functions in *Escherichia coli* to deliver *Pseudomonas syringae* Avr signals to plant cells and to secrete Avr proteins in culture. *Proc Natl Acad Sci U S A* 95 (17):10206-11.

Hansen-Wester, I., D. Chakravortty, and M. Hensel. 2004. Functional transfer of Salmonella pathogenicity island 2 to *Salmonella bongori* and *Escherichia coli*. *Infect Immun* 72 (5):2879-88.

McClelland, M., K. E. Sanderson, J. Spieth, S. W. Clifton, P. Latreille, L. Courtney, S. Porwollik, J. Ali, M. Dante, F. Du, S. Hou, D. Layman, S. Leonard, C. Nguyen, K. Scott, A. Holmes, N. Grewal, E. Mulvaney, E. Ryan, H. Sun, L. Florea, W. Miller, T. Stoneking, M. Nhan, R. Waterston, and R. K. Wilson. 2001. Complete genome sequence of *Salmonella enterica* serovar Typhimurium LT2. *Nature* 413 (6858): 852-6.

McDaniel, T. K., and J. B. Kaper. 1997. A cloned pathogenicity island from enteropathogenic *Escherichia coli* confers the attaching and effacing phenotype on *E. coli* K-12. *Mol Microbiol* 23 (2):399-407.

Pansegrau, W., E. Lanka, P. T. Barth, D. H. Figurski, D. G. Guiney, D. Haas, D. R. Helinski, H. Schwab, V. A. Stanisich, and C. M. Thomas. 1994. Complete nucleotide sequence of Birmingham IncP alpha plasmids. Compilation and comparative analysis. *J Mol Biol* 239 (5):623-63.

Quick, L. N., A. Shah, and J. W. Wilson. 2010. A series of vectors with alternative antibiotic resistance markers for use in lambda Red recombination. *J Microbiol Biotechnol* 20 (4):666-9.

Santiago, CP, LN Quick, and JW Wilson. 2011. Self-Transmissible IncP R995 Plasmids with Alternative Markers and Utility for Flp/FRT Cloning Strategies. *J Microbiol Biotechnol* 21 (11).

Thorsted, P. B., D. P. Macartney, P. Akhtar, A. S. Haines, N. Ali, P. Davidson, T. Stafford, M. J. Pocklington, W. Pansegrau, B. M. Wilkins, E. Lanka, and C. M. Thomas. 1998. Complete sequence of the IncPbeta plasmid R751: implications for evolution and organisation of the IncP backbone. *J Mol Biol* 282 (5):969-90.

Wilson, J. W., C. Coleman, and C. A. Nickerson. 2007. Cloning and transfer of the Salmonella pathogenicity island 2 type III secretion system for studies of a range of gram-negative genera. *Appl Environ Microbiol* 73 (18):5911-8.

Wilson, J. W., D. H. Figurski, and C. A. Nickerson. 2004. VEX-capture: a new technique that allows *in vivo* excision, cloning, and broad-host-range transfer of large bacterial genomic DNA segments. *J Microbiol Methods* 57 (3):297-308.

Wilson, J. W., and C. A. Nickerson. 2006. Cloning of a functional Salmonella SPI-1 type III secretion system and development of a method to create mutations and epitope fusions in the cloned genes. *J Biotechnol* 122 (2):147-60.

Wilson, J. W., and C. A. Nickerson. 2006. A new experimental approach for studying bacterial genomic island evolution identifies island genes with bacterial host-specific expression patterns. *BMC Evol Biol* 6:2.

Wilson, J. W., and C. A. Nickerson. 2007. *In vivo* excision, cloning, and broad-host-range transfer of large bacterial DNA segments using VEX-Capture. *Methods Mol Biol* 394:105-18.

Permissions

The contributors of this book come from diverse backgrounds, making this book a truly international effort. This book will bring forth new frontiers with its revolutionizing research information and detailed analysis of the nascent developments around the world.

We would like to thank David Figurski, for lending his expertise to make the book truly unique. He has played a crucial role in the development of this book. Without his invaluable contribution this book wouldn't have been possible. He has made vital efforts to compile up to date information on the varied aspects of this subject to make this book a valuable addition to the collection of many professionals and students.

This book was conceptualized with the vision of imparting up-to-date information and advanced data in this field. To ensure the same, a matchless editorial board was set up. Every individual on the board went through rigorous rounds of assessment to prove their worth. After which they invested a large part of their time researching and compiling the most relevant data for our readers. Conferences and sessions were held from time to time between the editorial board and the contributing authors to present the data in the most comprehensible form. The editorial team has worked tirelessly to provide valuable and valid information to help people across the globe.

Every chapter published in this book has been scrutinized by our experts. Their significance has been extensively debated. The topics covered herein carry significant findings which will fuel the growth of the discipline. They may even be implemented as practical applications or may be referred to as a beginning point for another development. Chapters in this book were first published by InTech; hereby published with permission under the Creative Commons Attribution License or equivalent.

The editorial board has been involved in producing this book since its inception. They have spent rigorous hours researching and exploring the diverse topics which have resulted in the successful publishing of this book. They have passed on their knowledge of decades through this book. To expedite this challenging task, the publisher supported the team at every step. A small team of assistant editors was also appointed to further simplify the editing procedure and attain best results for the readers.

Our editorial team has been hand-picked from every corner of the world. Their multi-ethnicity adds dynamic inputs to the discussions which result in innovative outcomes. These outcomes are then further discussed with the researchers and contributors who give their valuable feedback and opinion regarding the same. The feedback is then collaborated with the researches and they are edited in a comprehensive manner to aid the understanding of the subject.

Apart from the editorial board, the designing team has also invested a significant amount of their time in understanding the subject and creating the most relevant covers. They scrutinized every image to scout for the most suitable representation of the subject and create an appropriate cover for the book.

The publishing team has been involved in this book since its early stages. They were actively engaged in every process, be it collecting the data, connecting with the contributors or procuring relevant information. The team has been an ardent support to the editorial, designing and production team. Their endless efforts to recruit the best for this project, has resulted in the accomplishment of this book. They are a veteran in the field of academics and their pool of knowledge is as vast as their experience in printing. Their expertise and guidance has proved useful at every step. Their uncompromising quality standards have made this book an exceptional effort. Their encouragement from time to time has been an inspiration for everyone.

The publisher and the editorial board hope that this book will prove to be a valuable piece of knowledge for researchers, students, practitioners and scholars across the globe.

List of Contributors

Ahmed Chraibi and Stéphane Renauld
University of Sherbrooke, Québec, Canada

Kevin Hadi and Velpandi Ayyavoo
University of Pittsburgh, Pittsburgh, PA, USA

Oznur Tastan
Bilkent University, Ankara, Turkey

Alagarsamy Srinivasan
NanoBio Diagnostics, West Chester, PA, USA

M. Tang and K. Lai
University of Utah School of Medicine, USA

K.J. Wierenga
University of Oklahoma Health Sciences Center, USA

Christelle Bonod-Bidaud and Florence Ruggiero
Institut de Génomique Fonctionnelle de Lyon, ENS de Lyon, UMR CNRS 5242, University Lyon 1, France

Juanita Yazmin Damián-Almazo and Gloria Saab-Rincón
Departamento de Ingeniería Celular y Biocatálisis, Instituto de Biotecnología, Universidad Nacional Autónoma de México, Cuernavaca, Morelos, México
Silvio Alejandro López-Pazos
Facultad de Ciencias de la Salud, Universidad Colegio Mayor de Cundinamarca, Colombia
Biology & Rural Ecology research group, Corporación Ramsar Guamuez, El Encano (Pasto-Nariño), Colombia

Jairo Cerón
Instituto de Biotecnología, Universidad Nacional de Colombia, Santafé de Bogotá DC, Colombia

Nathan A. Sieracki and Yulia A. Komarova
University of Illinois – Chicago, USA

Shona A. Mookerjee
The Buck Institute for Research on Aging, USA

Elaine A. Sia
Department of Biology, The University of Rochester, USA

Joy Sturtevant
LSUHSC School of Medicine, USA

Roman G. Gerlach, Kathrin Blank and Thorsten Wille
Robert Koch-Institute Wernigerode Branch, Germany

Stathis Frillingos
University of Ioannina Medical School, Greece

James W. Wilson, Clayton P. Santiago and Jacquelyn Serfecz
Villanova University, USA

Laura N. Quick
Children's Hospital of Philadelphia, USA

Printed in the USA
CPSIA information can be obtained
at www.ICGtesting.com
JSHW011439221024
72173JS00004B/872